# Trends and Developments in Ordinary Differential Equations

Proceedings of the International Symposium

# Trends and Developments in Ordinary Differential Equations

Western Michigan University; Kalamazoo, Michigan, USA
20 – 22 May, 1993



*Editors*

**Yousef Alavi**
**Po-Fang Hsieh**
Western Michigan University; Kalamazoo, Michigan,
USA

**World Scientific**
*Singapore • New Jersey • London • Hong Kong*

*Published by*

World Scientific Publishing Co. Pte. Ltd.
P O Box 128, Farrer Road, Singapore 9128
*USA office:* Suite 1B, 1060 Main Street, River Edge, NJ 07661
*UK office:* 73 Lynton Mead, Totteridge, London N20 8DH

**TRENDS AND DEVELOPMENTS IN ORDINARY DIFFERENTIAL EQUATIONS**



ISBN 981-02-1530-4



Printed in Singapore.

# PREFACE

This volume constitutes the proceedings of the International Symposium on Ordinary Differential Equations and Applications, held at Western Michigan University, May 20 – May 22, 1993. Symposium participants included research mathematicians from colleges, universities and industry as well as graduate students. Altogether, there were representations from several states and eleven countries. The contributions in this volume include many topics in current research in both theory and applications in various areas of ordinary differential equations.

# DEDICATION

The organizers and participants dedicate the conference and its proceedings to Professors William Harris, Masahiro Iwano, and Yasutaka Sibuya for their outstanding and fundamental research contributions to the analytic theory of Differential and Functional Equations and its promotion throughout the world.

William Harris, Masahiro Iwano, and Yasutaka Sibuya wish to express their sincere gratitude to their teachers Masuo Hukuhara and Hugh Turrittin, and their mentor Wolfgang Wasow, for their valuable guidance and continuing encouragement.

William A. Harris, Jr.

Masahiro Iwano

Yasutaka Sibuya

## William Ashton Harris, Jr.

William Ashton Harris, Jr. was born December 18, 1930 in New Orleans, Louisiana, USA. He studied mathematics with minors in physics and applied mechanics (elasticity) at the University of Minnesota, receiving his Ph.D. in mathematics in 1958 under the direction of Professor H.L. Turrittin.

He has held two principle appointments: the University of Minnesota (1958-1970) and the University of Southern California (1970-present). He served as Chair, Department of Mathematics, University of Southern California, 1977-1984, and has been a consultant to Industry and Government.

Professor Harris has held visiting appointments at the University of Calgary (Canada), University of California (Los Angeles and San Diego), University of Dundee (Scotland), University of Edinburgh (Scotland), University of Groningen (Netherlands), Kyoto University (RIMS, Japan), University of Minnesota (IMA and AHPCRC), University of Nebraska, University of Nijmegen (Netherlands), University of Strasbourg (France), University of Texas (Arlington), University of Warwick (England), and the University of Wisconsin (MRC).

He has published more than 70 research articles. The major areas of his research are the behavior of solutions of ordinary differential and difference equations at a singular point, and singular perturbation problems, using analytic, algebraic, and asymptotic integration methods. His editorial duties include Associate Editorships for the SIAM Journal on Applied Mathematics and Nonlinear Analysis, TMA. With Yasutaka Sibuya, he was instrumental in establishing a United States - Japan Cooperative Science initiative in Differential and Functional Equations.

Professor Harris has given many invited addresses at national and international meetings. These include an invited hour address to the American Mathematical Society and the Mathematical Society of Japan. He has organized and been an invited participant in many special sessions of the American Mathematical Society. He has also been the Principle Lecturer for a CBMS-NSF Regional Conference in Applied Mathematics, and served as a SIAM Visiting Lecturer.

Professor Harris married Phyllis Marie Chatelle in 1953. They have a daughter, Anne Marie (Miller), a son William A. Harris III, and two granddaughters, Danielle Robyn Harris and Jessica Joyce Miller.

His hobbies include stained glass, stone sculpture, tennis, and ice dancing (Gold Dance Judge).

## Masahiro Iwano

Masahiro Iwano was born on February 1, 1931 in Chonju, Korea, to Japanese parents and returned to Japan before his first birthday. He studied mathematics at the University of Tokyo, receiving his DS in 1961 under the direction of Professor Masuo Hukuhara.

He had held two principle appointments: Tokyo Metropolitan Universtiy (1962 - 1984) and Chuo University (1984 - present). He is an emeritus professor of Tokyo Metropolitan University. He served as Chairman of the department at both places.

Professor Iwano has held visiting appointments at University of Tokyo, Kyoto University, Tohoku University, Yamanashi University , Tokyo University of Technology, Osaka City University, Keio University, Hiroshima University, Fukuoka University, Kumamoto University, University of Wisconsin (Mathematics Research Center), University of Minnesota and Western Michigan University.

He has published 45 research articles. The major area of his research is the study of a local problem of ordinary differential equations with an irregular type singularity or with a singularity with respect to parameters. The local problem includes the construction of the stable domain and analytic expressions of bounded solutions depending on several arbitrary parameters and, in particular, a general solution near an irregular type singular point of a system of nonlinear equations without Poincaré condition. A stable domain is certain product of a domain in the space of the independent variable and the domains of solution spaces of the simplified differential equations for the given equations. The stable domain consists of the product of circle bundles on the domain of the independent variable and is characterized by the property that, for arbitrary initial values in the stable domain, there is a curve, joining the initial point and the singular point, such that the graph of the general solution stays in the domain as the independent variable travels on this curve. The curves used by many authors required essentially the steadily decreasing of the modulus of the general solution on these curves. The stable domains constructed by Professor Iwano do not require such condition on the curves. Thus they provide wider regions of validity in the space of the independent variable and the domains of solutions.

Professor Iwano has given invited addresses at national and international meetings. These include two invited hour addresses to the Mathematical Society of Japan and five invited hour addresses to international symposiums.

Professor Iwano married Hiroko Aoki in 1959. They have three children: Koji (son), Motoko (daughter) and Yuji (son).

His hobby is growing many varieties of vegetables. He also likes to play Go.

## Yasutaka Sibuya

Yasutaka Sibuya was born on October 16, 1930 in Maizuru, Kyoto-fu, Japan. He studied mathematics at the University of Tokyo and the University of California, Los Angeles (as a Fulbright exchange student), receiving his PhD in mathematics at UCLA in 1959 under the direction of Professor E.A. Coddington and his DS at the University of Tokyo in 1961 under the direction of Professor M. Hukuhara.

He has held his principle academic appointment at the University of Minnesota from 1963 to present. He was also an assistant of Professor Hukuhara at the University of Tokyo (1955-58) and a lecturer and an assistant professor at Ochanomizu University, Tokyo (1958-1963).

Professor Sibuya had visiting appointments at Courant Institute at New York University, Keio University (Japan), Massachusetts Institute of Technology, Ulm University (Germany), University of California (Los Angeles), University of Edinburgh (Scotland), University of Groningen (Netherlands), University of Southern California, University of Strasbourg (Alsace), and the University of Wisconsin (MRC).

He was awarded a Senior Humboldt Research Scholarship in 1984. He is an editor of Asymptotic Analysis. He was also a member of organizing committees for meetings at Oberwolfach, Germany.

Professor Sibuya has published more than 100 research articles in various mathematical journals, and three books. The major area of his research was the study of formal power series solutions of ordinary differential and difference equations. This includes solutions at a singular point, solutions of problems of perturbations, and p-adic power series solutions. In these researches, he utilized analytic and algebraic methods extensively. He also studied ordinary and partial differential equations with almost periodic coefficients.

He has given many invited addresses at national and international meetings. These include an invited hour address to the American Mathematical Society and the Mathematical Society of Japan. He also gave many invited twenty minutes addresses in special sessions of the American Mathematical Society.

Professor Sibuya married his wife Yasuko in 1962. They have four children, Takasi (son), Atsuko (daughter), Maki (daughter) and Miki (son), and a grandchild, Stella A. Wiering (Maki's daughter).

He likes to read books on world history and collects small toys such as laughing ghosts of Halloween and dolls of Frenando Valenzuela and other athletes. He also goes fishing with his children.

# ACKNOWLEDGEMENT

The editors take pleasure in thanking the many people who contributed to the success of the International Symposium on Ordinary Differential Equations and Applications, as well as to the preparation of these Proceedings. In addition to the speakers, participants and contributors to this volume, we gratefully acknowledge

Western Michigan University;

WMU Department of Mathematics and Statistics
and the enthusiastic support of Dr. Ruth Ann Meyer, chairperson;

the financial support of several friends of the Department;

the fine support of the staff of the Department;

the valuable assistance of the Symposium Assistants:
Leslie Hayes, Feipeng Xie and Mei Yi.

Finally, we wish to thank the editors of World Scientific Publishing Co., for their valuable assistance with these Proceedings.

The editors apologize in advance for any oversight in acknowledgement or in any errors in the manuscripts.

YA
PFH

# CONTENTS

---

*indicates the presenter of a paper by multiple authors

# Computing Sturm-Liouville Eigenvalues for General Limit Circle Boundary Conditions

Paul B. Bailey
Department of Mathematics, Northern Illinois University
Dekalb, IL 60115, U.S.A.

## Abstract

We discuss the numerical computation of the eigen- values of Sturm-Liouville problems for formally self-adjoint differential equations of the form $-(py')' + qy = \lambda wy$ on $(a, b)$ (the interval $(a, b)$ may be infinite in either direction) when either $a$ or $b$ ( or both) is a singular point of Limit Circle type, either Oscillatory or Non-Oscillatory, and there is a general separated linear Boundary Condition there.

The work I'm going to talk about was done jointly with Prof. Norrie Everitt of The University, in Birmingham, England, and Prof. Tony Zettl of Northern Illinois University.

It concerns the Sturm-Liouville eigenvalue problem in the special case that one or the other of the endpoints is of the type called "Limit Circle".

Recall that the Sturm-Liouville Eigenvalue Problem is described in terms of a Second Order Ordinary Differential Equation of the form

$$-(py')' + qy = \lambda wy \quad \text{on} \quad (a, b),$$

where the coefficient functions $p, q, w$ will for convenience here be assumed to be continuous in the open interval $(a, b)$, and the two functions $p, w$ are strictly positive there.

In the event that the interval $(a, b)$ is finite and the just stated conditions on $p, q, w$ hold also at each of the two endpoints a and b, then the differential equation is Regular at both ends. In this case we also assume that a Boundary Condition is given at each end in the form

$$A_1 y(a) + A_2 p(a) y'(a) = 0$$

and

$$B_1 y(b) + B_2 p(b) y'(b) = 0,$$

where the constants $A_1, A_2$ (and $B_1, B_2$) are arbitrary except that at least one is not zero.

This is the most usual form of the Sturm-Liouville Eigenvalue problem when both endpoints are "Regular". When one or more of the endpoints is of Limit Circle type, the problem becomes more complicated in several ways, not the least of which is the matter of a suitable Boundary Condition at that endpoint.

Before I try to describe how we deal with a Limit Circle endpoint and Boundary Condition, let me very briefly recall how the Regular problem is handled.

We start with the Prufer transformation. That is, we transform from $y, py'$ to new unknowns $\rho, \theta$ by the equations

$$y = \rho \sin(\theta)$$

$$py' = \rho \cos(\theta),$$

or

$$\tan(\theta) = \frac{y}{py'}, \quad \rho = \sqrt{y^2 + (py')^2}.$$

This gives us a Differential Equation for $\theta$:

$$\theta' = \frac{1}{p}\cos(\theta)^2 + (\lambda w - q)\sin(\theta)^2 \qquad (*)$$

and the Boundary Conditions in terms of $y$, $py'$ give equivalent Boundary Conditions for $\theta$,

$$\theta(a) = \alpha, \quad \theta(b) = \beta + n\pi,$$

where

$$0 \le \alpha < \pi, \quad 0 < \beta \le \pi.$$

There is also a Differential Equation for $\rho$, but we won't be needing it right now.

To find a particular eigenvalue then, using the Prufer angle $\theta$, we integrate (*) for $\theta$ from $x = a$ to $x = M$ (where $M$ is some interior point), getting

$$\theta L(\lambda; M),$$

say, and then integrate (*) again from $x = b$ to $x = M$ to get

$$\theta R(\lambda; M).$$

Then find the zero, $\lambda(n)$, of the function of $\lambda$,

$$f(\lambda) = \theta L(\lambda; M) - \theta R(\lambda; M),$$

(there will be only one zero).

Now let us look at the situation when one of the endpoints is of Limit Circle type.

An example will show clearly why the Boundary Condition used for Regular problems will not do for Limit Circle problems. Consider Bessel's equation:

$$-y'' + (\nu\nu - 0.25)\frac{y}{xx} = \lambda y \qquad \text{on} \ \ (0,1).$$

The endpoint zero is Limit Circle whenever

$$0 < \nu < 1, \qquad \nu \neq \frac{1}{2}.$$

But notice that any solution near zero has the asymptotic form

$$y(x) \simeq Cx^{(0.5+\nu)} + Dx^{(0.5-\nu)}$$

so that

$$\begin{aligned} \frac{y}{py'}(x) &\simeq \frac{Cx^{\nu} + Dx^{-\nu}}{C(\nu + 0.5)x^{\nu} + D(-\nu + 0.5)x^{-\nu}}x \\ &\simeq \frac{x}{0.5 - \nu}, \end{aligned}$$

independent of the constants $C$, $D$ $(D \neq 0)$.

This means that a Boundary Condition

$$A_1 y(a) + A_2 y'(a) = 0$$

can be satisfied only if $A_2 = 0$. In other words, this kind of Boundary Condition does not enable one to distinguish one particular solution from all the others. So some other form of Boundary Condition is necessary.

The kind of Boundary Condition we have chosen to use is what may be called the Naimark Boundary Condition.

We assume that two functions $u$ and $v$ are known to be solutions of the original Differential Equation for some particular value of $\lambda$, say $\lambda^*$. (This condition can be relaxed somewhat: namely, it is only necessary that $u, v$ be functions in the "Maximal Domain", which is not so demanding as being "solutions" is. But we will assume for convenience here that $u$ and $v$ are actually solutions.)

Then, in terms of these functions $u$ and $v$, our Boundary Condition is that

$$A_1[y,u](x) + A_2[y,v](x) \to 0 \qquad \text{as } x \to a,$$

where

$$[y,u] = ypu' - upy'.$$

It turns out that this Naimark Boundary Condition does indeed enable one to select any particular solution from all the others.

The question now is: How can this Boundary Condition be employed in a numerical scheme?

The way we have chosen begins with a "Regularization" of the Differential Equation

$$-(py')' + qy = \lambda wy$$

given solutions $u$ and $v$ of the same D.E. when the parameter $\lambda$ is equal to some particular value, $\lambda^*$.

Let

$$y = z_1 u + z_2 v$$

and

$$py' = z_1 pu' + z_2 pv',$$

where $z_1$ and $z_2$ are two new dependent variables in place of $y$ and $py'$. Then the differential equation (*) is transformed into the system

$$z_1' = (\lambda - \lambda^*)w(uvz_1 + vvz_2)$$
$$z_2' = (\lambda - \lambda^*)w(-uuz_1 - uvz_2),$$

which is easily seen to be Regular (since $u$ and $v$ are each square integrable with the weight function $w$).

The Naimark Boundary Condition becomes

$$(A_1 z_2 - A_2 z_1)[u,v] \to 0,$$

or

$$\frac{z_1}{z_2} \to \frac{A_1}{A_2}.$$

Conceivably one could integrate for $z_1$, $z_2$ and then transform back to $y$, $py'$, then to $\theta$ and proceed as before. Alternatively, (and this is what we do), set

$$z_1 = \sigma\cos(\phi)$$
$$z_2 = \sigma\sin(\phi),$$

so that the differential system for $z_1$, $z_2$ is transformed into a differential equation for $\phi$,

$$\phi' = -(\lambda - \lambda^*)w(u\cos(\phi) + v\sin(\phi))^2$$

with Boundary Condition

$$\tan(\phi(a)) = \frac{A_2}{A_1}.$$

The relation between $\theta$ and $\phi$ is

$$\tan(\theta) = \frac{u\cos(\phi) + v\sin(\phi)}{pu'\cos(\phi) + pv'\sin(\phi)}.$$

In the case that $u$ and $v$ are not actually solutions of the Differential Equation, but only Maximal Domain Functions, the details are a little different from what was given above, but essentially the same sort of result is obtained.

So, as far as the Boundary Condition is concerned—the Naimark Boundary Condition, that is—we have at least one way of conceivably implementing it in a numerical scheme: Namely, integrate the Differential Equation for the "Regularization function", $\phi$, and then convert the result to our usual Prufer angle, $\theta$.

There is, to be sure, some ambiguity concerning the appropriate integer multiple of $\pi$ for $\theta$. When I said that we could get $\theta$ from $\phi$, or vice versa, from the relation above, I carefully avoided any reference to this ambiguity about multiples of $\pi$, until now, because how we deal with it depends upon which kind of Limit Circle problem we have. That is to say, whether the endpoint is

Limit Circle, Non-Oscillatory,

or,

Limit Circle, Oscillatory.

(Every Limit Circle endpoint is either the one kind or the other.)

It turns out that in the Non-Oscillatory case, there is no real difficulty in assigning the correct multiple of pi when converting from phi to theta, or vice versa. I won't bother you with the details here.

More interesting is the Oscillatory case. In this case all solutions of the original Differential Equation oscillate infinitely often near the endpoint. It also means that the Spectrum is unbounded below. Namely, there are arbitrarily large negative eigenvalues. In particular, there is no "lowest" eigenvalue. Consequently, if the eigenvalues be arranged in ascending order, we will have something like

$$\cdots, \lambda(-2), \lambda(-1), \lambda(0), \lambda(1), \cdots$$

(Whether or not this sequence is bounded above will depend upon the nature of the other endpoint.)

In any case, the indexing of this sequence is quite arbitrary, at least up to now. So to make matters definite, we might agree to have $\lambda(0)$ denote the first non-negative eigenvalue—if there is one. If there is not, we might alter the function $q$ by adding a constant to it so that the new problem does have a non-negative eigenvalue.

So we may as well assume that there is a first non-negative eigenvalue, $\lambda(0)$. Then we need to be able to compute, say, $\lambda(-2)$; or $\lambda(3)$ (if there is one). Recall that in the Regular case, the eigenfunction corresponding to lambda(n) has $n$ zeros in $(a, b)$, and the eigenfunction corresponding to $\lambda(0)$ has none. Similarly in the Oscillatory Limit Circle case it must be that the eigenfunction corresponding to $\lambda(n)$ has $n$ zeros "more" than does the eigenfunction corresponding to $\lambda(0)$. Since all solutions have infinitely many zeros, this sounds a little vague but can be made quite definite. To find $\lambda(0)$, for example, we can compare the solution $theta_0$ (obtained with eigen- parameter set equal to 0), to the solution $\theta$ (obtained with eigenparameter $\lambda$), and find the least positive value of $\lambda$ for which it is an eigenvalue. Then $\lambda(1)$ is the next positive eigenvalue, and so on.

Prem N. Bajaj

# Use of Operators in Difference Equations

## 1. Introduction:

The subject of difference equations got introduced to me while teaching a course in Mathematical Modelling. I stumbled upon a reference to difference equations in connection with population problems. Credit goes to Finizio and Laddas' book [1] for arousing my interest in the subject.

In the treatment of linear difference equations and linear differential equations , many similarities are noticable. Consider just the following two:

(1) The system $x_{k+1} = a_{11}\, x_k + a_{12}\, y_k + a_{13}\, z_k$ ,

$$y_{k+1} = a_{21}\, x_k + a_{22}\, y_k + a_{23}\, z_k,$$

$$z_{k+1} = a_{31}\, x_k + a_{32}\, y_k + a_{33}\, z_k.$$

of linear difference equations with constant coefficients can be solved, exactly as in ordinary differential equations, by finding eigen – values and eigen – vectors of the coefficient matrix of the system.

(2) Consider the system:

$$x_{k+1} = a_{11}\, x_k + a_{12}\, y_k + b_k,$$

$$y_{k+1} = a_{21}\, x_k + a_{22}\, y_k + c_k.$$

Introducing operator $D$ by $Dx_k = x_{k+1}$ (and $Dy_k = y_{k+1}$) the system can be written as

$$(D - a_{11})\, x_k - a_{12}\, y_k = b_k$$

$$-a_{21}\,x_k + (D - a_{22})\,y_k = c_k.$$

Eliminating $y_k$, in the natural way, we obtain

$$[(D-a_{11})(D-a_{22}) - a_{12}\,a_{21}]\;\;x_k = (b_{k+1} - a_{22}\,b_k + \;c_k\,a_{12}.$$

in one unknown $x_k$.

This resembles the familar method in ODE where operator $D$ is the differential operator.

The question arises if the methods, using differential operator $D$, for finding particular integral for linear ordinary differential equations extend to difference equations also. See [3] for some of these methods. To be specific, we are interested in examining if the following formulae from ordinary differential equations will carry over to difference equations also. Let $f$ be an $n$th degree polynomial with constant coefficients. Let $D$ be a differential operator. Let $y_p$ be a particular integral of $f(D)y = g(x)$.

(F – 1). Let $g(x) = e^{ax}$. Then

(i) $y_p = \dfrac{1}{f(a)}\; e^{ax}$ , if $f(a) \neq 0$.

(ii) $y_p = x^m \dfrac{1}{f^{(m)}(a)}\; e^{ax}$ if $f$ has a zero of order $m \geq 1$ at $a$.

(F – 2). Let $g(x) = sin\ ax$. Then

(i) $y_p = \dfrac{1}{f(D)|_{D^2=-a^2}}\; sin\ ax$ if $(D^2 + a^2)$ is not a factor of $f(D)$.

(ii) $y_p = x\dfrac{1}{f\,'(D)|_{D^2=-a^2}}\; sin\ ax$ if $(D^2 + a^2)$ is a factor of $f(D)$.

Odd exponents of $D$ are treated in the natural way.

The case of $\cos ax$ is similar.

(F – 3). Let $g(x)$ be a polynomial in $x$. Let $f(D) = D^m F(D)$ , $m \geq 0$, $F(0) \neq 0$. Then

$$y_p = \left(\frac{b_0}{D^m} + \frac{b_1}{D^{m-1}} + \ldots + b_m + b_{m+1}D + \ldots\right) g(x) \quad \text{where } b_0 + b_1 D + b_2 D^2 + \ldots$$

is the Taylor series expansion of $\frac{1}{F(D)}$ at $D = 0$.

(F – 4). Let $V = V(x)$ be of class $C^m$. Then

$$f(D)(e^{ax}V) = e^{ax}\, f(D + a)(V)$$

## 2. Main results

We claim that (F – 1) and (F – 3) carry over, elegantly, to difference equations. Extension of (F – 2) to difference equations, though natural and direct, is not neat. We find no clue for (F – 4) and invite the reader to deal with it.

To give the idea of our results, we extend (F – 1) by the following:

## 3. Theorem

Define operator $D$ by $Dy_k = y_{k+1}$ Let $f$ be a polynomial with constant coefficients. Then a particular solution $\overset{p}{y}_k$ of $f(D)y_k = a^k$, $a \neq 1$ is given by

(i) $\overset{p}{y}_k = \dfrac{1}{f(a)}\, a^k$ if $f(a) \neq 0$.

(ii) $\overset{p}{y}_k = \dfrac{k^n}{a^n} \dfrac{1}{f^n(a)}\, a^k$ if $f$ has a zero of order $n$ at $a$.

Outline of proof

Part (i) Observe that $f(D)a^k = f(a)a^k$

Part(ii) We need the following lemma.

**4 Lemma**

Let $n$ be a natural number. Then for all $k$,

$(k+n)^n - \binom{n}{1}(k+n-1)^n + \binom{n}{2}(k+n-2)^n + \ldots + (-1)^n k^n = n!$

Proof of lemma. Define operator $\Delta$ by $\Delta = D - 1$.

We compute $\Delta^n k^n$ in two ways.

$\Delta k^n = (k+1)^n - k^n = nk^{n-1}$ + lower degree terms of $k$

$\Delta^2 k^n = \Delta(\Delta k^n) = n(n-1)k^{n-2}$ + lower degree terms of $k$.

Continuing, we obtain $\Delta^n k^n = n!$

Next $\Delta k^n = (k+1)^n - k^n$

$\Delta^2 k^n = \Delta(\Delta k^n) = (k+2)^n - (n+1)^n - [(k+1)^n - k^n]$

$= (k+2)^n - 2(k+1)^n + k^n.$

We claim that:

$$\Delta^i k^n = \sum_{j=0}^{i} (-1)^j \binom{i}{j} (k+i-j)^n, \quad 1 \le i \le n \qquad (A)$$

Assuming that (A) holds for some $i$, $1 \le i < n$, we have

$$\Delta^{i+1} k^n = \sum_{j=0}^{i} (-1)^j \binom{i}{j} (k+1+i-j)^n$$

$$-\sum_{j=0}^{i} (-1)^j \binom{i}{j} (k+i-j)^n$$

Writing the second term on the right as

$$\sum_{j=1}^{i+1} (-1)^j \binom{i}{j-1} (k+i-j+1)^n ,$$

it, easily , follows that

$$\Delta^{i+1} k^n = \sum_{j=0}^{i+1} (-1)^j \binom{i+1}{j} (k+i+1-j)^n ,$$

Now, (A) follows from induction. In particular (A) gives another expression for $\Delta^n k^n$; equating this expression for $\Delta^n k^n$ with $n!$ , the proof of lemma follows.

To prove part (*ii*) , first show (by using above lemma) that $\frac{1}{n!} \frac{k^n}{a^n} a^k$ is a particular solution of the difference equation.

$$(D-a)^n y_k = a^k.$$

## REFERENCES


[1] Finizio N. and Laddas G. (1982), An introduction to differential equations, Wadsworth.

[2] Lakshmikantham V. and Trigiante D. (1988), Theory of difference equations, Academic Press.

[3] Rainville E.D. and Bedient P.E. (1981), Elementary differential equations, sixth Edition, Macmillan.



The Wichita State University,

Wichita, Kansas 67260-0033

# Recent Developments in Complex Oscillation Theory

Steven B. Bank
Department of Mathematics, University of Illinois
Urbana, IL 61801, U.S.A.

This talk is a survey of results over the past ten years concerning the frequency of zeros in the plane of solutions of linear differential equations whose coefficients are entire functions. The research has been carried out by a variety of mathematicians from all over the world, and many of the results can now be found in the exellent book by Ilpo Laine, entitled "Nevanlinna Theory and Complex Differential Equations", published in 1993 by Walter de Gruyter and Company.

# Multisummability and Stokes phenomenon for linear meromorphic differential equations

B.L.J. Braaksma

*Department of Mathematics, University of Groningen,*
*P.O.Box 800, 9700 AV, Groningen, The Netherlands*

**Abstract**

Formal solutions of meromorphic differential equations can be summed uniquely by the method of multisummability developed by Ecalle, Martinet and Ramis. We give an outline of this method using generalized Borel transforms. We apply it to linear differential equations and indicate how it yields information on the Stokes phenomenon.

## 1 Introduction

A formal power series solution of a meromorphic differential equation with an irregular singularity is in general divergent. On sectors with sufficiently small opening one may associate a holomorphic solution with the formal solution as asymptotic expansion on that sector. However, this solution is not uniquely determined by this requirement, even not if the sector is chosen as large as possible, the so-called maximal asymptotic sectors. This is due to the occurrence of exponentially small solutions. To get unicity we need to add requirements on the solution outside these maximal asymptotic sectors. An example of such uniquely determined solutions associated with formal solutions are the normal solutions of Jurkat [Ju], where certain conditions are imposed on functions on a sequence of overlapping sectors.

This is an example of a so-called multisum of a formal power series solution. In general a formal power series is multisummable on a set of overlapping sectors if on every of these sectors functions exist which have the given series as asymptotic expansion and such that the differences of these functions are exponentially small of certain orders on the intersections of these sectors. This form of the definition of multisummability has been given by Malgrange and Ramis [Mal Ra]. It will be given in detail in section 3. This concept has its origin in the work of Ecalle [Ec] who introduced the more general notion of accelero-summability. Next Martinet and Ramis developed in [Mar Ra] the notion of multisummability. Equivalent definitions were given subsequently by Balser [Ba] and Malgrange and Ramis.

A useful method to prove multisummability for differential equations is based on the translation to convolution equations via generalized Borel transforms (another cohomological method has been used in [BBRS]). In section 2 we give the definition and some results on such transforms due to Malgrange [Mal]. In section 4 we give multisummability results for linear differential equations and in section 5 we show that the accelerates of the formal solutions of the distinguished levels contain all information concerning the Stokes phenomenon for that level. Furthermore we give a relation for the Stokes multipliers from which these multipliers can be determined. This concerns work in progress with G.K. Immink and Y. Sibuya.

# 2 Borel and Laplace transforms

## 2.1 Notation and some definitions

We denote by $\mathbf{C}_\infty$ the Riemann surface of $\log x$. Unless stated otherwise a sector will always be an open sector with vertex 0 on $\mathbf{C}_\infty$, so a set $S := \{x \in \mathbf{C}_\infty | x \neq 0, a < \arg x < b\}$ for some $a < b, a, b \in \mathbf{R}$. We also denote this sector by $S(a, b)$ and define the opening $|S|$ of this sector as $|S| := b - a$. A neighborhood of 0 in $S$ is an open subset $U$ of $S$ such that for any sector $S'$ which is relatively closed in $S$ there exists an $r > 0$ such that $\Delta(0, r) \cap S' \subset U$. Here $\Delta(0, r)$ denotes the open disc with center 0 and radius $r$.

If $S$ is a sector then we denote by $\mathcal{A}(S)$ the set of functions $f$ which are holomorphic in some neighborhood $U$ of 0 in $S$ and which have an asymptotic expansion $\sum_{j=0}^{\infty} a_j x^j$ as $x \to 0$ in $U$.

If $S$ is a sector and $k$ is a positive number, then $\mathcal{E}(S, k)$ denotes the set of holomorphic functions $f : S \to \mathbf{C}$ which belong to $\mathcal{A}(S)$ and which are of **exponential order at most $k$ at $\infty$** in $S$, i.e. for every sector $S'$ which is relatively closed in $S$ there exist positive constants $A, B$ and $R$ such that $|f(x)| \leq A \exp(B|x|^k)$ for all $x \in S'$ with $|x| > R$.

Furthermore, we denote by $\mathcal{A}^{\leq -k}(S)$ the subset of functions $f \in \mathcal{A}(S)$ with asymptotic expansion 0 and such that for every sector $S'$ which is relatively closed in $S$ there exist positive constants $A$ and $B$ with the property that

$$|f(x)| \leq A \exp(-B|x|^{-k})$$

for all $x$ in a neighborhood of 0 in $S'$. One may define sheaves $\mathcal{A}$ and $\mathcal{A}^{\leq -k}$ using the sets $\mathcal{A}(S)$ and $\mathcal{A}^{\leq -k}(S)$.

## 2.2 Definition of $k$-precise quasi-functions.

We now consider certain sets of holomorphic functions. Let $S$ be a sector and let $\{S_j\}_{j \in J}$ be an open covering of $S$, i.e. $S_j$ is a subsector of $S$ and $\cup_{j \in J} S_j = S$. Suppose that $f_j$ is holomorphic in a neighborhood of 0 in $S_j$ and $f_j - f_h \in \mathcal{A}^{\leq -k}(S_j \cap S_h)$ if

$S_j \cap S_h \neq \emptyset$. Then $\{f_j\}_{j \in J}$ is said to be a **$k$-precise quasi-function on** $S$(cf. [Ra]). Two $k$-precise quasi-functions $\{f_j\}$ and $\{g_h\}$ on a sector $S$ are said to be equivalent and will be identified if $f_j - g_h \in \mathcal{A}^{\leq -k}(S')$ if $S'$ is the intersection of the sectors on which $f_j$ and $g_h$ are defined near the origin.

The concept of $k$-precise quasi-function formalizes the notion of holomorphic function known with a precision up to functions which are exponentially small of order $k$. The set of $k$-precise quasi-functions $\{f_j\}$ on a covering $\{S_j\}$ of $S$ as above with the extra property that $f_j \in \mathcal{A}(S_j)$ is denoted by $\mathcal{A}/\mathcal{A}^{\leq -k}(S)$. They are sections of the quotient-sheaf $\mathcal{A}/\mathcal{A}^{\leq -k}$ over $S$.

## 2.3 Definition of Borel transforms

Let $S := S(a, b)$ be a sector with opening $b - a > \pi$ and let $U$ be a neighborhood of 0 in $S$. Assume that $f$ is a holomorphic function in $S$ which is $\mathcal{O}(x^p)$ as $x \to 0$ in $U$ where $p$ is some real number. Let $\gamma$ be a smooth contour in $U$ which runs from 0 to 0 and which in its starting point is tangent to $\arg x = b_1$ and in its endpoint is tangent to $\arg x = a_1$ with $a < a_1 < b_1 < b, b_1 - a_1 > \pi$. Then we define the Borel transform of $f$ by

$$\mathcal{B}f(\xi) := \frac{1}{2\pi i} \int_\gamma \exp(\xi/x) f(x) d(x^{-1}),$$

if $\xi \in S(a + \pi/2, b - \pi/2) =: S^*$. Then $\mathcal{B}f$ is holomorphic on the complete sector $S^*$ and is of exponential order at most 1 at $\infty$ in $S^*$. For example we have

$$\mathcal{B}x^\lambda(\xi) = \frac{\xi^{\lambda - 1}}{\Gamma(\lambda)}.$$

Therefore the **formal Borel transform** of a formal power series $\hat{f} := \sum_{j=1}^\infty c_j x^j$ is defined by

$$\hat{\mathcal{B}}\hat{f}(\xi) := \sum_{j=1}^\infty c_j \frac{\xi^{j-1}}{(j-1)!}.$$

If $f \in x\mathcal{A}(S)$ then there exists a formal power series $\hat{f}$ as above such that $f \sim \hat{f}$ as $x \to 0$ in $S$. Then

$$\mathcal{B}f(\xi) \sim \hat{\mathcal{B}}\hat{f}(\xi),$$

as $\xi \to 0$ in $S^*$. Hence the Borel transform maps $x\mathcal{A}(S)$ into $\mathcal{E}(S^*, 1)$. Moreover, if $f, g \in x\mathcal{A}(S)$ then $\mathcal{B}(fg) = (\mathcal{B}f) * (\mathcal{B}g)$. Note that the definition of Borel transform which we use here differs from the usual one since we start with functions in a neighborhood of 0 instead of $\infty$.

## 2.4 Definition of Laplace transforms

Let $S^* := S(a^*, b^*)$ be a sector and $\phi$ be a holomorphic function in $S^*$ which is of exponential order at most one in $S^*$ and which is $\mathcal{O}(\xi^p)$ as $\xi \to 0$ in $S^*$ where $p > -1$.

Then the Laplace transform of $\phi$ is defined by

$$\mathcal{L}\phi(x) := \int_0^{\infty(\theta)} \exp(-\xi/x)\phi(\xi)d\xi,$$

where the path of integration is in $S^*$ from 0 to $\infty\exp(i\theta)$ with $a^* < \theta < b^*$. Thus one defines a holomorphic function $f := \mathcal{L}\phi$ in a neighborhood of 0 in $S(a,b)$ where $a := a^* - \pi/2$ and $b := b^* + \pi/2$.

Moreover, if $\phi \in \mathcal{E}(S^*, 1)$ then there exists $\hat{\phi} \in \mathbf{C}[[\xi]]$, say $\hat{\phi}(\xi) = \sum_{j=0}^{\infty} \gamma_j \xi^j$ such that $\phi(\xi) \sim \hat{\phi}(\xi)$ as $\xi \to 0$ in $S^*$ and one has $f(x) \sim \sum_{j=0}^{\infty} \gamma_j j! x^{j+1}$ as $x \to 0$ in $S$. Here the righthandside defines the formal Laplace transform $\hat{\mathcal{L}}\hat{\phi}$ of $\hat{\phi}$. It follows that the Laplace transform maps $\mathcal{E}(S^*, 1)$ into $x\mathcal{A}(S)$. Here we used a modification of the usual Laplace transform similar to the modification of the Borel transform.

The following result is well known.

**Theorem 1** *Let $S = S(a,b)$ be a sector with opening $b - a > \pi$ and $S^* := S(a^*, b^*)$ where $a^* := a + \pi/2$ and $b^* := b - \pi/2$. Then the Borel transform defines an isomorphism of $x\mathcal{A}(S)$ onto $\mathcal{E}(S^*, 1)$ with the Laplace transform as its inverse.*

## 2.5 Extension to 1-precise quasi-functions

Let $f$ be a 1-precise quasi-function on a sector $S = S(a,b)$ with opening $b - a > \pi$. Let $T$ be any closed subsector $a_1 \le \arg x \le b_1$ of $S$, where $b_1 - a_1 > \pi$. Then $f|_T$ is represented by a 0-cochain $\{f_j\}_{j=1}^m$ on an open covering $\{S_j\}_{j=1}^m$ of $T$ such that $S_j$ only intersects $S_{j-1}$ and $S_{j+1}$, $S_1$ intersects the ray $\arg x = a_1$. So $f_j$ is holomorphic in a neighborhood $U_j$ of 0 in $S_j$ and $f_{j+1} - f_j \in \mathcal{A}^{\le -1}(S_{j+1} \cap S_j)$. Suppose $f_j = \mathcal{O}(x^p)$ as $x \to 0$ in $S_j$ for some real number $p$, $j = 1$ and $j = m$ .
Choose $x_j \in U_j \cap U_{j+1}$ and smooth contours $\gamma_j, j = 1, ..., m$ in $U_j$ which run from $x_{j-1}$ to $x_j$ where we put $x_0 := 0, x_m := 0$. As above we require that $\gamma_1$ is tangent to $\arg x = b_1$ in 0 and $\gamma_m$ is tangent to $\arg x = a_1$ in 0 where $a < a_1 < b_1 < b$ and $b_1 - a_1 > \pi$. Then we define

$$\mathcal{B}f(\xi) := \frac{1}{2\pi i}[\sum_{j=1}^{m} \int_{\gamma_j} \exp(\xi/x) f_j(x) d(x^{-1}) +$$

$$\sum_{j=1}^{m-1} \int_0^{x_j} \exp(\xi/x)(f_{j+1} - f_j)(x) d(x^{-1})].$$

Here the last sum is holomorphic at 0 since $f_{j+1} - f_j \in \mathcal{A}^{\le -1}$. In the first sum all terms are entire in $\xi$ except for the first and last term which are holomorphic for $\arg \xi \in (a_1 + \pi/2, b_1 - \pi/2)$. This definition of Borel transform extends the one from sect. 2.3 and it will be denoted by the same symbol. One may show that if $f \in \mathcal{A}/\mathcal{A}^{\le -1}(S)$ then $\mathcal{B}f \in \mathcal{A}(S^*)$ where $S^* := S(a + \pi/2, b - \pi/2)$.

The inverse of the Borel transform for quasi-functions appears to be the finite Laplace transform, which is defined as follows: Let $\phi$ be holomorphic in a neighborhood $U$ of 0 in some sector $S^* := S(a^*, b^*)$. Assume $\phi(\xi) = \mathcal{O}(\xi^p)$ as $\xi \to 0$ on $U$ for some $p > -1$. Choose $\xi_j \in U$ and define

$$f_j(x) := \int_0^{\xi_j} \exp(-\xi/x)\phi(\xi)d\xi.$$

Then $f_j$ is holomorphic on $S(a^* - \pi/2, b^* + \pi/2)$. Moreover, it is easy to see that $f_j - f_{j+1} \in \mathcal{A}^{\leq -1}(S_j \cap S_{j+1})$, where $S_j := S(\arg \xi_j - \pi/2, \arg \xi_j + \pi/2)$. Hence we get a 1-precise quasi-function on $S(a^* - \pi/2, b^* + \pi/2)$ which will be denoted by $\mathcal{L}\phi$.

One may extend the previous isomorphism theorem to the case of 1-precise quasi-functions as follows:

**Theorem 2** *Let $S := S(a, b)$ be a sector with opening $> \pi$ and $S^* := S(a + \pi/2, b - \pi/2)$. Then the Borel transform is an isomorphism of $x\mathcal{A}/\mathcal{A}^{\leq -1}(S)$ onto $\mathcal{A}(S^*)$ with as inverse the finite Laplace transform as defined in the previous paragraph. Moreover, if $f, g \in x\mathcal{A}/\mathcal{A}^{\leq -1}(S)$ then $\mathcal{B}(fg) = (\mathcal{B}f) * (\mathcal{B}g)$.*

## 2.6 Extension by ramification of the variables

For the treatment of multisummability we also use the above transforms with $x$ replaced by $x^k$ where $k$ is a positive number. Define:

$$\rho_k f(x) := f(x^{1/k}),$$

$$\mathcal{B}_k := \rho_k^{-1} \circ \mathcal{B} \circ \rho_k,$$

$$\mathcal{L}_k := \rho_k^{-1} \circ \mathcal{L} \circ \rho_k,$$

$$\phi *_k \psi := \rho_k^{-1}\{(\rho_k\phi) * (\rho_k\psi)\}.$$

If $\hat{f} := \sum_{h=1}^{\infty} c_h x^h$ is a formal series then one defines the formal Borel transform of order $k$ of $\hat{f}$ by

$$\hat{\mathcal{B}}_k \hat{f}(\xi) := \sum_{h=1}^{\infty} c_h \frac{\xi^{h-k}}{\Gamma(h/k)}.$$

Malgrange showed

**Theorem 3 (Isomorphism theorem)** *Let $k$ and $l$ be positive numbers with $k < l$ and let $S := S(a, b)$ be a sector with opening $> \pi/k$. Let $S^* := S(a + \pi/(2k), b - \pi/(2k))$. Define $\kappa$ by $\kappa^{-1} = k^{-1} - l^{-1}$.*

*Then the Borel transform $\mathcal{B}_k$ of order $k$ is an isomorphism from $x\mathcal{A}(S)$ onto $\mathcal{E}(S^*, k)$, from $x\mathcal{A}/\mathcal{A}^{\leq -k}(S)$ onto $\xi^{1-k}\mathcal{A}(S^*)$ and from $x\mathcal{A}/\mathcal{A}^{\leq -l}(S)$ onto $\xi^{1-k}\mathcal{E}(S^*, \kappa)$.*

*The inverse is the corresponding (finite) Laplace transform of order $k$. In the cases above one has $\mathcal{B}_k f \sim \hat{\mathcal{B}}_k \hat{f}$ as $\xi \to 0$ in $S^*$ if $f \sim \hat{f}$ as $x \to 0$ in $S$.*

# 3 Multisummability and exact asymptotics

Let $\hat{f} := \sum_{h=1}^{\infty} c_h x^h$ be a formal power series. Let $k_1, k_2, ..., k_r$ be an increasing sequence of positive numbers. Define $k_{r+1} := +\infty$. Let $S_1, S_2, ..., S_r$ be a decreasing set of sectors (so $S_j \supset S_{j+1}$) with $|S_j| > \pi/k_j$.

Assume that the formal Borel transform $\hat{\phi}_1$ of order $k_1$ of $\hat{f}$ is convergent in a neighborhood of the origin with sum $\phi_1$. So

$$\phi_1(\xi) = \sum_{h=1}^{\infty} c_h \frac{\xi^{h-k_1}}{\Gamma(h/k)}. \tag{1}$$

Let $f_0$ be its finite Laplace transform of order $k_1$ on any sector $S_0 \supset S_1$ in the sense of sect. 2.5 and 2.6.

Assume that there exist $f_j \in \mathcal{A}/\mathcal{A}^{\leq -k_{j+1}}(S_j), j = 1, ..., r$ such that $f_j|_{S_{j+1}} = f_{j+1}$ mod $\mathcal{A}^{\leq -k_{j+1}}, j = 0, ..., r-1$. Then $\hat{f}$ **is said to be** $(k_1, ..., k_r)$**-summable on** $(S_1, ..., S_r)$ **with multisum** $(f_1, ..., f_r)$ (cf. [Mal Ra]).

Then $f_j \sim \hat{f}, j = 0, ..., r$. Since $k_{r+1} = \infty$ we have $f_r \in \mathcal{A}(S_r)$, so $f_r$ is a function on $S_r$. Furthermore, at stage $j$ we have a $k_{j+1}$-precise quasifunction on $S_j$ which refines on $S_{j+1}$ to a $k_{j+2}$- precise quasifunction. One can show that if it exists **the multisum is uniquely determined** by $\hat{f}, k_j$ and $S_j, j = 1, ..., r$. So we have here a case of **exact asymptotics**. In case $r = 1$ we have ordinary $k$-summability. One may extend the definition trivially to the case that $\hat{f}$ also contains a constant term $c_0$ by adding $c_0$ to $f_j$. The numbers $k_1, ..., k_r$ are the so-called **levels** of the multisum.

By means of the isomorphism theorem for Borel transforms Malgrange gave an **equivalent form of the definition of multisummability** as follows:

Let $\hat{f}, k_j$ and $S_j$ be as before. Let $S_j^*$ and $S_j^{**}$ be sectors with the same bisector as $S_j$ and with opening which is $\pi/k_j$ and $\pi/k_{j+1}$ smaller than that of $S_j$ respectively. Let $\kappa_j$ be defined by $\kappa_j^{-1} = k_j^{-1} - k_{j+1}^{-1}$ for $j = 1, ..., r$. So $\kappa_r = k_r$. Define $\phi_1$ by (1) as before.

**Assumptions:**

1) **Assume** $\phi_1$ can be analytically continued to an element $\phi_1 \in \xi^{1-k_1}\mathcal{E}(S_1^*, \kappa_1)$. Then define $f_1 := \mathcal{L}_{k_1}\phi_1$. Hence $f_1 \in x\mathcal{A}/\mathcal{A}^{\leq -k_2}(S_1)$ by theorem 3.

2)Define for $j = 2, ..., r$ recursively $\phi_j := \mathcal{B}_{k_j} f_{j-1}$. Thus $\phi_j \in \xi^{1-k_j}\mathcal{A}(S_{j-1}^{**})$ by theorem 3. **Assume** that $\phi_j$ can be analytically continued to an element $\phi_j \in \xi^{1-k_j}\mathcal{E}(S_j^*, \kappa_j)$. Then define $f_j := \mathcal{L}_{k_j}\phi_j$. Thus by theorem 3 we have $f_j \in \mathcal{A}/\mathcal{A}^{\leq -k_{j+1}}(S_j)$.

Then $\hat{f}$ has the multisum $(f_1, ..., f_r)$ on $(S_1, ..., S_r)$. This definition is equivalent to the previous one. The function $\phi_j$ is called the $j^{th}$ accelerate of $\phi_1$. The acceleration of $\phi_j$ to $\phi_{j+1}$ by means of $\mathcal{B}_{k_{j+1}}\mathcal{L}_{k_j}$ may be obtained directly utilizing the acceleration operator of Ecalle (cf. [Mar Ra]). Let

$$C_\alpha(t) := \sum_{h=1}^{\infty} \frac{(-t)^h}{h!\Gamma(-h/\alpha)} \text{ if } \alpha > 1.$$

Then

$$\phi_{j+1}(\xi) = \xi^{-k_{j+1}} \int_0^\infty C_\alpha((t/\xi)^{k_j})\phi_j(t)d(t^{k_j})$$

where $\alpha := k_{j+1}/k_j$ and the path of integration is chosen in $S_j^*$. So after calculating the formal Borel transform of order $k_1$ of $\hat{f}$ one may determine successively all accelerates $\phi_j$ by means of the acceleration operators and from these one determines $f_j$ by Laplace transforms. By means of these acceleration operators Martinet and Ramis defined multisummability in [Mar Ra]. We will see that the accelerates $\phi_j$ contain all the information for the Stokes phenomenon of "level $k_j$".

Suppose $\hat{f}$ has $(k_1, ..., k_r)$-sum $(f_1, ..., f_r)$ on $(S_1, ..., S_r)$. Let $0 < k_1' < ... < k_p'$ and $S_1' \supset ... \supset S_p'$ be a nested set of sectors with $|S_j'| > \pi/k_j'$. Assume that for every $j \in \{1, ..., r\}$ there exists a $h \in \{1, ..., p\}$ such that $k_j = k_h', S_j = S_h'$. Then $\hat{f}$ is also $(k_1', ..., k_p')$-summable on $(S_1', ..., S_p')$ with multisum $(g_1, ..., g_p)$ where $f_j = g_h$ if $k_j = k_h'$ (cf. [Mar Ra]).

# 4 Application to linear differential equations

## 4.1 A multisummability theorem

We consider a linear meromorphic differential equation with an irregular singularity. Such a differential equation may be transformed to a normal form by a holomorphic linear transformation of the dependent variable and a translation and ramification of the independent variable. The normal form reads

$$x\frac{dy}{dx} - (\oplus_{h=0}^{r} x^{-k_h} A_h)(I + xA_+(x))y = g(x) \tag{2}$$

where $k_0 = 0 < k_1 < ... < k_r$ are integers, $A_h$ is a constant $(n_h \times n_h)$-matrix, $n_0 + ... + n_r = n$; $A_h$ is diagonal and invertible if $h > 0$; $A_+(x)$ is a $(n \times n)$-matrix holomorphic at the origin and $g(x)$ is an $n$-vector holomorphic at the origin. $I$ denotes the identity matrix.

A direction $d_h$, i.e. a half line starting at the origin, is said to be a **singular direction of level** $k_h$ for (2) if $h > 0$ and $d_h$ contains a point $\xi$ where $A_h - \xi^{k_h} I_h$ is not invertible. With such a direction we associate two **Stokes directions** $d_h \pm \pi/(2k_h)$ **of level** $k_h$ which form a **Stokes pair of level** $k_h$. Then one may show (cf. [BBRS], [Br 1])

**Theorem 4** *Let $\hat{y}$ be a formal power series solution of (2). Let $S_1 \supset S_2 \supset ... \supset S_r$ be sectors such that $|S_h| > \pi/k_h$ and $S_h$ does not contain a Stokes pair of level $k_h$.*

*Then $\hat{y}$ is $(k_1, ..., k_r)$-summable on $(S_1, ..., S_r)$ with multisum $(y_1, ..., y_r)$ where $y_r$ is a solution of (2) whereas $y_h$ is a solution mod $\mathcal{A}^{\leq -k_{h+1}}$.*

The proof may be given via Borel transforms of order $k_h$ of (2) which gives a convolution equation for $\phi_h$.These equations are of the form $\phi_h = M_h \mathcal{G}_h(\phi_h)$ where

$\mathcal{G}_h$ is a "nice" integral operator and $M_h$ is a block-diagonal matrix whose blocks are "harmless" except one which equals $(A_h - k_h\xi^{k_h}I_h)^{-1}$. The latter gives rise to singularities of $\phi_h$ and the singular directions of level $k_h$.

By a contraction argument one shows that the formal Borel transform of order $k_1$ of $\hat{y}$ is a convergent solution $\phi_1$ of the convolution equation of level $k_1$. Then one shows that the assumptions of the second definition in sect. 3 are satisfied: from the convolution equation it follows that $\phi_h$ may be continued analytically outside the singularities mentioned above and that they possess exponential estimates such that $\phi_h \in \xi^{1-k_h}\mathcal{E}(S_h^*, \kappa_h)$ (cf. [Br 1]). The transition from a solution of the equation of level $k_j$ to one of level $k_{j+1}$ is performed by acceleration: $\mathcal{B}_{k_{j+1}}\mathcal{L}_{k_j}$.

An extension to nonlinear equations has been given in [Br 2].

## 4.2 Fundamental matrices

We now consider fundamental matrices of

$$x\frac{dy}{dx} = (P(x^{-1}) + L + xR(x))y, P(x) = \oplus_{h=1}^m p_h(x)I_h, L = \oplus_{h=1}^m L_h. \tag{3}$$

Here $p_h(x) \in x\mathbf{C}[x]$, $I_h$ and $L_h$ are constant $(n_h \times n_h)$-matrices, $I_h$ is an identity matrix, $n_1 + ... + n_m = n$ and $R(x)$ is an $(n \times n)$-matrix which is holomorphic at 0.

Then (3) has a formal fundamental matrix

$$\hat{Y}(x) = \hat{H}(x)x^L \exp Q(x^{-1})$$

where

$$Q(x) := \oplus_{h=1}^m q_h(x)I_h,\ q_h(x) := -\int_0^x p_h(t)\frac{dt}{t}.$$

Furthermore, $\hat{H}(x) \in GL(n; \mathbf{C}[[x]])$ is a formal power series solution of

$$x\frac{dH}{dx} = (P(x^{-1}) + L)H - H(P(x^{-1}) + L) + xR(x)H. \tag{4}$$

Let $H_j$ denote the $(n \times n_j)$-matrix consisting of the columns $n_1 + ...n_{j-1} + 1, ..., n_1 + ... + n_j$ of $H$. Then $\hat{H}_j$ is a formal solution of

$$x\frac{dH_j}{dx} = (P(x^{-1}) - p_j(x^{-1})I)H_j + LH_j - H_jL_j + xR(x)H_j. \tag{5}$$

The leading term in the coefficient of $H_j$ involves $(p_h - p_j)(x^{-1})$ where if $p_h \neq p_j$:

$$p_h(x) - p_j(x) = k_{hj}(\rho_{hj}x)^{k_{hj}}(1 + \mathcal{O}(x^{-1})) \tag{6}$$

as $x \to \infty$ for some integer $k_{hj} > 0$ and some $\rho_{hj} \neq 0$.

Hence (5) is of the form (2) with as levels the degrees $k_{hj} > 0$, singular directions $\arg \rho_{hj}$ of level $k_{hj}$ and corresponding Stokes directions $\arg \rho_{hj} \pm \pi/(2k_{hj})$. Let $m_{1j} < ... < m_{r(j)j}$ be the numbers $k_{hj}$ arranged in increasing order with $j$ fixed. By variation of $j$ we get the levels $m_1 < ... < m_r$ of (4). The points $\rho_{hj}$ are called singular values of level $k_{hj}$. Then we have

**Theorem 5** *Let $S_{1j} \supset S_{2j} \supset \ldots \supset S_{r(j)j}$ be sectors with the property that $S_{hj}$ does not contain any pair of Stokes directions of level $m_{hj}$ and $|S_{hj}| > \pi/m_{hj}$. Then $\hat{H}_j$ is $(m_{1j}, \ldots, m_{r(j)j})$-summable on $(S_{1j}, \ldots, S_{r(j)j})$.*

*If $S_1 \supset \ldots \supset S_r$ are sectors such that $S_h$ does not contain any pair of Stokes directions of level $m_h$ and $|S_h| > \pi/m_h$ then $\hat{H}$ is $(m_1, \ldots, m_r)$-summable on $(S_1, \ldots, S_r)$. The last element $\tilde{H}$ of the multisum satisfies (4), has asymptotic expansion $\hat{H}(x)$ on $S_r$ and $\tilde{H}(x)x^L \exp Q(x^{-1})$ is a fundamental matrix of (3).*

# 5 Stokes phenomenon

Here we consider the Stokes phenomenon for fundamental matrices of differential equation (3). We will use the notation of sect. 4.2. We assume that the entries of $L$ have positive real part which is no real restriction.

Choose $j \in \{1, \ldots, m\}$, $i \in \{1, \ldots, r\}$ such that $m_i = m_{gj}$ for some $g \in \{1, \ldots, r(j)\}$. Keep $j$ and $i$ fixed. Let $d$ be a singular direction of level $m_i$ and let $d^-, d, d^+$ be three consecutive singular directions of level $m_i$. The arguments of these directions will also be denoted by $d^-, d, d^+$. Consider two multi-sectors $\mathbf{S}^\pm := (S_1^\pm, \ldots, S_r^\pm)$ satisfying the conditions of theorem 5 where $S_h^+ = S_h^- =: S_h$ if $h \neq i$, and $S_i^+ := S(d - \pi/(2m_i), d^+ + \pi/(2m_i))$ and $S_i^- := S(d^- - \pi/(2m_i), d + \pi/(2m_i))$. Here and in the following the upper (lower) signs belong together. Let $\mathbf{H}_h^\pm = (H_{1h}^\pm, \ldots, H_{rh}^\pm)$ be multisum of $\hat{H}_h$ on $\mathbf{S}^\pm$. Define

$$F_{lh}^\pm(x) := H_{lh}^\pm(x) x^{L_h} \exp q_{hj}(x^{-1}), \tag{7}$$

where $q_{hj}(x) := (q_h - q_j)(x) \sim -(\rho_{hj} x)^{k_{hj}}$ as $x \to \infty$ with $\rho_{hj}$ singular value of level $k_{hj}$ if $k_{hj} > 0$. Then $F_{rh}^\pm$ is solution of (3) with $P(x^{-1})$ replaced by $P(x^{-1}) - p_j(x^{-1})I$.

We consider the connection problem or Stokes phenomenon for $H_{rj}^\pm$, so for $F_{rj}^\pm$ near $d$. We have

$$(F_{rj}^+ - F_{rj}^-)(x) = \sum_{h \in J} F_{rh}^-(x) C_h, \tag{8}$$

where $C_h$ is a constant $(n_h \times n_j)$-matrix, so a connection or Stokes coefficient. The sum is extended over the set $J$ of indices $h$ such that $k_{hj} > 0$ and $\exp q_{hj}(x^{-1})$ is exponentially small in the neighborhood of 0 on $S_r$. Because $H_{lh}^\pm$ in (7) are elements of multisums one may determine the connection coefficients from relations for accelerates.

From the definition of multisum it follows that $H_{lh}^+ = H_{lh}^-$ if $l < i$ since then $S_l^+ = S_l^-$. Define

$$\phi_{ih} := \mathcal{B}_{m_i}\{H_{(i-1)h}^\pm(x) x^{L_h}\}. \tag{9}$$

Then $\phi_{ih}$ is holomorphic in a neighborhood of 0 in $S(d^-, d^+)$ and can be continued analytically in this sector except for singularities at the singular values $\rho_{lh} \in d$ of level $k_{lh} = m_i$ (cf. sect. 4.2).

Define

$$g_h(\xi) := \frac{1}{2\pi i}\int_0^{\xi_0} \exp(\xi/x)^{m_i} F_{rh}^-(x) d(x^{-m_i}), h \in J, \tag{10}$$

where the path of integration is along $d$, $\xi_0 \in d$ such that $F_{rh}^-$ exists on $(0, \xi_0)$ for all $h \in J$. Then $g_h(\xi)$ is holomorphic for $\Re(\xi^{m_i} - \rho_{hj}^{m_i})\xi_0^{-m_i} < 0$ if $k_{hj} = m_i$ and entire in $\xi^{m_i}$ if $k_{hj} > m_i$. We have

**Theorem 6** *Let $J_0 := \{h | k_{hj} = m_i, \rho_{hj} \in d\}$. Then the connection coefficients $C_h$ in the connection formula (8) vanish if $h \notin J_0$. The coefficients $C_h$ with $h \in J_0$ can be determined from the property that*

$$\phi_{ij}(\xi) + \sum_{h \in J_0} g_h(\xi) C_h \tag{11}$$

*can be continued analytically from a neighborhood of 0 on $d$ to the entire half-line $d$. Here $g_h$ with $h \in J_0$ has an analytic continuation on $S(d, d^+)$, which can be analytically continued across $d$ except for a singularity in $\rho_{hj}$. The same holds for $\phi_{ij}$ except that now all singular values $\rho_{hj}, h \in J_0$, are singularities.*

**Remark** This result remains valid if $\hat{H}_h, h = 1, ..., m$, is multisummable but not necessarily a solution of a differential equation. Instead of this one now assumes that (8) holds with (7) and some conditions on analytic continuation of the accelerates $\phi_{lh}$. In (11) the accelerate $\phi_{ij}$ may be calculated from $\hat{H}_j x^{L_j}$ by means of the acceleration operators. The analytic continuation of $g_h$ can be obtained using the properties of $H_{ih}$ (cf. below) and in several cases its behavior at $\rho_{hj}$ only depends on $\hat{H}_h(0)$.

We give a rough sketch of the proof of the theorem in case $i < r$.
From the definition of a multisum it follows that $H_{lj}^{\pm}|_{S_r} = H_{rj}^{\pm} \bmod \mathcal{A}^{\leq -m_{l+1}}$. Since $H_{(i-1)j}^+ = H_{(i-1)j}^-$ (see above) we have $C_h = 0$ in (8) if $k_{hj} < m_i$.

Choose $z_0$ on $d$ with $|z_0|$ so large that all singular values of level $m_i$ on $d$ are lying between 0 and $z_0$. Choose a path $\Gamma$ which runs from $z_0$ to $z_0$ in $S(d^-, d^+)$ and which goes in the negative sense around those singularities. Let

$$G(x) := \int_\Gamma \exp -(\xi/x)^{m_i} \phi_{ij}(\xi) d(\xi^{m_i}), \tag{12}$$

which is the difference of two Laplace integrals of $\phi_{ij}$. So

$$G(x) = (F_{rj}^+ - F_{rj}^-)(x) \bmod \mathcal{A}^{\leq -m_{i+1}}(S)$$

where $S$ is a sector which contains $d$ and has sufficiently small opening. Combining this with (8) we get

$$G(x) = \sum_{h \in J_1} F_{rh}^-(x) C_h \bmod \mathcal{A}^{\leq -m_{i+1}}(S), \tag{13}$$

where $J_1$ is the set of $h \in J$ with $k_{hj} = m_i$.

Analogous to (10) we define

$$g(\xi) := \frac{1}{2\pi i} \int_0^{\xi_0} \exp(\xi/x)^{m_i} G(x) d(x^{-m_i}). \tag{14}$$

Substitute the integral which defines $G$ in this formula, interchange the order of integration and vary $\Gamma$ in $S(d^-, d^+)$. Then one may show that $\alpha := g + \phi_{ij}$ is holomorphic in $\tilde{S} := S(d^-, d) \cup S(d, d^+) \cup (0, z_0)$.

From (14), (13) and (10) it follows that $\alpha_1 := g - \sum_{h \in J_1} g_h C_h$ is entire in $\xi^{m_i}$. Hence

$$\phi_{ij} + \sum_{h \in J_1} g_h C_h = \alpha - \alpha_1 \tag{15}$$

can be analytically continued in $\tilde{S}$.

From the multisum properties of $\mathbf{H}_h^-$ one may deduce that if $h \in J_1$ then $g_h(\xi)$ (cf. (10)) has a singularity in $\rho_{hj}$ and that it can be analytically continued for $\zeta := (\xi^{m_i} - \rho_{hj}^{m_i})^{1/m_i} \in S(d^-, d + 2\pi/m_i)$. If $h \in J_1 \backslash J_0$ then $\rho_{hj} \notin d$, so $\rho_{hj} \notin S(d^-, d^+)$ since $d^-, d, d^+$ are consecutive singular directions. So in this case $g_h$ is holomorphic in $S(d^-, d^+)$. Since the lefthand side of (15) is holomorphic on $\tilde{S}$ it follows that the expression (11) can be continued analytically on the entire sector $S(d^-, d^+)$.

Now suppose $h \in J_0$. Then we have direct analytic continuations $g_h^+$ and $g_h^-$ of $g_h$ on $S(d, d^+)$ and $S(d^-, d)$ respectively, and there exists a direct analytic continuation of $g_h^+$ across $d$ for $|\xi| > |\rho_{hj}|$ which will also be denoted by $g_h^+$. Then one may show that $\mathcal{B}_{m_i} F_{ih}^- = g_h^- - g_h^+$ if $\zeta \in S(d^-, d)$ and $h \in J_0$.

Let $\phi_{ih}^+$ and $\phi_{ih}^-$ denote the analytic continuations of $\phi_{ih}$ on $S(d, d^+)$ and $S(d^-, d)$ respectively. Since the expression (11) is holomorphic on $S(d^-, d^+)$ we have

$$\phi_{ij}^+ + \sum_{h \in J_0} g_h^+ C_h = \phi_{ij}^- + \sum_{h \in J_0} g_h^- C_h, \tag{16}$$

on $d$ and therefore

$$\phi_{ij}^+ - \phi_{ij}^- = \sum_{h \in J_0} \mathcal{B}_{m_i} F_{ih}^- C_h. \tag{17}$$

Applying acceleration operators to (16) with the fact that $S_l^+ = S_l^-$ if $l > i$ we may show that $C_h = 0$ if $h \in J \backslash J_0$. □

**Corollary.** *Let*

$$\gamma_{hj}(s) = \frac{1}{2\pi i} \int_d \exp[(\frac{s}{x})^{m_i} + q_{hj}(x^{-1})] d(x^{-m_i}), \tag{18}$$

*if* $Re\{(s^{m_i} - \rho_{hj}^{m_i}) x^{-m_i}\} < 0$ *for* $x \in d$. *We also denote by* $\gamma_{hj}$ *the analytic continuation of this function on* $S(d^-, d)$. *Then the Stokes coefficients* $C_h, h \in J_0$, *also may be determined from the property that*

$$\phi_{ij}(\xi) + \sum_{h \in J_0} \int_0^{s_0} \phi_{ih}(s) \gamma_{hj}((\xi^m - s^m)^{\frac{1}{m}}) d(s^m) C_h \tag{19}$$

*can be continued analytically from* $S(d^-, d)$ *to* $d$. *Here* $m := m_i, s_0 \in S(d^-, d)$ *and the path of integration is in* $S(d^-, d)$.

*If* $h \in J_0$ *and* $k_{lh} = m_i$ *implies* $\rho_{lh} \neq \rho_{hj}$, *then we may replace* $s_0$ *by* $\xi$ *in (18) (cf. [Br 1, Theorem 3]).*

This follows by replacing $F_{rh}^-$ in (10) by a finite Laplace transform of $\phi_{ih}$, which gives a new function which differs from $g_h$ by an entire function.

**Remark** The Stokes phenomenon for two multisums on multisectors which have more than one pair of different sectors can be studied similarly or by decomposition of these multisums into multisums where only one pair of different sectors occurs and using the result obtained above. Details and other approaches to the Stokes problem will be given in a forthcoming paper with G.K. Immink and Y. Sibuya.

## References


[Ba] W. Balser, *A different characterization of multi-summable power series,* Analysis **12** (1992), 57-65.

[BBRS] W. Balser, B.L.J. Braaksma, J.-P. Ramis and Y. Sibuya, *Multisummability of formal power series solutions of linear ordinary differential equations*, Asymptotic Anal. **5** (1991), 27-45.

[Br 1] B.L.J. Braaksma, *Multisummability and Stokes multipliers of linear meromorphic differential equations*, J. Diff. Eqns. **92** (1991), 45-75.

[Br 2] B.L.J. Braaksma, *Multisummability of formal power series solutions of nonlinear meromorphic differential equations*, Ann. Inst. Fourier **42** (1992), 517-540.

[Ec] J. Ecalle, *Introduction à l'accélération et à ses applications*, book submitted to Travaux en Cours (1992).

[Ju] W.B. Jurkat, *Meromorphe Differentialgleichungen*, Lecture Notes in Math., vol. 637, Springer, 1978.

[Mal] B. Malgrange, *Equations différentielles à coefficients polynomiaux*, Progress in Math., vol. 96, Birkhäuser, 1991.

[Mal Ra] B. Malgrange, J.-P. Ramis, *Fonctions multisommables*, Ann. Inst. Fourier **42** (1992), 353-368.

[Mar Ra] J.Martinet, J.-P. Ramis, *Elementary acceleration and multisummability*, Ann. Inst. H. Poincaré **54** (1991), 331-401.

[Ra] J.-P. Ramis, *Séries divergentes et théories asymptotiques*, Publ. I.R.M.A. Strasbourg, 478/P-274 (1991).

# ON A REPRESENTATION OF A FUNDAMENTAL SYSTEM OF SOLUTIONS FOR THE STURM-LIOUVILLE EQUATION AND ITS APPLICATIONS

N. CHERNYAVSKAYA
*Dept. of Math. & Comp. Sci.*
*Ben Gurion University of the Negev, P.O.B. 653*
*Beer-Sheva 84105, Israel*

and

L. SHUSTER
*Dept. of Math. & Comp. Sci.*
*Bar-Ilan University, Ramat-Gan 52900, Israel*

ABSTRACT

The following results are obtained: necessary and sufficient conditions of asymptotic equivalence of fundamental systems of solutions for Sturm-Liouville equations in the nonoscillatory case; asymptotics on the diagonal of Green's function and, as a consequence, asymptotic distribution of the spectrum for the semibounded Sturm-Liouville operator.

## 1. Introduction

In this work we find:

(I) conditions of asymptotic equivalence of fundamental systems of solutions for two Sturm-Liouville equations

$$y''(x) = q(x)y(x), \quad q(x) \in L_1^{\text{loc}}(\mathbb{R}_+), \quad x \in \mathbb{R}, \tag{1.1}$$

$$z''(x) = q_1(x)z(x), \quad q_1(x) \in L_1^{\text{loc}}(\mathbb{R}_+), \quad x \in \mathbb{R}, \tag{1.2}$$

in the case when (1.2) is a nonoscillatory equation;

(II) asymptotics on the diagonal of Green's function and asymptotic distribution of the spectrum for the semibounded Sturm-Liouville operator.

These problems were studied in numerous works (see, for example, [4,5]) and the methods used for investigating these two problems were quite different.

In this work we offer a single approach, which may be applied to the treatment of both of them. We obtain new effective results concerning problems I and II on the basis of the following facts:

(1) the representation of the basic fundamental system of solutions $\{u_1(x),\ v_1(x)\}$ of equation (1.2) (see Definition 2.1.3 below)

$$u_1(x) = \rho_1^{1/2}(x)\exp\left(-\frac{1}{2}\int_{x_1}^{x}\frac{dt}{\rho_1(t)}\right),\ v_1(x) = \rho_1^{1/2}(x)\exp\left(\frac{1}{2}\int_{x_1}^{x}\frac{dt}{\rho_1(t)}\right), \quad (1.3)$$

where $\rho_1(x) = u_1(x)v_1(x)$ , $x \geq x_1$ and $x_1$ is a certain point in $\mathbb{R}_+$;

(2) estimates [1] of the function $\rho_1(x)$

$$\frac{1}{4}q_1^*(x)^{-1/2} \leq \rho_1(x) \leq \frac{3}{2}\,q_1^*(x)^{-1/2},\ x \geq x_1, \quad (1.4)$$

which hold under the single requirement $1 \leq q_1(x) \in L_1^{\mathrm{loc}}(\mathbb{R}_+)$. Here $q_1^*(x)$ is a specific average of the function $q_1(x)$ (see (3.5-6) below).

Note that one more argument on behalf of the proposed method is that our results do not contain any requirements on differential properties of the functions $q(x)$ and $q_1(x)$ (cf. [3,5]).

## 2. Asymptotic equivalence of fundamental systems of solutions for the Sturm-Liouville equations in the nonoscillatory case

### *2.1. Preliminaries. Definition and properties of the basic fundamental system of solutions (BFSS)*

For the convenience of the reader, we cite below some definitions and results from [4].

Consider the equation

$$z''(x) = q_1(x)z(x), \quad x \in \mathbb{R}_+, \quad q_1(x) \in L_1^{\mathrm{loc}}(\mathbb{R}_+). \quad (2.1.1)$$

*Definition 2.1.1.* [4] Equation (2.1.1) is said to be nonoscillatory as $x = \infty$ if there exists an FSS $\{z_1(x),\ z_2(x)\}$ and a point $x_0 \in \mathbb{R}_+$ such that $z_1(x) > 0$, $z_2(x) > 0$ as $x \geq x_0$.

**Theorem 2.1.1.** [4] *If (2.1.1) is a nonoscillatory equation as $x = \infty$ then it has a solution $u_1(x)$ that is defined up to a constant factor by one of two equivalent conditions*

$$\lim_{x\to\infty}\frac{u_1(x)}{v_1(x)} = 0 \quad (2.1.2)$$

*or*

$$\int_{x_0}^{\infty}\frac{dt}{u_1^2(t)} = \infty\ ,\ \int_{x_0}^{\infty}\frac{dt}{v_1^2(t)} < \infty\ , \quad (2.1.3)$$

*where $v_1(x)$ is an arbitrary solution of* (2.1.1) *such that $u_1(x)$ and $v_1(x)$ are linearly independent.*

*Definition 2.1.2.* [4] The solution $u_1(x)$ defined in Theorem 2.1.1 is called the principal solution of equation (2.1.1). A solution of equation (2.1.1) which is linearly independent to $u_1(x)$ is called a nonprincipal solution.

**Theorem 2.1.2.** [4] *Let* (2.1.1) *be a nonoscillatory equation as $x = \infty$ with a principal solution $u_1(x)$, $u_1(x) > 0$ as $x \geq x_0$. Then the function*

$$v_1(x) = u_1(x) \int_{x_0}^{x} \frac{d\xi}{u_1^2(\xi)}, \quad x \geq x_0 \tag{2.1.4}$$

*is a nonprincipal solution of equation* (2.1.1), *moreover*

$$v_1'(x)\, u_1(x) - u_1'(x)\, v_1(x) = 1, \quad x \geq x_0.$$

Note that if $\{u_1(x),\ \tilde{v}_1(x)\}$ is an FSS of equation (2.1.1) normalized by the condition $\tilde{v}_1'(x)\, u_1(x) - u_1'(x)\, \tilde{v}_1(x) = 1$, $x \geq x_0$, then the solution $\tilde{v}_1(x)$ has the form $\tilde{v}_1(x) = v_1(x) + c u_1(x)$, $c = \text{const}$, and $\tilde{v}_1(x) v_1^{-1}(x) \to 1$ as $x \to \infty$. Taking into account this simple fact, we introduce the following definition.

*Definition 2.1.3.* Let $u_1(x)$ be a principal solution of equation (2.1.1), $u_1(x) > 0$ as $x \geq x_0$ and $v_1(x)$ is defined by formula (2.1.4). The FSS $\{u_1(x),\ v_1(x)\}$ is called the basic FSS (BFSS) of equation (2.1.1).

We will use the notation $\{u_1(x),\ v_1(x)\}$ below for the BFSS of equation (2.1.1). Denote

$$\rho_1(x) = u_1(x)\, v_1(x) = u_1^2(x) \int_{x_0}^{x} \frac{d\xi}{u_1^2(\xi)}. \tag{2.1.5}$$

The following statement contains an important property of the BFSS.

**Theorem 2.1.3.** [3] (see also [1]) *Let $x_1$ be the interection point of the graphs of the functions $u_1(x)$ and $v_1(x)$ (it exists by Theorem 2.1.1). For all $x \geq x_1$ the representations*

$$v_1(x) = \rho_1^{1/2}(x) \exp\left\{\frac{1}{2}\int_{x_1}^{x} \frac{dt}{\rho_1(t)}\right\}, \quad u_1(x) = \rho_1^{1/2}(x) \exp\left\{-\frac{1}{2}\int_{x_1}^{x} \frac{dt}{\rho_1(t)}\right\} \tag{2.1.6}$$

*hold.*

*2.2. Formulation of the results*

Consider the equations

$$y''(x) = q(x)y(x), \quad q(x) \in L_1^{\text{loc}}(\mathbb{R}_+), \quad x \in \mathbb{R}_+, \tag{2.2.1}$$

$$z''(x) = q_1(x)z(x), \quad q_1(x) \in L_1^{\text{loc}}(\mathbb{R}_+), \quad x \in \mathbb{R}_+. \tag{2.2.2}$$

In the following we assume that (2.2.2) is a nonoscillatory equation as $x = \infty$ with the BFSS $\{u_1(x),\ v_1(x)\}$. Various absolute constants are denoted by $c$.

We require the following definitions.

*Definition 2.2.1.* We say that equations (2.2.1) and (2.2.2) have asymptotically equivalent FSS's as $x \to \infty$ if there exists an FSS $\{u(x),\ v(x)\}$ of equation (2.2.1) for which the equalities

$$\lim_{x\to\infty} \frac{v(x)}{v_1(x)} = \lim_{x\to\infty} \frac{u(x)}{u_1(x)} = 1 \tag{2.2.3}$$

hold. We call FSS's of equations (2.2.1), (2.2.2) completely asymptotically equivalent as $x \to \infty$ if, in addition, the equalities

$$\lim_{x\to\infty} \frac{u'(x)}{u_1'(x)} = \lim_{x\to\infty} \frac{v'(x)}{v_1'(x)} = 1 \tag{2.2.4}$$

hold.

Finally, we define regular functions.

*Definition 2.2.2.* Let $q_1(x)$ be a function which satisfies the condition

$$1 \le q_1(x) \in L_1^{\text{loc}}(\mathbb{R}_+). \tag{2.2.5}$$

Denote $q_1^-(x) = d_-(x)^{-2}$, $q_1^+(x) = d_+(x)^{-2}$, where for every fixed $x \in \mathbb{R}_+$, $d_-(x)$ and $d_+(x)$ are the positive solutions of the equations

$$1 = \int_0^{\sqrt{2}d} \int_{x-t}^{x} q_1(\xi)d\xi dt, \quad 1 = \int_0^{\sqrt{2}d} \int_x^{x+t} q_1(\xi)d\xi dt,$$

respectively.

We say that $q_1(x)$ is a regular function if for $x \ge x_1$ the inequalities $c^{-1} \le q_1^-(x)(q_1^+(x))^{-1} \le c$ hold.

*Remark 2.2.1.* The functions $q_1^+(x)$ and $q_1^-(x)$ are local one-sided integral averages of the function $q_1(x)$; they were introduced in [2].

Denote by $\mathbb{C}^{\text{loc}}(\mathbb{R}_+)$ a set of continuous (not necessarily bounded) functions at $\mathbb{R}_+$.

Next we present a theorem which reduces the problem of asymptotic equivalence of FSS's for equations (2.2.1) and (2.2.2) to the study of a certain integral equation.

**Theorem 2.2.1.** *Let $q_1(x)$ be a regular function and $(q(x) - q_1(x)) \in \mathbf{C}^{loc}(\mathbf{R}_+)$. Equations (2.2.1) and (2.2.2) have asymptotically equivalent FSS's if and only if the integral equation*

$$\beta(x) = 1 + \int_x^\infty (q(t) - q_1(t))\, \rho_1(t) \left\{1 - \exp\left(-\int_x^t \frac{d\xi}{\rho_1(\xi)}\right)\right\} \beta(t)dt \qquad (2.2.6)$$

*has a solution $\beta(x)$ such that*

$$\lim_{x\to\infty} \beta(x) = 1. \qquad (2.2.7)$$

In the general case of the nonoscillatory equation (2.2.2), the corresponding result looks as follows.

**Theorem 2.2.2.** *Let $(q(x) - q_1(x)) \in \mathbf{C}^{loc}(\mathbf{R}_+)$. If there exists a solution $\beta(x)$ of equation (2.2.6) such that (2.2.7) holds then equations (2.2.1) and (2.2.2) have asymptotically equivalent FSS's; moreover,*

$$u(x) = u_1(x)\eta_1(x), \quad \eta_1(x) = \beta(x), \quad x \ge x_1, \qquad (2.2.8)$$

$$v(x) = v_1(x)\eta_2(x), \quad x \ge x_1, \quad \lim_{x\to\infty} \eta_2(x) = 1. \qquad (2.2.9)$$

*If, in addition, for all $x \ge x_1$ the inequality*

$$|\rho_1'(x)| \le 1 - \delta, \quad 0 < \delta \le 1 \qquad (2.2.10)$$

*holds, then the FSS's of equations (2.2.1) and (2.2.2) are completely asymptotically equivalent and*

$$u'(x) = u_1'(x)\eta_3(x), \quad x \ge x_1, \quad \lim_{x\to\infty} \eta_3(x) = 1, \qquad (2.2.11)$$

$$v'(x) = v_1'(x)\eta_4(x), \quad x \ge x_1, \quad \lim_{x\to\infty} \eta_4(x) = 1. \qquad (2.2.12)$$

*Furthermore, for $x \ge x_1$*

$$\eta_3(x) = \beta(x) + \varepsilon(x), \quad |\varepsilon(x)| \le c_1|\beta'(x)\rho_1(x)|, \quad \lim_{x\to\infty} \beta'(x)\rho_1(x) = 0. \qquad (2.2.13)$$

Note that under some additional conditions formulae, (2.2.9) and (2.2.12) of Theorem 2.2.2 may be revised. Namely, the following statement holds.

**Corollary 2.2.1.** *Suppose that the conditions of Theorem 2.2.2 are fulfilled and, moreover,* $\|\beta'(x)\|_{C(\mathbb{R}_+)} < \infty$, $\lim_{x\to\infty} \rho_1(x) = 0$.
*Then the functions* $\eta_2(x)$ *and* $\eta_4(x)$ *in Theorem 2.2.2 (see (2.2.9), (2.2.12)) have the form*

$$\eta_2(x) = \frac{1}{\beta(x)} + \varphi_1(x), \quad \eta_4(x) = \frac{1}{\beta(x)} + \varphi_2(x); \quad |\varphi_{1,2}(x)| \le c\rho_1(x). \qquad (2.2.14)$$

It turns out that Theorems 2.2.1 and 2.2.2 allow to obtain effective sufficient conditions of asymptotic equivalence in terms of the function $\rho_1(x)$.

**Theorem 2.2.3.** *Let (2.2.2) be a nonoscillatory equation as* $x = \infty$. *Introduce the functions*

$$A_1(x) = \int_x^\infty |q(t) - q_1(t)|\rho_1(t)dt, \quad x \ge x_1,$$
$$A_2(x) = \sup_{t\ge x}\left|\int_t^\infty (q(\xi) - q_1(\xi))\rho_1(\xi)d\xi\right|, \quad x \ge x_1,$$
$$A_3(x) = \int_x^\infty \rho_1^{-1}(t)A_2^2(t)dt, \quad x \ge x_1.$$

*If either condition*

(I) $A_1(x_1) < \infty$

*or condition*

(II) $(q(x) - q_1(x)) \in \mathbb{C}^{\text{loc}}(\mathbb{R}_+)$, $A_3(x_1) < \infty$

*is fulfilled, problem (2.2.6-7) has the solution* $\beta(x)$ *such that*

$$\beta(x) = 1 + \mu(x), \quad |\mu(x)| \le c\begin{cases} A_1(x) & \text{in case I,} \\ A_2(x) + A_3(x) & \text{in case II} \end{cases}$$

*and, therefore, equations (2.2.1) and (2.2.2) have asymptotically equivalent FSS's.*

*If, in addition, (2.2.10) holds, then the FSS's of equations (2.2.1) and (2.2.2) are completely asymptotically equivalent, relations (2.2.11-13) hold and*

$$|\beta'(x)\rho_1(x)| \le c\gamma(x), \quad \text{where} \quad \gamma(x) = \begin{cases} A_1(x) & \text{in case I,} \\ A_2(x) & \text{in case II.} \end{cases}$$

**Corollary 2.2.2.** *Assume that the conditions of Theorem 2.2.3 are fulfilled,* $\rho_1(x) \to 0$ *as* $x \to \infty$ *and, furthermore,* $\rho_1(x) \ge c\gamma(x)$ *as* $x \ge x_1$. *Then the conclusion of Corollary 2.2.1 holds.*

It is easy to obtain from Theorem 2.2.3 various generalizations of the JWKB-method. As an example, we cite below the following statement.

**Theorem 2.2.4.** *Consider equation (2.2.1). Assume that there exists a function $q_1(x)$ such that* $0 < q_1(x) \in \mathbb{C}^{\mathrm{loc}}(\mathbb{R}_+)$, $q_1''(x) \in \mathbb{C}^{\mathrm{loc}}(\mathbb{R}_+)$ *and* $\int_0^\infty q_1^{1/2}(x)\ A^2(x)\ dx < \infty$, *where*

$$A(x) = \sup_{t\geq x}\left|\int_t^\infty \left[(q(\xi) - q_1(\xi))q_1^{-1/2}(\xi) + q_1^{-1/4}(\xi)\left(q_1^{-1/4}(\xi)\right)''\right] d\xi\right|.$$

*Then equation (2.2.1) has an FSS of the form (2.2.8-9). where*

$$u_1(x) = q_1^{-1/4}(x)\exp\left(-\int_{x_0}^x q_1^{1/2}(t)\ dt\right), \quad v_1(x) = q_1^{-1/4}(x)\exp\left(\int_{x_0}^x q_1^{1/2}(t)\ dt\right),$$

$$|\beta(x) - 1| \leq c\left(\int_x^\infty q_1^{1/2}(t)A^2(t)dt + A(x)\right).$$

*If, in addition, the inequality* $\frac{1}{4}q_1'(x)q_1^{-3/2}(x) \leq 1 - \delta$, $\delta \in (0, 1]$ *is fulfilled, then formulae (2.2.11-12) hold and* $|\eta_3(x) - \beta(x)| \leq cA(x)$. *Finally, if* $A(x) \leq cq_1^{-1/2}(x)$ *and* $q_1(x) \to \infty$ *as* $x \to \infty$ *then (2.2.14) holds with* $\rho_1(x) = q_1^{-1/2}(x)$.

In conclusion, we emphasize that the proofs of all the above-mentioned results are based only on representation (2.1.6) of the BFSS of equation (2.2.2).

## 3. Asymptotics on the diagonal of Green's function and asymptotic distribution of the spectrum for the Sturm-Liouville operator

Consider the boundary value problem

$$-y''(x) + q(x)y(x) = f(x), \qquad \lim_{|x|\to\infty} y(x) = 0, \tag{3.1}$$

where $x \in \mathbb{R}$, $f(x) \in L_2(\mathbb{R})$ and

$$1 \leq q(x) \in L_1^{\mathrm{loc}}(\mathbb{R}). \tag{3.2}$$

It is well-known [5] that problem (3.1) has a unique solution. By the Sturm-Liouville operator $L$ we mean the differential expression $\ell = -d^2/dx^2 + q(x)$ defined on the set of solutions of the boundary value problem (3.1) for all $f(x) \in L_2(\mathbb{R})$. Denote by $G(x,t)$ Green's function of problem (3.1). We also call $G(x,t)$ "Green's function of the Sturm-Liouville operator $L$."

It was proved in [3] (see also [1]) that the following representation for Green's function $G(x,t)$ holds for all $x, t \in \mathbb{R}$

$$G(x,t) = \rho^{1/2}(x)\rho^{1/2}(t)\exp\left\{-\frac{1}{2}\left|\int_x^t \frac{d\xi}{\rho(\xi)}\right|\right\}; \quad \rho(x) \equiv G(x,t)|_{t=x}. \tag{3.3}$$

Moreover, we showed in [1] that for the diagonal function $\rho(x)$ the inequalities

$$\frac{1}{4}q^*(x)^{-1/2} \le \rho(x) \le \frac{3}{2}q^*(x)^{-1/2}, \quad x \in \mathbb{R} \tag{3.4}$$

hold. Here

$$q^*(x) = d^{-2}(x) \tag{3.5}$$

and for a fixed $x \in \mathbb{R}$, $d(x)$ is the positive solution of the equation

$$\frac{1}{d} = \frac{1}{2}\int_{x-d}^{x+d} q(t)dt. \text{ [1]} \tag{3.6}$$

In this section we present a result on asymptotics of the diagonal function $\rho(x)$ as $|x| \to \infty$. On the basis of the asymptotics of $\rho(x)$ as $|x| \to \infty$ we find the asymptotic distribution of the spectrum for the Sturm-Liouville operator. It is interesting to mention that in the proofs we use only representation (3.3) and estimates (3.4).

*Definition 3.1.* Suppose that condition (3.2) is fulfilled. We say that the function $q(x)$ belongs to the class $H$ $(q(x) \in H)$ if there exists a positive continuous function $K(x)$ such that

(1) $K(x) \to \infty$ as $|x| \to \infty$,

(2) $c^{-1} \le K(t)K^{-1}(x) \le c$ as $|x - t| \le K(x)q^*(x)^{-1/2}$, $x \in \mathbb{R}$,

(3) $K(x)q^*(x)^{-1/2} \sup\limits_{z \in w(x)} \left|\int_0^z (q(x+t) - q(x-t))dt\right| \le c < \infty$, $x \in \mathbb{R}$.

Here $w(x) = [0, K(x)q^*(x)^{-1/2}]$.

Before we turn to the formulation of Theorem 3.1 let us adopt the following notation:

$$F(x) = K^{1/2}(x)q^*(x)^{-1/2} \sup_{z \in \tilde{w}(x)} \left|\int_0^z (q(x+t) - q(x-t))dt\right|,$$

$$\tilde{w}(x) = [0, K^{1/2}(x)q^*(x)^{-1/2}];$$

$$\alpha(x) = \begin{cases} \sup_{t \ge x} F(t) + \exp(-cK^{1/2}(x)) & \text{if } x \ge 0, \\ \sup_{t \le x} F(t) + \exp(-cK^{1/2}(x)) & \text{if } x \le 0. \end{cases}$$

---

[1] $q^*(x)$ was originally introduced by M. Otelbaev (see [6]); $q^*(x)$ is an average of the function $q(x)$ of the Steklov type (see [1]).

**Theorem 3.1.** *If $q(x) \in H$ then for all $x \in \mathbb{R}$ the relations*

$$\rho(x) = \frac{1+\varepsilon_1(x)}{2q^*(x)^{1/2}}, \quad |\varepsilon_1(x)| \le c\alpha(x); \ |\rho'(x)| \le c\alpha(x)$$

*hold.*

Next we establish asymptotics of the function $d(x) = q^*(x)^{-1/2}$.

**Theorem 3.2.** *Suppose that condition (3.2) is fulfilled and that the function $q(x)$ may be expressed in the form $q(x) = q_1(x) + q_2(x)$, where $1 \le q_1(x) \in \mathbb{C}^{\mathrm{loc}}(\mathbb{R})$ and $q_2(x) \in L_1^{\mathrm{loc}}(\mathbb{R})$.*

*If the conditions*

$$h_1(x) \to 0\,, \quad h_2(x) \to 0 \ \text{ as } \ |x| \to \infty$$

*hold, then*

$$d(x) = \frac{1+\varepsilon_2(x)}{q_1^{1/2}(x)}\,, \qquad |\varepsilon_2(x)| \le c[h_1(x) + h_2(x)].$$

Here

$$h_1(x) = \frac{1}{2}q_1^{-1/2}(x) \sup_{t\in[0,2q_1^{-1/2}(x)]} \left| \int_0^t (q_1(x+s) - 2q_1(x) + q_1(x-s))ds \right|,$$

$$h_2(x) = \frac{1}{2}q_1^{-1/2}(x) \sup_{t\in[0,2q_1^{-1/2}(x)]} \left| \int_{x-t}^{x+t} q_2(s)ds \right|.$$

Let $\lambda \ge 0$. Consider the problem

$$(L+\lambda E)y = -y''(x) + (q(x)+\lambda)y(x) = f(x); \qquad \lim_{|x|\to\infty} y(x) = 0. \tag{3.7}$$

Denote by $\rho_\lambda(x)$ the diagonal value of Green's function $G(x,t,\lambda)$ of problem (3.7), $\rho_\lambda(x) = G(x,t,\lambda)\,\big|_{t=x}$, $x \in \mathbb{R}$. As a rule, one can take $K(x) = (q_1(x)+\lambda)^\alpha$, $0 < \alpha < 1/2$ in Theorems 3.1 and 3.2. In this case, for all $\lambda \gg 1$, the relations

$$\rho_\lambda(x) = \frac{1+O(\lambda^{-\varepsilon})}{2(q_1(x)+\lambda)^{1/2}}, \quad |\rho'_\lambda(x)| \le \frac{c}{\lambda^\varepsilon}, \quad \varepsilon > 0 \tag{3.8}$$

hold uniformly with respect to $x \in \mathbb{R}$. As a consequence, we obtain the following result.

**Theorem 3.3.** *Suppose that condition (3.2) is fulfilled. Assume that for the function $q(x)$ there exists a function $q_1(x)$ such that $1 \le q_1(x) \in \mathbb{C}^{\mathrm{loc}}(\mathbb{R})$, $q_1^{-3/2}(x) \in L_1(\mathbb{R})$ and for all $\lambda \gg 1$ the relations (3.8) hold uniformly with respect to $x \in \mathbb{R}$. Then the operator $L$ has a discrete spectrum. Denote $N(\lambda) = \sum_{\lambda_k < \lambda} 1$, where $\lambda_k$, $k = 1, 2, \ldots$ are eigenvalues of the operator $L$. If the function*

$$\psi(\lambda) = \int_{q_1(x) \le \lambda} (\lambda - q_1(x))^{1/2} dx$$

*satisfies the inequalities* $c^{-1}\psi(\lambda) \le \psi'(\lambda) \le c\psi(\lambda), \quad \lambda \gg 1,$ *then*

$$N(\lambda) = (1 + o(1))\frac{1}{\pi}\psi(\lambda).$$

*Example.* In the case $q(x) = 1 + \exp(|x|) + \exp(|x|)\sin(\exp(\gamma|x|))$, $\gamma > 1/2$, consider problem (3.1). Then the choice $q_1(x) = 1 + \exp(|x|)$, $K(x) = (q_1(x) + \lambda)^{\delta}$, $\delta = \min\{1/6, \gamma - 1/2\}$ in Theorems 3.1 and 3.2 leads to the formulae (3.8) with $\varepsilon = \delta$, and we obtain $N(\lambda) = \frac{4}{\pi}\sqrt{\lambda}\ln\lambda(1 + o(1))$.

## References


1. N. Chernyavskaya and L. Shuster, *Estimates for Green's function of the Sturm-Liouville operator*, J. Diff. Eq. (to appear).
2. N. Chernyavskaya and L. Shuster, *WKB-approximations from the perturbation theory viewpoint*, Operator Theory: Advances and Applications, Birkhäuser Verlag, Basel, Vol. **46**, 1990, pp. 119-123.
3. E.B. Davies and E.M. Harrell, *Conformally flat Riemannian metrics, Schrödinger operators and semiclassical approximation*, J. Diff. Eq. **66, 2** (1987), 165-188.
4. P. Hartman, *Ordinary Differential Equations*, Wiley, New York, 1964.
5. B.M. Levitan and J.S. Sargsjan, *Introduction to Spectral Theory: Selfadjoint Ordinary Differential Operators*, Transl. Math. Monographs, vol. **39**, Amer. Math. Soc., Providence, RI, 1975.
6. K. Mynbaev and M. Otelbaev, *Weighted Functional Spaces and Differential Operator Spectrum*, Nauka, Moskva, 1988.

# THE SINGULARITY DEFERRED APPROACH TO THE LIMIT

ELISE DE DONCKER
*Department of Computer Science,*
*Western Michigan University,*
*Kalamazoo, MI 49008*
elise@cs.wmich.edu

ABSTRACT

We consider the asymptotic behavior as the step size $h \to 0$, of numerical solutions of second order differential equations. We focus, in particular, on singular cases where the error is not $\mathcal{O}(h^2)$, and the standard deferred approach to the limit fails to work. The validity and effectivity of other potential extrapolation schemes is assessed, for the successful elimination of error terms from the expansion. These include more general linear extrapolations, Brezinski's E-algorithm, Levin and Sidi's $d$-transformation and the $\varepsilon$-algorithm (Wynn). In particular, the $d$-transformation handles terms in $\mathcal{O}(h^\alpha \log h)$, where $\alpha$ is not necessarily integer and a non-geometric progression of the step size is used.

## 1. Asymptotic Error

In this section we will review the error functional expansion satisfied by

$$p(x)y'' + q(x)y' = f(x, y) \tag{1}$$

for the simple finite difference approximation

$$h^{-2} p_n \delta^2 y_n + h^{-1} q_n \mu \delta y_n = f(x_n, y_n), \tag{2}$$

as given by Mayers [11]. His analysis disproves the validity of $h^2$-extrapolation for some types of problems. In the next section, we will explore possible alternative extrapolation techniques, valid under more general circumstances. We will report results of some numerical experiments in Section 3.

Assume for now that the error can be expressed in the form of an expansion in odd and even powers of the stepsize $h$,

$$Y(x) - y(x,h) = A_1(x)h + A_2(x)h^2 + A_3(x)h^3 + \cdots, \quad \text{as } h \to 0, \tag{3}$$

and that furthermore the difference operators can be replaced in Eq. (2) by their series expansions of the form

$$\begin{aligned}\delta^2 &= h^2(D^2 + \frac{1}{12}h^2D^4 + \frac{1}{360}h^4D^6 + \cdots)\\ \mu\delta &= h(D + \frac{1}{6}h^2D^3 + \frac{1}{120}h^4D^5 + \cdots).\end{aligned} \tag{4}$$

Equating terms in equal powers of $h$ then yields a system of differential equations, the $j+1$-th of which defines $A_j$ in terms of $A_0 = Y$ for $j = 1$ and in terms of $A_0, A_1, \cdots, A_{j-1}$ for $j \geq 2$ (and the first of which states Eq. (1) for $y = Y$).

Solutions for the $A_j$ depend on the boundary conditions. For example,

$$\begin{aligned} y(a,h) &= Y(a) = \alpha \\ y(b,h) &= Y(b) = \beta \end{aligned} \tag{5}$$

generally lead to $A_{2i+1} = 0$, $i \geq 0$, so that the error satisfies an expansion in even powers of $h$. The same result is found if the second boundary condition in Eq. (5) is replaced by $Y'(a) = \gamma$ and the derivative is approximated as $h^{-1}\mu\delta y(a,h)$. However, if instead the forward difference $h^{-1}\Delta y(a,h)$ is used, the odd power terms in Eq. (3) are also present. Likewise, replacing the second boundary condition in Eq. (5) by $y(a+h,h) = Y(a+h)$ obtained via a series expansion, yields that only $A_1$ necessarily vanishes. Evidently, in these cases, a version of the Neville algorithm can be applied to eliminate terms from the error expansion.

This process breaks down in singular cases as the ones given by Fox [5] and Mayers [11], and leads to differential equations for the coefficients $A_j$ of Eq. (3) which are found to have no solutions (satisfying the boundary conditions). Eq. (3) is then replaced by

$$Y(x) - y(x,h) = \psi(x,h). \tag{6}$$

By expressing $y_n$ according to (6) in the difference equation Eq. (2), where the difference operators are expanded using Eq. (4), a second order equation is produced for the error $\psi(x,h)$,

$$p\psi'' + q\psi' = \frac{1}{12}h^2pY^{iv} + \frac{1}{6}h^2qY''' + \psi\frac{\partial f}{\partial Y}, \tag{7}$$

where terms of orders higher than $h^2$ are neglected. Note that the values $x$ where Eq. (7) is valid may be restricted by requirements involving the existence of the derivatives on the right hand side and the convergence of the series expansions (4) for the difference operators.

Provided with the necessary boundary conditions, Eq. (7) can be solved to determine the dominant error behavior. In Section 3, we will test the efficiency of alternative extrapolation methods for an example with error behavior of $\mathcal{O}(h^2 \log h)$.

## 2. Numerical Extrapolation

Extrapolation processes are generally used to accelerate sequence convergence. The process can be viewed as successive sequence-to-sequence transformations where, under certain conditions, a newly generated sequence converges faster than its predecessor. Considering a sequence of quantities $T(h)$ depending on a decreasing stepsize $h$, such that $S = \lim_{h\to 0} T(h)$, extrapolation may yield an efficient means of approximating the unknown quantity $S$. The transformations may be linear or non-linear. Inspired by their success in numerical quadrature, it is appropriate to present them as they effect the elimination of terms from an Euler-Maclaurin type expansion for the error of $T(h)$.

For example, Romberg integration is based on a sequence of trapezoidal rule sums $T(h)$, for a geometric progression with ratio $\frac{1}{2}$ of the stepsize. The extrapolated results are organized in a triangular table, with as its columns the above-mentioned transformed sequences. Romberg integration is successful on the condition that the approximation error of $T(h)$ has an Euler Maclaurin asymptotic expansion in even powers of $h$, i.e.,

$$T(h) \sim S + \sum_{k>0} C_{2k} h^{2k}, \quad \text{as } h \to 0, \tag{8}$$

where the $C_{2k}$ are independent of $h$. It may be noted that the table elements are linear combinations of the $T(h)$, and the combinations are such that terms of Eq. (8) in successive powers of $h$ are eliminated in successive columns, making the error of the $\ell$-th column of $\mathcal{O}(h^{2\ell+2})$, $\ell = 0, 1, \cdots$. As $C_{2k}$ involves an integral of the $2k$-th derivative of the integrand, the asymptotic expansion in Eq. (8) may have to be replaced by a finite sum with a remainder term of $\mathcal{O}(h^p)$, if the derivatives of orders $\leq p$ are known to be integrable. In this case the extrapolation is known to be valid up to that point.

In [3, 2], a survey on the treatment of certain quadrature expansions by extrapolation is given, and the $d$-transformation [9] is validated for an important class of problems. We will put this information to use for the numerical solution of differential equations yielding a similar error expansion of non-standard form.

If the basic rules are not symmetric, then generally also terms in odd powers of $h$ appear in the expansion, and the methods classified as linear extrapolation provide the table elements by implicitly or explicitly solving a linear system of order $\ell + 1$ for the approximate $C_k$, $k = 0, 1, \ldots, \ell$ with $C_0 \approx S$. In its more general form, linear extrapolation can also be applied when the error satisfies an expansion in terms involving more general functions of $h$, as long as these functions are known. The expansion Eq. (8) is then replaced by

$$T(h) \sim S + \sum_{k>0} C_k \varphi_k(h), \qquad \text{as } h \to 0, \tag{9}$$

where

$$\lim_{h \to 0} \frac{\varphi_{i+1}(h)}{\varphi_i(h)} = 0.$$

Denoting the table elements by $T_j^\ell$, with $T_0^i = T(h_i)$, the systems solved are

$$T_0^i = T_\ell^j + \sum_{k=1}^{\ell} \tilde{C}_k \varphi_k(h_i), \qquad i = j, j+1, \cdots, j+\ell. \tag{10}$$

Note that the $E$-algorithm [8, 1] is a recursive method for calculating the $T_\ell^j$ elements satisfying Eq. (10).

Truncating Eq. (9) and replacing the constants $C_k$ by functions $\beta_k(h)$ which are assumed to satisfy an expansion of the form

$$\beta_k(h) \sim h^{\alpha_k}(\beta_{k0} + \beta_{k1} h + \beta_{k2} h^2 + \cdots)$$

yields

$$T(h) \sim S + \sum_{k=0}^{m-1} \beta_k(h)\varphi_k(h). \tag{11}$$

The $W^{(m)}$ algorithm [12, 4] then calculates $A_N^j = T_{n,m}^j \approx S$ determined by

$$T(h_i) = T_{n,m}^j + \sum_{k=0}^{m-1} \varphi_k(h_i) \sum_{\ell=0}^{n_k} \tilde{\beta}_{k\ell} {h_i}^{\ell+\alpha_k}, \quad i = j, j+1, \ldots, j+N, \tag{12}$$

with $N = \sum_{k=0}^{m-1}(n_k+1)$, $n = (n_0, n_1, \ldots, n_{m-1})$ and $h_j > h_{j+1} > \cdots > h_{j+N}$.

Letting

$$\varphi_k(h)h^{\alpha_k} = \phi_{k+1}(h), \qquad k = 0, 1, \ldots, m-1,$$

the application of the $W^{(m)}$ algorithm is justified [3] for

$$T(h) \sim S + \sum_{s>0}(A_s + C_s \log h)h^{\beta+s} + \sum_{k=1}^{M}\sum_{s>0} E_{ks}h^{\gamma_k+s}, \tag{13}$$

with

$$\begin{aligned} \phi_1(h) &= h^{\beta+1} \\ \phi_2(h) &= h^{\beta+1}\log h \\ \phi_{k+2}(h) &= h^{\gamma_k+1}, \quad k = 1, 2, \ldots, M \end{aligned}$$

and $m = M+2$. It is assumed that the values of $\beta$ and $\gamma_k$ are such that $m$ corresponds to the number of distinct subseries in Eq. (13).

The $d$-transformation [9] alleviates the specification of the functions $\phi_k(h)$ under conditions given in [9, 3], by replacing

$$\phi_k(h_i) = \Delta^k u_i {h_i}^{\alpha_k}, \qquad \text{as } h_i \to 0, \tag{14}$$

where

$$u_i = \Delta T(h_i) = T(h_{i+1}) - T(h_i). \tag{15}$$

Principally, the $u_i$ are required to satisfy a linear difference equation of order $m$,

$$u_i = \sum_{k=1}^{m} \Delta^k u_i q_k(h_i), \qquad \text{as } h_i \to 0, \tag{16}$$

where $q_k(h) \in \mathcal{B}^{(\alpha_k)}$ and $\mathcal{B}^{(\kappa)}$ is the set of functions

$$p(x) \sim x^{\kappa}(p_0 + p_1 x + p_2 x^2 + \cdots) \quad \text{as } x \to 0. \tag{17}$$

Note that by replacing $\phi_k(h)$ as given by Eq. (14) in Eq. (12) and using (15-17), the double summation on the right of Eq. (10) reduces to $u_i$, which is a measure of the error in $T(h_i)$.

Shanks' transformation for the sequence $T(h_i), i = 0, 1, \ldots$, is calculated recursively by the $\varepsilon$-algorithm [13] and allows the approximation of $S$ in

$$T(h_i) \sim S + \sum_{\ell=1}^{k} \beta_\ell(i)\zeta_\ell^i \tag{18}$$

where $\beta_\ell(i)$ is a polynomial of degree $d_\ell$ in i. The $\varepsilon$-algorithm calculates the entries $\varepsilon_p^{(j)}$ according to

$$\begin{aligned} \varepsilon_{-1}^{(i)} &= 0, \\ \varepsilon_0^{(i)} &= T(h_i), \\ \varepsilon_p^{(j)} &= \varepsilon_{p-2}^{(j+1)} + \frac{1}{\varepsilon_{p-1}^{(j+1)} - \varepsilon_{p-1}^{(j)}}. \end{aligned}$$

If Eq. (18) is an equality, then [6]

$$\varepsilon_{2k'}^{(0)} = S,$$

where

$$k' = d_1 + d_2 + \cdots + d_k + k.$$

As an example, for

$$T(h) = S + Ah^2 + Ch^2 \log_2 h,$$

let $h = 2^{-i}, i = 0, 1, \ldots$. Then $T(2^{-i}) = S + 4^{-i}A - 4^{-i}Bi$, which is of the form of Eq. (18).

## 3. Numerical experiments

For the example [5, 11]

$$xy'' - y' - x = 0, \qquad y(0) = y(1) = 0,$$

satisfied by

$$Y(x) = \frac{1}{2}x^2 logx,$$

the error function $\psi(x,h)$ given by Eq. (7) is

$$x\psi'' - \psi' = -\frac{h^2}{4x}, \qquad h \leq x \leq 1. \tag{19}$$

Note that the solution function $Y(x)$ has singular derivatives of orders $\geq 2$ at $x = 0$ and the series expansions of the difference operators given by Eq. (4) do not converge for $0 < x < h$. The general solution of Eq. (19) is

$$\psi(x,h) = -\frac{h^2}{8}\log x + px^2 + q.$$

Since $\psi(1,h) = 0$, it follows that $p = -q$. The difference equation Eq. (2) at $x = h$,

$$p(h)[y(2h) - 2y(h) + y(0)] + hq(h)[y(2h) - y(0)] = h^2 f(h, y(h)), \tag{20}$$

produces a second boundary condition,

$$2\psi(h) - \frac{1}{2}\psi(2h) = h^2(1 - \log 2).$$

This leads to $p = -\frac{h^2}{8}\log h + h^2(\frac{17}{24}\log 2 - \frac{2}{3})$, so that

$$\psi(x,h) \sim -\frac{h^2}{8}[(x^2 - 1)\log h + \log x].$$

Note that the error is of $\mathcal{O}(h^2 \log h)$, thus invalidating $h^2$-extrapolation.

Solving the difference equation Eq. (2) for this problem yields, for mesh size $h = \frac{1}{r}$,

$$y_n = \alpha n^2 + \frac{1}{4}h^2(4n^2 - 1)\sigma_n,$$

where $\alpha = -(4r^2 - 1)\frac{\sigma_r}{4r^4}$ and $\sigma_\ell = \sum_{j=1}^{\ell}\frac{1}{2j-1}$, or

$$y_n = \frac{1}{4}(\sigma_r n^2 h^2 - \sigma_n)h^2 - (\sigma_r - \sigma_n)n^2 h^2.$$

For example, with $r = 2n$ this gives

$$y_n = \frac{1}{4}(\frac{1}{4}\sigma_{2n} - \sigma_n)h^2 - \frac{1}{4}(\sigma_{2n} - \sigma_n).$$

*The Singularity Deferred Approach to the Limit*

Use of [7],

$$\sigma_\ell = \frac{1}{2}(C + \log \ell) + \log 2 + \frac{B_2}{8\ell^2} + \frac{(2^3 - 1)B_4}{64\ell^4} + \ldots,$$

$C$ representing Euler's constant and $B_k$ the $k$-th Bernoulli number, leads to an expansion of the form

$$Y(x_n) - y_n = Ch^2 \log h + \sum_{k>0} C_{2k} h^{2k}.$$

In Figure 1, we show the results of direct extrapolation at $x = \frac{1}{2}$, by supplying the $h$-dependencies of successive terms and solving a linear system. The bullets indicate the results obtained with a harmonic mesh sequence, the triangles those obtained with a geometric sequence.

Figure 2 gives results of the $d$-transformation, the stars for $m = 2$ and a geometric mesh sequence, and the other symbols for a harmonic mesh sequence and various values of $m$, apart from the circles which correspond to the values of the original (input) sequence. The actual number of subseries in the expansion is $m = 2$, corresponding to the fact that the $u_i$ satisfy a linear difference equation of order $m \geq 2$. For $m = 2$ and the harmonic sequence, high accuracy is reached early on, but there is a loss of accuracy due to roundoff at later extrapolation stages. With the larger values of $m$, higher accuracies are obtained at a later stage. The calculations were performed in double precision, using the program from [9], which was modified for our examples. The values plotted are the outputs $APPROX(J, P)$ from the program, for $J = 0$ and for the mesh sequences listed in Figure 2.

## 4. Conclusions

Methods such as the $E$-algorithm and the $W^{(m)}$ algorithm can be used effectively if the nature of a number of terms in the error expansion is known or can be guessed. However, determining the structure of the part of the expansion needed may be too lengthy or impossible, even if one would resort to symbolic manipulation.

The $\epsilon$-algorithm does not require this type of knowledge. For the experiments described in the previous section, it would be effective on the condition that a geometric

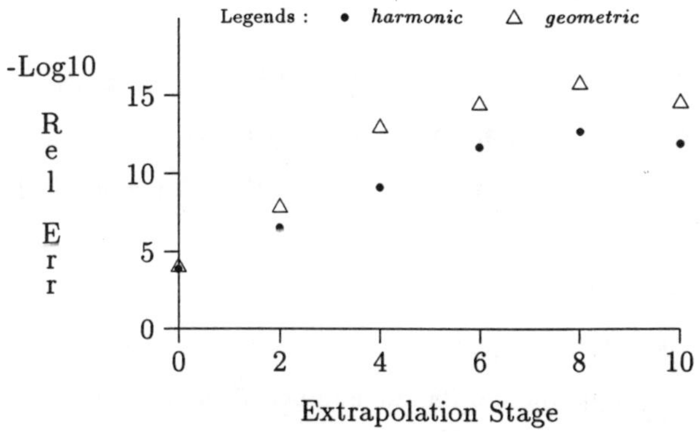


Figure 1: Direct extrapolation using harmonic mesh sequence $r = 2, 4, 6, \ldots$ vs. geometric mesh sequence $r = 2, 4, 16, \ldots$.

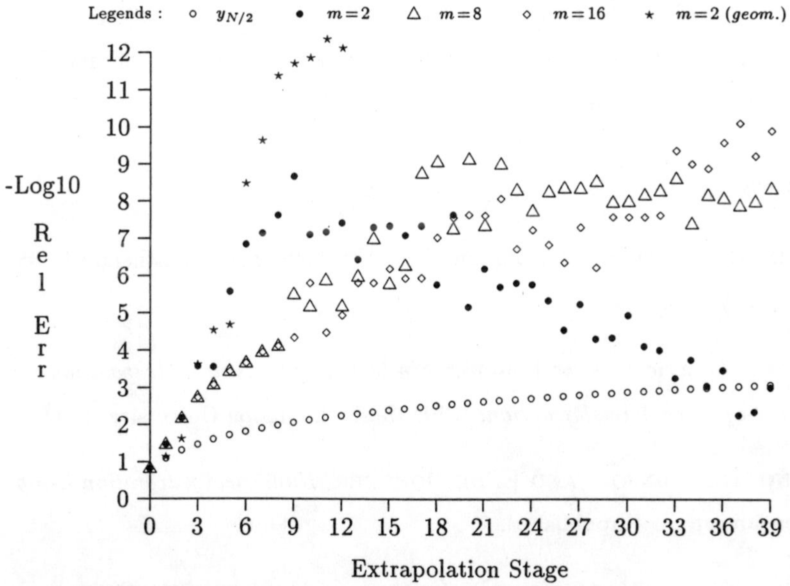


Figure 2: d-transformation using $m = 2, 8, 16$, and mesh sequence $r = 4, 6, 8, \ldots$, or $m = 2$ and mesh sequence $r = 4, 8, 16, \ldots$.

progression is employed for the mesh size. Other sequences allow evaluating the $T(h)$ more economically, with for example a linear increase in the number of points for harmonic sequences, versus an exponential increase for geometric sequences. Note that the use of more slowly decreasing progressions of $h$ as they affect the stability of the method was studied in the framework of linear extrapolation for numerical quadrature in [10].

It is our hope that the $d$-transformation could be used as a black-box method for a range of problems in ODEs, as it does not require explicit knowledge of the type of terms in the error expansion, and an upperbound for the number of subseries can be used. Furthermore it is also not restricted to geometric grid size progressions.

Work is needed to establish a basis for the above, by specifying problem characteristics which give rise to certain terms or a specific structure of the error. The validity of the $W^{(m)}$ algorithm and $d$-transformation should then be verified for that structure, if it is not covered by the results given for example by Carino [3, 2]. Another topic for investigation is the combined use of higher order ODE numerical methods with a suitable extrapolation technique.

# References

[1] C. Brezinski, *A general extrapolation algorithm*, Numerische Mathematik, 35 (1980), pp. 175–187.

[2] R. Cariño, *Numerical Integration over Finite Regions, using Extrapolation by Non-linear Sequence Transformations*, PhD thesis, La Trobe University, 1992.

[3] R. Cariño, I. Robinson, and E. de Doncker, *Numerical integration using the d-transformation.* Unpublished.

[4] W. Ford and A. Sidi, *An algorithm for the generalization of the Richardson extrapolation process*, SIAM Journal on Numerical Analysis, 24 (1987), pp. 1212–1232.

[5] L. FOX, *Numerical Solution of Ordinary and Partial Differential Equations*, Pergamon Press, Oxford, 1962.

[6] A. GENZ, *The Approximate Calculation of Multidimensional Integrals Using Extrapolation Methods*, PhD thesis, University of Kent at Canterbury, 1975.

[7] I. S. GRADSHTEYN AND I. M. RYZHIK, *Table of Integrals, Series, and Products*, Academic Press, 1980.

[8] T. HÄVIE, *Generalized Neville-type extrapolation schemes*, BIT, 19 (1979), pp. 204–213.

[9] D. LEVIN AND A. SIDI, *Two classes of non-linear transformations for accelerating the convergence of infinite integrals and series*, Appl. Math. Comp., 9 (1981), pp. 175–215.

[10] J. N. LYNESS, *Applications of extrapolation techniques to multidimensional quadrature of some integrand functions with a singularity*, Journal of Computational Physics, 20 (1976), pp. 346–364.

[11] D. F. MAYERS, *The deferred approach to the limit in ordinary differential equations*, Computer Journal, 7 (1964), pp. 54–57.

[12] A. SIDI, *Some properties of a generalization of the Richardson extrapolation process*, JIMA, 24 (1979), pp. 327–346.

[13] P. WYNN, *On a device for computing the $e_m(s_n)$ transformation*, Mathematical Tables and Aids to Computing, 10 (1956), pp. 91–96.

# MATHIEU FUNCTIONS WITH LARGE PARAMETERS: THE EIGENVALUE PROBLEM

T. M. DUNSTER
*Department of Mathematical Sciences, San Diego State University, San Diego, CA 92182-0314, U.S.A.*

ABSTRACT

Uniform asymptotic solutions of Mathieu's equation

$$d^2w/dz^2 = \{2q\cos(2z) - a\}w,$$

are derived for the case $q \to \infty$. These are uniformly valid for $a$ lying in the interval $-2q \le a \le (2-d)q$ $(d > 0)$, for complex values of $z$. The solutions involve both elementary functions (Liouville-Green) and a Whittaker function. The latter solution is derived by an application of a recent asymptotic theory of a coalescing turning point and simple pole in the complex plane. The new asymptotic approximations are then used in a connection formula to derive a uniform asymptotic approximation between the characteristic exponent $\nu$ and the parameters $a$ and $q$. Error bounds are available for all approximations.

## 1. Introduction

In this paper we shall employ uniform asymptotic solutions of Mathieu's equation

$$\frac{d^2w}{dz^2} = \{2q\cos(2z) - a\}w, \tag{1.1}$$

for $z$ complex, and $q$ and $a$ real with $q \to \infty$, which have recently been obtained in [3]. These asymptotic solutions are uniformly valid for

$$-2q \le a \le (2-d)q, \tag{1.2}$$

($d$ an arbitrary small positive constant), with $z$ lying in the semi-infinite strip $0 \le \mathrm{Re}(z) \le \pi$, $\mathrm{Im}(z) \ge 0$.

It is well-known that there always exists one or more *pseudo-periodic* (or *Floquet*) Mathieu functions $w(z)$ having the property that $w(z+\pi) = M^{\pm 1}w(z)$, for some number $M$ that depends on the parameters. We shall use the standard notation of $M = e^{\nu\pi i}$, where $\nu$ is the so-called *characteristic exponent*. In [3] uniform asymptotic approximations were derived for Floquet and other Mathieu functions. The focus of the present paper is the derivation, via a connection formulae involving two asymptotic solutions of Eq. (1.1), of a uniform asymptotic relationship between the parameters $q$, $a$ and the characteristic exponent $\nu$. This relationship will be uniformly valid for $q \to \infty$ with Eq. (1.2) holding, and error bounds are readily available from referenced sources.

## 2. Asymptotic Solutions Involving Elementary Functions

To describe the asymptotic solutions of [3], we begin by converting the trigonometric form Eq. (1.1) of Mathieu's equation into an algebraic form, via a new independent variable $t$ and a new dependent variable $\tilde{w}(t)$ defined by

$$t=\left[-\cos(z)\right]^2,\quad \tilde{w}(t)=\left\{\cos(z)\sin(z)\right\}^{1/2}w(z). \tag{2.1a,b}$$

As a result Mathieu's equation takes the algebraic form

$$\frac{d^2\tilde{w}}{dt^2}=\left\{u^2 f(\tilde{a},t)+g(t)\right\}\tilde{w}, \tag{2.2}$$

where

$$f(\tilde{a},t)=\frac{t-\tilde{a}}{t(1-t)},\quad g(t)=-\frac{4t^2-4t+3}{16t^2(1-t)^2}. \tag{2.3a,b}$$

Here we have introduced the new parameters

$$\tilde{a}=\frac{1}{4}\left(\frac{a}{q}+2\right),\quad u=\sqrt{q}. \tag{2.4a,b}$$

The semi-infinite strip $0\leq \mathrm{Re}(z)\leq \pi$, $\mathrm{Im}(z)\geq 0$ is mapped to the cut $t$-plane as depicted in figs. 1 and 2, and we denote this $t$-domain by $\tilde{\mathbf{D}}$; we prescribe that $0\leq \arg(t)\leq 2\pi$ for this domain.

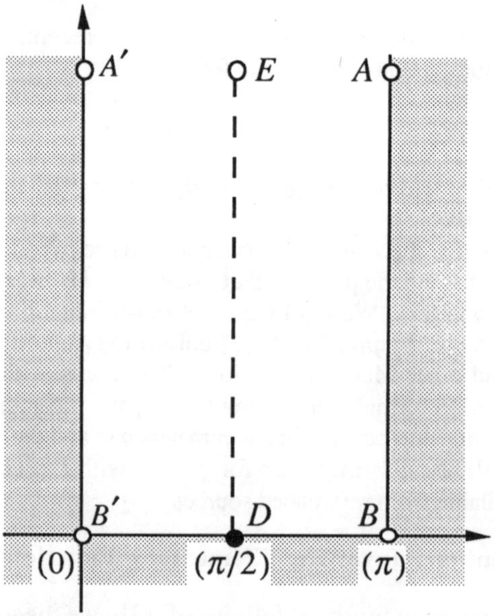


*Figure* 1. *z* plane

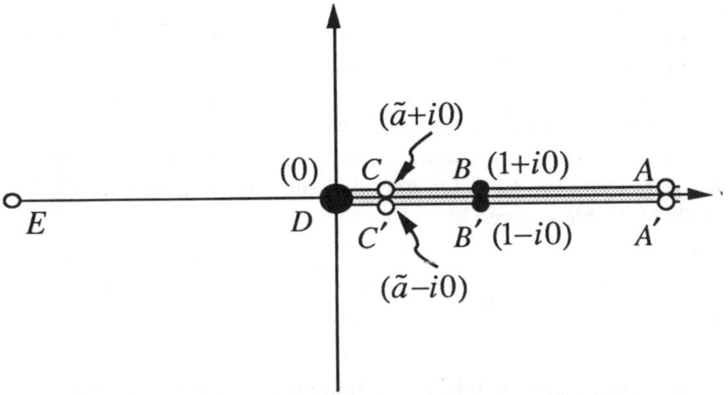


*Figure* 2. *t* plane

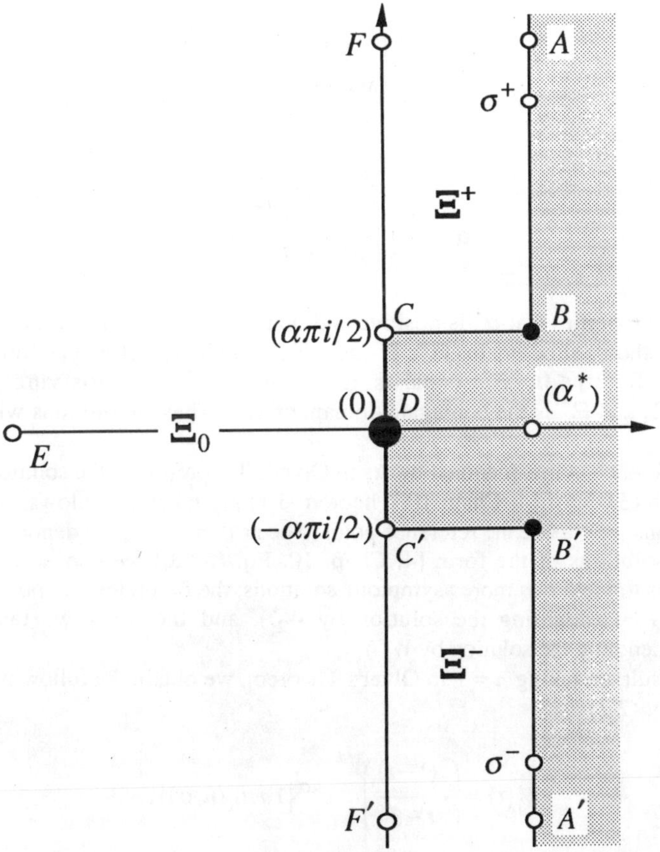


*Figure* 3. σ plane

Eq. (2.2) is characterized by $f(\tilde{a},t)$ having simple poles at $t=0,1$, and a turning point at $t=\tilde{a}$. The turning point at $t=\tilde{a}$ is located on the cut along the positive real axis between the two poles, and coalesces with the pole at the origin in the critical case $\tilde{a}=0$. We present asymptotic solutions of Eq. (2.2) in $\tilde{\mathbf{D}}$, which are uniformly valid for $0 \le \tilde{a} \le 1-d$ $(d>0)$ as $u \to \infty$; see Eq. (1.2) above.

In [3] Liouville-Green (L-G) solutions were constructed by applying a Theorem of Olver [4, Chap. 10, Theorem 3.1]. On denoting the independent variable $\xi$ of Olver's Theorem by $\sigma$, this variable is given by

$$\sigma = \int_0^t \left\{ \frac{s-\tilde{a}}{s(1-s)} \right\}^{1/2} ds. \tag{2.6}$$

The branches for the integrand are taken so that the $\sigma$-domain corresponding to $\tilde{\mathbf{D}}$, which we denote by $\Xi$, is as depicted in fig. 3. The points $t=0, \tilde{a}\pm i0, 1\pm i0$ are mapped to $\sigma = 0, \pm \alpha\pi i/2, \alpha^* \pm \alpha\pi i/2$, respectively, where $\alpha^*$ and $\alpha$ are defined by

$$\alpha^* = \int_{\tilde{a}}^1 \left\{ \frac{s-\tilde{a}}{s(1-s)} \right\}^{1/2} ds, \tag{2.7}$$

and

$$\alpha = \frac{2}{\pi}\int_0^{\tilde{a}} \left\{ \frac{\tilde{a}-s}{s(1-s)} \right\}^{1/2} ds. \tag{2.8}$$

These have the properties that $\alpha^*$ is positive and bounded away from zero, and $\alpha \to 0$ as $\tilde{a} \to 0$. Fig. 3 shows subdomains $\Xi^+$, $\Xi^-$, and $\Xi_0$, which are defined as follows. $\Xi_0$ is the half-plane $\mathrm{Re}(\sigma) \le 0$, $\Xi^+$ consists of all points in $\Xi$ satisfying $\mathrm{Re}(\sigma) \ge 0$, $\mathrm{Im}(\sigma) \ge \alpha\pi/2$, and $\Xi^-$ is the conjugate domain of $\Xi^+$. These definitions will be of use below.

The reference points (denoted by $\alpha_j$ in Olver's Theorem) for the solutions given by Eqs. (3.02) and (3.03) of [4, Chap. 10, Theorem 3.1] are taken as follows. For the first of these solutions we choose the reference point to be at $\sigma = \infty e^{\pi i}$ (and denote the solution by $W_1$). For solutions of the form [4, Chap. 10, Eq. (3.03)] we choose two different reference points to yield *two* more asymptotic solutions; the first reference point is taken to be at $\sigma = \alpha^* + i\infty$ (denoting the solution by $W_2^+$), and the other we take to be at $\sigma = \alpha^* - i\infty$ (denoting the solution by $W_2^-$).

As a result, on taking $n=1$ in Olver's Theorem, we obtain the following three L-G solutions of Eq. (2.2)

$$W_1(t) = \left( \frac{t(1-t)}{t-\tilde{a}} \right)^{1/4} e^{u\sigma}\{1+\eta_1(u,\sigma)\}, \tag{2.9}$$

$$W_2^{\pm}(t) = \left(\frac{t(1-t)}{t-\tilde{a}}\right)^{1/4} e^{-u\sigma}\left\{1+\eta_2^{\pm}(u,\sigma)\right\}. \qquad (2.10a,b)$$

The error terms are bounded by [4, Chap. 10, Eq. (3.04)] for $\alpha_j$ taken at the respective reference points at infinity. The bound for $\eta_1(u,\sigma)$ holds uniformly for $\sigma$ in $\Xi$ with neighbourhoods of the points $\sigma = \pm\alpha\pi i/2$ excluded, the bound for $\eta_2^+(u,\sigma)$ is uniformly valid for $\sigma$ in $\Xi_0 \cup \Xi^+$ with a neighbourhood of all points on the imaginary axis below $\sigma = \alpha\pi i/2$ excluded, and the bound for $\eta_2^-(u,\sigma)$ is uniformly valid for $\sigma$ in $\Xi_0 \cup \Xi^-$ with a neighbourhood of all points on the imaginary axis above $\sigma = -\alpha\pi i/2$ excluded.

The error terms $\eta_1(u,\sigma)$, $\eta_2^+(u,\sigma)$ and $\eta_2^-(u,\sigma)$ are $O\left(u^{-1}\right)$ as $u \to \infty$, and $O\left(\sigma^{-1}\right)$ as $\sigma \to \infty$, in their respective regions of asymptotic validity. The importance of the three L-G solutions $W_1$, $W_2^+$, and $W_2^-$ is that they are recessive in $\Xi_0$, $\Xi^+$, and $\Xi^-$, respectively.

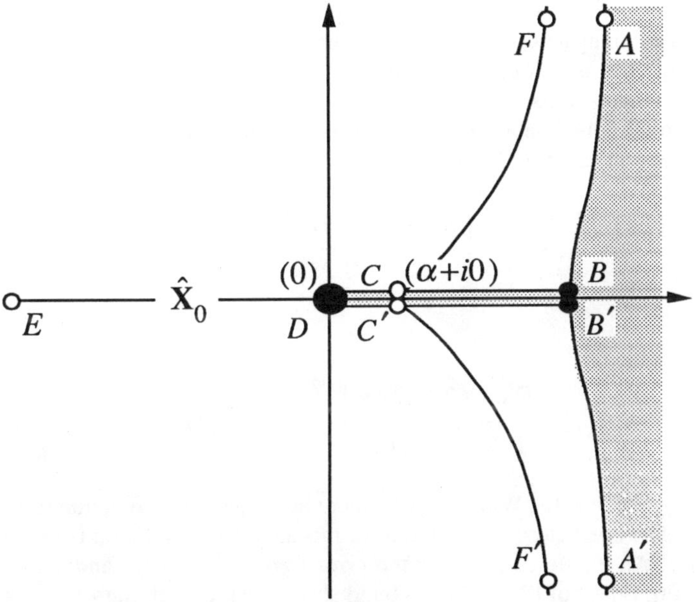


*Figure* 4. $\xi$ plane

## 3. A Uniform Asymptotic Solution Involving a Whittaker Function

The three L-G solutions of the previous section are not valid in a neighbourhood of the turning point. An asymptotic solution to Eq. (2.2) which *is* uniformly valid in a domain which contains the coalescing turning point and pole at the origin is given as follows. It involves a Whittaker confluent hypergeometric function of argument $\xi$ defined next, and comes from an application of the general theory of [2, Theorem II]. The independent variable $\xi$ is defined by [2, Eq. (4.4)]

$$\int_{\alpha}^{\xi}\left\{\frac{\tau-\alpha}{\tau}\right\}^{1/2}d\tau=\int_{\tilde{a}}^{t}\left\{\frac{s-\tilde{a}}{s(1-s)}\right\}^{1/2}ds, \tag{3.1}$$

where the branches on both sides are chosen so that the $\xi$ region $\hat{\mathbf{D}}$ corresponding to $t$ region $\tilde{\mathbf{D}}$ is as depicted in fig. 4. We specify that $0 \le \arg(\xi) \le 2\pi$ for $\xi \in \hat{\mathbf{D}}$. The points $t = 0, \tilde{a}$ are mapped to $\xi = 0, \alpha$, respectively, and we denote the point corresponding to $t = 1$ by $\xi = \xi_1$. The bounding curve $AB$ emanating at an angle $\pi/2$ from $\xi_1 + i0$ is the map of the upper part of the cut from $t = 1$ to $t = \infty$ ($\arg(t) = 0$), and the curve $A'B'$ is the conjugate of $AB$.

The curve $CF$ emanating at an angle $\pi/3$ from $\alpha + i0$ is defined by

$$\mathrm{Re}\int_{\alpha+i0}^{\xi}\left\{\frac{\tau-\alpha}{\tau}\right\}^{1/2}d\tau=0, \quad \left(\mathrm{Im}(\xi)\ge 0\right), \tag{3.2}$$

and $C'F'$ is the conjugate curve. These two curves form a boundary of a subdomain $\hat{\mathbf{X}}_0$ as depicted in fig. 4, *which is defined to include all its boundary points*. The $\xi$-domain $\hat{\mathbf{X}}_0$ corresponds to the $\sigma$-domain $\Xi_0$.

We now apply Theorem II of [2] to obtain the following asymptotic solution of Mathieu's algebraic equation (2.2)

$$\hat{w}_0(t)=\left(\frac{t(\alpha-\xi)(1-t)}{\xi(\tilde{a}-t)}\right)^{1/4}\hat{W}^{(0)}(u,\alpha,\xi), \tag{3.3}$$

where

$$\hat{W}^{(0)}(u,\alpha,\xi)=\frac{\Gamma(u\alpha/2+3/4)e^{u\alpha/2}}{(u\alpha/2)^{u\alpha/2}\Gamma(3/2)}M_{u\alpha/2,1/4}(2u\xi)+\hat{\varepsilon}^{(0)}(u,\alpha,\xi). \tag{3.4}$$

Here $M_{u\alpha/2,1/4}(2u\xi)$ is the Whittaker confluent hypergeometric function (see, for example [4, p. 260]). The characteristic behaviour of this asymptotic solution is that it is recessive at the origin. An explicit bound for the error term $\hat{\varepsilon}^{(0)}(u,\alpha,\xi)$ and its derivatives is supplied by Eq. (4.80) of [2], and this bound shows that, for arbitrary fixed positive $\delta$,

$$\hat{\varepsilon}^{(0)}(u,\alpha,\xi)=\frac{\Gamma(u\alpha/2+3/4)e^{u\alpha/2}}{(u\alpha/2)^{u\alpha/2}\Gamma(3/2)}M_{u\alpha/2,1/4}(2u\xi)O\left((u\tilde{a}+1)^{1/3}u^{-1+\delta}\right), \tag{3.5}$$

uniformly for $\xi \in \hat{\mathbf{X}}_0$ as $u \to \infty$, except near the zeroes of $M_{u\alpha/2,1/4}(2u\xi)$. In fact, the bound can be extended to a certain larger region (which depends on the values of the parameters), but such an extension is not required in the present investigation.

Our next task is the determination of constants $\lambda_1^+$, $\lambda_1^-$, and $\lambda_2$ such that

$$\hat{w}_0(t)=\lambda_1^{\pm}W_1(t)+\lambda_2 W_2^{\pm}(t). \tag{3.6a,b}$$

To determine these constants, we first use Eqs. (2.6) and (3.1) to relate $\sigma$ to $\xi$, to observe that

$$\sigma = \xi - \tfrac{1}{2}\alpha \ln(4\xi) - \tfrac{1}{2}\alpha + \tfrac{1}{2}\alpha \ln(\alpha) + \tfrac{1}{2}\alpha\pi i + O(\xi^{-1}), \tag{3.7}$$

as $\xi \to \infty$ in $\hat{\mathbf{D}}$. Consequently, on letting $\xi \to \infty e^{\pi i}$ in Eqs. (3.6a,b), and invoking Eqs. (2.9), (2.10a,b), (3.3), (3.7) and the well-known behaviour of $M_{u\alpha/2,1/4}(2u\xi)$ at infinity, it can be shown that

$$\lambda_2 = -e^{-\pi i/4}\{1 + \delta_0\}. \tag{3.8}$$

Here $\delta_0$ is some constant which can be explicitly bounded (see [3, Eq. (4.22)]), having the asymptotic property

$$\delta_0 = O\left((u\tilde{a} + 1)^{1/3} u^{-1+\delta}\right), \tag{3.9}$$

uniformly for $0 \le \tilde{a} \le 1 - d$, $(d > 0)$ as $u \to \infty$, for arbitrary fixed positive $\delta$.

Having found $\lambda_2$ we can determine $\lambda_1^+$ by letting $\xi \to \infty$ along the curve $CF$, since this curve is contained in the common region of asymptotic validity of the three asymptotic solutions in Eq. (3.6a), and $L_1(u,\sigma)$ does not vanish as $\xi \to \infty$ along this curve. It can be shown that

$$\lambda_1^+ = \left(\frac{2e}{u\alpha}\right)^{u\alpha} \frac{e^{-u\alpha\pi i/2}\Gamma(u\alpha/2 + 3/4)}{\Gamma(3/4 - u\alpha/2)} + \delta_0^+, \tag{3.10}$$

where $\delta_0^+$ is a constant which is also $\delta_0 = O\left((u\tilde{a} + 1)^{1/3} u^{-1+\delta}\right)$ uniformly for $0 \le \tilde{a} \le 1 - d$, $(d > 0)$ as $u \to \infty$. Again, for further details see [3].

The determination of $\lambda_1^-$ is similar, and by letting $\xi \to \infty$ on $C'F'$ one can show from the relation (3.6b) that

$$\lambda_1^- = -i\left(\frac{2e}{u\alpha}\right)^{u\alpha} \frac{e^{u\alpha\pi i/2}\Gamma(u\alpha/2 + 3/4)}{\Gamma(3/4 - u\alpha/2)} + \delta_0^-, \tag{3.11}$$

where $\delta_0^-$ is the constant with the same asymptotic behaviour as $\delta_0^+$.

## 4. The Eigenvalue Problem

We now are in a position to determine the relationship between the characteristic exponent $\nu$ and the parameters $q$ and $a$. We shall achieve this by solving for $M = e^{\nu\pi i}$ as an eigenvalue of the matrix

$$\mathbf{A} = \begin{pmatrix} a_{11} & a_{12} \\ a_{21} & a_{22} \end{pmatrix}, \tag{4.1}$$

whose coefficients are given by the connection problem

$$w_0(z+\pi) = a_{11}w_0(z) + a_{12}w_1(z), \tag{4.2}$$

$$w_1(z+\pi) = a_{21}w_0(z) + a_{22}w_1(z), \tag{4.3}$$

where $w_0(z)$ and $w_1(z)$ are *any* linearly independent Mathieu functions (solutions of Eq. (1.1)). For a proof that $M$ (and $M^{-1}$) is a solution of $\det\{\mathbf{A} - M\mathbf{I}\} = 0$ see [1, p. 30].

We shall use the asymptotic solutions of the previous two sections in Eqs. (4.2) and (4.3); specifically, we utilize the following two solutions of Mathieu's (trigonometric) equation

$$w_0(z) = \left(t(t-1)\right)^{-1/4}\hat{w}_0(t), \tag{4.4}$$

and

$$w_1(z) = (\tilde{a}-t)^{-1/4}e^{u\sigma}\{1+\eta_1(u,\sigma)\}. \tag{4.5}$$

We take $z$ in Eqs. (4.2) and (4.3) to lie on the positive imaginary axis, and denote the corresponding value of $\sigma$ by $\sigma^-$, and the value of $\sigma$ corresponding to $z+\pi$ by $\sigma^+$. Then from the definition Eq. (2.6) it is not difficult to show that

$$\sigma^- = -\sigma^+ + 2\alpha^*, \tag{4.6}$$

where $\alpha^*$ is defined by Eq. (2.8).

The solution $w_1(z)$ can be directly evaluated at both $\sigma^+$ and $\sigma^-$, since both these points are included in its domain of asymptotic validity. Moreover, since we have solved the connection problem

$$w_0(z) = (\tilde{a}-t)^{-1/4}\left[\lambda_1^{\pm}e^{u\sigma}\{1+\eta_1(u,\sigma)\} + \lambda_2 e^{-u\sigma}\left\{1+\eta_2^{\pm}(u,\sigma)\right\}\right], \tag{4.7a,b}$$

we can also evaluate $w_0(z)$ at these two points, by using Eq. (4.7a) at $\sigma^+$ and Eq. (4.7b) at $\sigma^-$.

To determine $a_{11}$ and $a_{12}$ we use Eq. (4.2) and its differentiated form

$$\frac{\partial w_0(z+\pi)}{\partial \sigma^-} = a_{11}\frac{\partial w_0(z)}{\partial \sigma^-} + a_{12}\frac{\partial w_1(z)}{\partial \sigma^-}. \tag{4.8}$$

Now, from Eqs. (2.6) and (4.4)

$$\frac{\partial w_0(z)}{\partial \sigma^-} = \frac{1}{4}\left(\tilde{a}-t^-\right)^{-1}\left(\frac{t^-\left(1-t^-\right)}{t^- - \tilde{a}}\right)^{1/2} w_0(z) + \left(\tilde{a}-t^-\right)^{-1/4}$$

$$\left[\lambda_1^- u e^{u\sigma^-}\left\{1+\eta_1\left(u,\sigma^-\right)+\frac{\partial\eta_1\left(u,\sigma^-\right)}{u\partial\sigma^-}\right\}-\lambda_2 u e^{-u\sigma^-}\left\{1+\eta_2^-\left(u,\sigma^-\right)-\frac{\partial\eta_2^-\left(u,\sigma^-\right)}{u\partial\sigma^-}\right\}\right]. \quad (4.9)$$

The corresponding expression for $\partial w_0(z+\pi)/\partial\sigma^-$ comes from replacing the minus suffices on the RHS by plus suffices, and multiplying by the factor $-1$ $(= d\sigma^+/d\sigma^-)$. Similarly, we find that

$$\frac{\partial w_1(z)}{\partial\sigma^-}=\frac{1}{4}\left(\tilde{a}-t^-\right)^{-1}\left(\frac{t^-\left(1-t^-\right)}{t^- -\tilde{a}}\right)^{1/2} w_1(z)$$

$$+\left(\tilde{a}-t^-\right)^{-1/4} u e^{u\sigma^-}\left\{1+\eta_1\left(u,\sigma^-\right)+\frac{\partial\eta_1\left(u,\sigma^-\right)}{u\partial\sigma^-}\right\}. \quad (4.10)$$

Therefore, on inserting Eqs. (4.9) and (4.10) into (4.2), and letting $\sigma^\pm \to \alpha^* \pm i\infty$, we obtain

$$e^{\pi i/2}\left[\lambda_1^+ e^{u\sigma^+}+\lambda_2 e^{-u\sigma^+}\right]=a_{11}\left[\lambda_1^- e^{u\sigma^-}+\lambda_2 e^{-u\sigma^-}\right]+a_{12}e^{u\sigma^-}+o(1). \quad (4.11)$$

Likewise, from Eq. (4.8) we arrive at

$$-e^{\pi i/2}\left[\lambda_1^+ e^{u\sigma^+}-\lambda_2 e^{-u\sigma^+}\right]=a_{11}\left[\lambda_1^- e^{u\sigma^-}-\lambda_2 e^{-u\sigma^-}\right]+a_{12}e^{u\sigma^-}+o(1). \quad (4.12)$$

On solving this pair of equations we find that

$$a_{11}=i\frac{\lambda_1^+}{\lambda_2}e^{2u\alpha^*}, \qquad a_{12}=-i\frac{\lambda_1^-\lambda_1^+}{\lambda_2}e^{2u\alpha^*}+i\lambda_2 e^{-2u\alpha^*}. \quad (4.13)$$

By a similar method, using Eq. (4.3) and its differentiated form we find that

$$a_{21}=\frac{i}{\lambda_2}e^{2u\alpha^*}, \qquad a_{22}=-i\frac{\lambda_1^-}{\lambda_2}e^{2u\alpha^*}. \quad (4.14)$$

We can now solve for $M$ from the equation $\det\{\mathbf{A}-M\mathbf{I}\}=0$; using Eqs. (3.8), (3.10), (3.11), (4.1), (4.13), and (4.14) we see that $M$ satisfies the quadratic equation

$$M^2-bM+1=0, \quad (4.15)$$

where

$$b=a_{11}+a_{22}=\left\{1+\hat{\delta}_0\right\}^{-1}e^{2u\alpha^*}\left[G(u\alpha/2)\cos(u\alpha\pi)-e^{3\pi i/4}\hat{\delta}_0^+ +e^{3\pi i/4}\hat{\delta}_0^-\right]. \quad (4.16)$$

and

$$G(k) = \frac{1}{\pi}\left(\frac{e}{k}\right)^{2k} \Gamma(k+3/4)\Gamma(k+1/4). \tag{4.17}$$

From Stirling's formula we observe that $G(k)$ is non-vanishing and bounded for $0 \le k < \infty$. Now, on writing $M = e^{\nu\pi i}$ in Eq. (4.15) we arrive at our main result

$$\cos(\nu\pi) = \tfrac{1}{2}\left\{1+\hat{\delta}_0\right\}^{-1} e^{2u\alpha^*}\left[G(u\alpha/2)\cos(u\alpha\pi) - e^{3\pi i/4}\hat{\delta}_0^+ + e^{3\pi i/4}\hat{\delta}_0^-\right]. \tag{4.18}$$

Since $\alpha = O(\tilde{a})$ as $\tilde{a} \to 0$, the constants $\delta_0$, $\delta_0^\pm$ are $O\left((u\tilde{a}+1)^{1/3}u^{-1+\delta}\right)$ uniformly for $0 \le \tilde{a} \le 1-d$ as $u \to \infty$, for arbitrary fixed positive $d$ and $\delta$.

When $\nu$ is real we have a special simplification in Eq. (4.18): $\cos(\nu\pi)$ is bounded (which includes the special case when there exists a basically-periodic solution, that is, when $\cos(\nu\pi) = \pm 1$), and therefore

$$\cos(u\alpha\pi) = \frac{1}{G(u\alpha/2)}\left[e^{3\pi i/4}\left\{\hat{\delta}_0^+ - \hat{\delta}_0^-\right\} + O(1)e^{-2u\alpha^*}\left\{1+\hat{\delta}_0\right\}\right], \tag{4.19}$$

when $\nu$ is real. On solving for $\alpha$ and then returning to the original parameters, and recalling that $\alpha^*$ is positive and bounded away from zero, we arrive at the following relationship

$$\tilde{a}\int_0^1 \left\{\frac{1-s}{s(1-\tilde{a}s)}\right\}^{1/2} ds = \frac{(2m+1)\pi}{4\sqrt{q}} + O\left(\left(\sqrt{q}\tilde{a}+1\right)^{1/3} q^{-1+\delta}\right) \qquad (m = 0,1,2,\cdots), \tag{4.20}$$

which is uniformly valid for $0 \le \tilde{a} \le 1-d$ as $q \to \infty$. Eq. (6.13) can be solved for $\tilde{a}$, from which $a$ can be determined by the relation $a = 4\tilde{a}q - 2q$.

**Acknowledgment.** This research was supported by the National Science Foundation under grant DMS-9102834.

## References

[1] F. M. ARSCOTT (1964), *Periodic Differential equations*, Pergamon Press, Oxford.

[2] T. M. DUNSTER (1993), *Uniform asymptotic solutions of second-order linear differential equations having a simple pole and a coalescing turning point in the complex plane*, SIAM J. Math. Anal., [To appear].

[3] T. M. DUNSTER (1993), *Uniform asymptotic approximations for Mathieu functions*, [Submitted to Methods and Applications of Analysis].

[4] F. W. J. OLVER (1974), *Asymptotics and Special Functions*, Academic Press, New York.

# Linear Differential Equations with Operator-Valued Coefficients and Asymptotic Parameters

SERGEI EDELSTEIN

*Mechanical and Mathematical Department*
*Rostov State University*
*Ul. Zorge 5 Rostov-on-Don 344104, Russia*

## Abstract

Boundary value problems for differential equations in the domain $(-\infty, \infty)$ with operator-valued coefficients are considered. The main results are the theorems of well-posedness and an asymptotic formulae for the solutions. The principal idea is the asymptotic decomposition of the boundary value problem into two evolution problems. This idea is a basis of one-directed models of waveguides and our results give the proof of some of these models.

## 1. Introduction.

We consider the boundary value problems

$$\frac{d^2y}{dt^2} - k^2Q(t)y = F(t), \qquad (t \in \mathcal{R}), \tag{1}$$

$$\lim_{t\to\pm\infty}\left(\frac{dy}{dt} \pm kQ^{\frac{1}{2}}(\pm\infty)y\right) = 0, \tag{2}$$

where $y$ and $F$ are vector-functions (v-f), whose values are elements of some Banach space, $Q$ is the operator-function (o-f) and $k \gg 1$ is the asymptotic parameter. In some cases we shall set problem for Eq. (1) by means of limiting absorption principle. It should be noteed that both approaches are equal in the situation considered, but in the present paper we do not discuss this problem.

The problems with operator-valued coefficients occur, for example, in the theory of waveguide, when the environment is slowly range-dependent ($t = \epsilon x$ is the "slow coordinate" with $\epsilon = k^{-1}$). In this case the boundary conditions are so-called radiation conditions and $Q(t)$ are differential operators. For instance $Q(t)$

may be Sturm-Liouville operators in the domain $[0,\infty)$ or $[0,h]$ with Dirichlet boundary conditions (see Examples 1 and 2 below).

Before studying the general situation, it is convenient to remind the well known WKB estimates for scalar equation with the nonzero coefficient.

Let $Q : \mathcal{R} \to \mathcal{C}$ be a function which satisfies the following conditions:

(i) $Q', Q'' \in L_1(\mathcal{R})$, (in particular $Q(\pm\infty)$ exist);

(ii) $Q(t) \neq 0$ for all $t \in \mathcal{R}$ and moreover, there exists $\delta > 0$ such that $|Q(t)| > \delta$;

(iii) there exists the continuous branch of the root $M(t) = Q^{\frac{1}{2}}(t)$ such that $\Re Q^{\frac{1}{2}}(t) > 0$.

Then, the homogeneous scalar Eq. (1) (i.e. $F \equiv 0$) has two solutions $y_\pm(t)$, whose asymptotic behavior for large $k$ are as follows

$$y_\pm(t) = Q^{-\frac{1}{4}}(t) \exp\left[\mp k \int_{t_0}^{t} Q^{\frac{1}{2}}(s)\, ds\right](1 + k^{-1}\epsilon_\pm(t)).$$

The functions $\epsilon_\pm(t)$ are bounded and $\epsilon_\pm(t) \to 0$ as $t \to \pm\infty$. By means of $y_\pm(t)$, we can define the Green function $G(t,s,k)$ of the boundary value problem in the scalar case and for $\pm t \geq \pm s$,

$$G(t,s,k) = -(2k)^{-1}Q^{-\frac{1}{4}}(t)Q^{-\frac{1}{4}}(s) \exp\left[\mp k \int_{s}^{t} Q^{\frac{1}{2}}(s)\, ds\right](1 + O(k^{-1})),$$

uniformly in $t$ and $s$. The present paper show that a result like this asymptotic formula is valid for problems with operator-valued coefficients, but the exponents $\exp\left[\pm k \int_s^t Q^{\frac{1}{2}}(s)\, ds\right]$ should be replaced by the evolution operators for the Cauchy problems:

$$\frac{du_\pm}{dt}(t) = \pm k Q^{\frac{1}{2}}(t) u_\pm(t), \quad (\pm t > \pm s) \quad u(s) = f. \tag{3}$$

Of course we consider situations when these problems are correct and we remind that evolution operators are defined by formulae $u_\pm(t) = U_\pm(t,s,k)f$.

## 2. Principal set of conditions.

We should explain now in which way the conditions (i), (ii) and (iii) are transformed when $Q$ is an operator-function. We describe two sets of conditions in the present paper. One of them is a direct generalization and we call it the principal set of condtions.

When $B$ is the Banach space, $-\infty \leq a < b \leq \infty$ and $1 \leq p \leq \infty$, we denote by $L_p((a,b),B)$ the space consisting of those strong measurable v-f $f$

which satisfy the condition $\|f(\cdot)\|_B \in L_p((a,b))$. We denote by $L((a,b),B)$ the space of local integrable v-f $f:(a,b)\to B$, and $W^m((a,b),B)=\{f|f\in L((a,b),B)\}$, the generalized derivatives $f^{(k)}\in L((a,b),B)$ for $k=1,2\cdots,m\}$; $W_p^m((a,b),B)=\{f|f\in W^m((a,b),B), f^{(k)}\in L_p((a,b),B), k=0,1,2\cdots,m\}$. If $B_1$ and $B_2$ are two Banach spaces. We shall denote by $\mathcal{L}(B_1,B_2)$ the space of bounded linear operators acting from $B_1$ to $B_2$, and $\mathcal{L}(B)=\mathcal{L}(B,B)$.

Let us assume that the following Assumptions are true:

**Assumption 1.** $Q(t)=Q_0+Q_1(t)$, where $Q_0$ is some closed operator, $Q\in W^2(\mathcal{R},\mathcal{L}(B)), Q'\in W_1^1(\mathcal{R},\mathcal{L}(B))$.

**Assumption 2.** There exist $\zeta\in\mathcal{C}$, $|\zeta|=1$, $c>0$ such that $\{\zeta\tau:\tau\geq 0\}\cap\sigma(Q(t))=\emptyset$, for all $t\in\mathcal{R}$, and $\|(Q(t)-\zeta\tau)^{-1}\|\leq c(1+\tau)^{-1}$ ($\sigma(\cdot)$ : the spectra).

**Assumption 3.** For all $t\in\mathcal{R}$, the operator $-M(t)$ generates a semigroup $\exp(-\tau M(t))$ which satisfies the estimate $\|\exp(-\tau M(t))\|\leq 1$ for all $\tau>0$.

From Assumptions 1 and 2, we know that $\|Q(t)^{-1}\|\leq c$ and there exists a strong continuous branch of the root $M(t)=Q^{\frac{1}{2}}(t)$ and the limits $M(\pm\infty)=\lim_{t\to\pm\infty}M(t)$ exist.

Let us consider norms $\|f\|_{-1}=\|(M(t_0)-\lambda_1)^{-1}f\|_B$ ($\Im\lambda_1<0$), $\|f\|_{-2}=\|(Q(t_0)-\lambda_2)^{-1}f\|_B$ ($\lambda_2=\zeta\tau$, $\tau<0$), and the spaces $B^{-1}$, $B^{-2}$, which are completions of $B$ with these norms. When $F\in L((a,b),B)$ we shall consider the solution of Eq. (1) as v-f from $W^2((a,b),B^{-2})\cap W^1((a,b),B^{-1})\cap L((a,b),B)$ and the solution of the equations $\dfrac{du_\pm}{dt}=\mp kM(t)u_\pm+F(t)$ as a v-f from $W^1((a,b),B^{-1})\cap L((a,b),B)$. Then both sides of the equation can be considered as v-f from $W^2((a,b),B^{-2})$ or from $W^1((a,b),B^{-1})$.

From Assumption 3 we obtain that the Cauchy problems (Eq. (3)) are correct and $\|U_\pm(t,s,k)\|\leq 1$.

Besides the operator $M(t)=Q^{\frac{1}{2}}(t)$, we need $M^{\frac{1}{2}}(t)=Q^{\frac{1}{4}}(t)$ which can be defined to satisfy $\sigma(M^{\frac{1}{2}}(t))\subset\{\lambda\in\mathcal{C}:|\arg(\lambda)|<\frac{\pi}{4}\}$.

Let us consider the integral operators

$$(G_0(k)F)(t)=-(2k)^{-1}\left[\int_{-\infty}^{t}M^{-\frac{1}{2}}(t)U_+(t,s,k)M^{-\frac{1}{2}}(s)F(s)ds\right.$$
$$\left.+\int_t^\infty M^{-\frac{1}{2}}(t)U_-(t,s,k)M^{-\frac{1}{2}}(s)F(s)ds\right].$$

It is obvious that $G_0(k)F\in\mathcal{L}(L_1(\mathcal{R},B),L_\infty(\mathcal{R},B))$, and $\|G_0(k)\|=O(k^{-1})$.

THEOREM 1. *If Assumptions 1 – 3 are valid then for large $k$ and for any $F\in L_1(\mathcal{R},B)$ the problem Eqs. (1) and (2) has the unique (generalized) solution*

*$y = G(k)F \in L_\infty(\mathcal{R}, B)$ which satisfies the following estimated*

$$\|G_0(k)\| = O(k^{-1}), \quad \|G(k) - G_0(k)\| = O(k^{-2}) \tag{4}$$

*(all norms are in the space $\mathcal{L}(L_1(\mathcal{R}, B), L_\infty(\mathcal{R}, B))$).*

One of the most important tools of proving this theorem was the technique of asymptotic decomposition of abstract differential equation with small parameter which was created (for Cauchy problems) by S. G. Krein [1] and Yu. L. Daletskii [2].

The approximate reduction of a boundary value problem to evolution problems is the main idea of one-directed models of waveguide which are widely practiced in the theory of underwater sound propagation [3], in seismicity and radio physics. But we should note that the reduction (it is called "parabolic equation method") was not proved mathematically previously. Our results give the proof of this reduction in some cases.

Theorem 1 allows us to consider waveguides with bounded cross sections.

**Example 1.** (Waveguide strip) Let $\mathcal{D}(Q(t)) = \{f \in W_2^2(0,h) : f(0) = f(h) = 0\}$, $Q(t)f = -\dfrac{df^2}{dz^2} - a(t,z)f$. If we make the change of variables $t = \xi x$, $\xi = k^{-1}$, then Eq. (1) is transformed into Helmholtz equation with the coefficient slowly depending on $x$:

$$\Delta u + a(\xi x, z)u = F_1(x,z) \qquad (\xi \ll 1) \tag{5}$$

and the boundary conditions $u(x,0) = 0, u(x,h) = 0$.

For waveguides of this type the spectra of $Q(t)$ consist (as a rule) only of eigenvalues $\lambda_n(t)$, so if $\lambda_n(t) \neq 0$ we can use this theorem.

The majority of natural waveguides are open waveguides, their cross sections should be considered as unbounded sets.

**Example 2.** (The simplest model of range-depending ocean waveguide) Let $\mathcal{D}(Q(t)) = \{f \in W_2^2(0,\infty) : f(0) = 0\}$, $Q(t)f = -\dfrac{df^2}{dz^2} - a(t,z)f$. After the change of variables $t = \xi x$, $\xi = k^{-1}$, Eq. (1) is transformed into (5) in half-plane with condition $u(x,0) = 0$. We suppose that $a(t,z) = \dfrac{\omega^2}{c^2(t,z)} > \delta > 0$ ($\omega$ is the frequency and $c(\xi x, z)$ is the sound speed), $a(t,z) \to a_0$ when $z \to \infty$ and $a_0$ does not depend on $t$. In this situation $0 \in \sigma(Q(t))$ for all $t \in \mathcal{R}$, so Theorm 1 can not be applied.

**3. Analytic deformations.**

Let $B$ and $E$ be Hilbert spaces, $\{0\} \neq E \subseteq B$ and the inclusion map $j : E \to B$ is continuous. Let $N$ be a self-adjoint operator in $B$, $NB \subseteq E$ and $N \in \mathcal{L}(B, E)$. We define the map $j^* : B \to E^*$ by the formula $(j^*j)g = (j, g)_B$ foa all $j, g \in B$.

We shall assume that the following assumptions are true:

**Assumption 4.** $Q(t) = Q_0 + NK(t)N$, where $Q_0$ is a closed operator, $K \in L_1(\mathcal{R}, \mathcal{L}(B))$, $K' \in W_1^1(\mathcal{R}, \mathcal{L}(B))$ and all $Q(t)$ are self-ajoint operators in $B$.

Let us denote by $B_{ac}(t)$ the absolutely continuous subspace connected with $Q(t)$ and let $B_{[a,b]}(t)$ be the invariant subspace of $Q(t)$ corresponding to the spectral subset $[a, b]$.

**Assumption 5.** There exists $\delta > 0$ such that for all $t \in \mathcal{R}$ $B_{[-\delta,\delta]}(t) \in B_{ac}(t)$ and $j^*(Q(t) - \lambda)^{-1}j$ can be continued from $\mathcal{C}_+$ and from $\mathcal{C}_-$ in the neighborhood $\mathcal{O}_\delta = \{\lambda \in \mathcal{C} : |\lambda| < \delta\}$ and for the corresponding o-f $R^\pm(\lambda, t)$ the following estimate holds: $\sup\left\{\|R^\pm(\lambda, t)\|_{E\to E^*} : \lambda \in \mathcal{O}_\delta, t \in \mathcal{R}\right\} < \infty$.

In this situation it is convenient to use the limiting absorption principle to set up problem for Eq. (1). The operator $\frac{d^2}{dt^2} - k^2Q(t)$ is self-adjoint in $L_2(\mathcal{R}, \mathcal{L}(B))$, so the following equation $\frac{d^2y_\epsilon}{dt^2} - (k^2Q(t) - i\epsilon I)y_\epsilon = F(t)$ $(F \in L_2(\mathcal{R}, \mathcal{L}(B)))$ is uniquely solvable in $L_2(\mathcal{R}, \mathcal{L}(B))$ for all $\epsilon > 0$.

THEOREM 2. *If Assumptions 4 and 5 hold then for large $k$ and for all $F \in L_1(\mathcal{R}, E) \cap L_2(\mathcal{R}, B)$ there exists $y_0 \in L_\infty(\mathcal{R}, E^*)$ such that for any $a < b$, $\|y_0(t) - j^*y_\epsilon(t)\|_{E^*} \to 0$, uniformly on $[a, b]$, as $\epsilon \to +0$.*

The estimates like Eq. (4) is valid too (in norms of the space $\mathcal{L}(L_1(\mathcal{R}, E)$, $L_\infty(\mathcal{R}, E^*)))$. This theorem can be used when we consider Example 2 and $a(t, z) = a_0 + q(t, z)$, where $q(t, \cdot) \in E$ with $E = L_{\infty,\gamma}(0, \infty) = \left\{\exp(\gamma z)f(z) : f \in L_{\infty,\gamma}(0, \infty)\right\}$ and $\gamma < 0$.

Some results related to this paper were published in [4], [5], [6] and [7].

The author is very grateful to Professor P. Hsieh for his kind invitation to participate in the International Symposium on Ordinary Differential Equations and Applications and his suggestions to improve this paper. I gratefully acknowledge the travel grant of the International Science Foundation.

## References

[1] S. G. Krein, *Linear Differential Equations in a Banach Space*, Moscow, Nauka, 1967 (Russian).

[2] Yu. L. Daletzkii and S. G. Krein, *Stability of Solutions of Differential Equations in a Banach Space*, Moscow, Nauka, 1970 (Russian).

[3] F. D. Tappert, The parabolic approximation method, in *Wave Propagation and Underwater Acoustics*, edited by J. B. Keller and J. S. Papadakis, Lecture Notes in Physics, Vol. 70, Springer-Verlag, New York, 1977.

[4] S. L. Edelstein, Estimates and asymptotics of solutions of two-dimensional boundary value problem for the Helmholtz equation with the coefficient weakly depending on horizontal coordinate, *Differentsialnye Uravnenia (Differential Equations)*, 26 (1990), No 4, 727 – 729 (Russian).

[5] S. L. Edelstein, Asymptotic behavior of modes of the slowly depending on distance wave-guide with penetrable boundaries, *Akusticheskii zhurnal (Acoustic Journal)*, 36 (1990), No 2, 355 – 359 (Russian).

[6] S. L. Edelstein, Operator analogs of WKB-estimates and solvability of boundary value problems, *Matematichaeskie zametki (Mathematical Notes)*, 51 (1992), No 4, 124 – 131 (Russian).

[7] S. L. Edelstein, Asymptotic decomposition of boundary value problems for differential equations in Banach spaces, *Doklady Akademii Nauk (Journal Russian Academy Science. Daklady, Vol. 45)*, 323 (1992), No 3, 427 – 429 (Russian).

# On a Generalization of Bessel Functions Satisfying Higher-Order Differential Equations

by

W.N. Everitt and C. Markett

Abstract: Bessel-type functions $\{J_\lambda^{\alpha,M}(x)\}_{\lambda\geq 0}$ with two parameters $\alpha \geq -1/2$ and $M \geq 0$, which include the classical normalized Bessel function for $M = 0$, are introduced as a certain confluent limit of Koornwinder's Laguerre-type polynomials $\{L_n^{\alpha,N}(x)\}_{n\in\mathbb{N}_0}$. For any $\alpha \in \mathbb{N}_0$, they arise as solutions of a spectral $\lambda$-dependent differential equation of order $2\alpha + 4$. A necessary and sufficient condition to transform the differential equation into symmetric form is given in terms of an overdetermined system of linear equations. It is shown that for $\alpha = 0,1,2$, a solution of this problem exists and leads to a (symmetric) fourth-, sixth-, and eighth-order differential equation, respectively. For any $\alpha \geq -1/2$, we also derive a second-order differential equation, but with coefficients depending nonlinearly on the eigenvalue parameter $\lambda$. Finally, two differential expressions of order $2\alpha + 2$, $\alpha \in \mathbb{N}_0$, are constructed which map the classical Bessel functions onto their non-classical counterparts, and vice versa. This result may be used to establish an orthogonality relation for the Bessel-type functions (in a distributional sense) with respect to the distributional weight function $(2/\Gamma(\alpha+1))x^{2\alpha+1} + M\delta(x)$, $\delta$ denoting the point mass centered at the origin.

## Results

It is a characteristic feature of the classical orthogonal polynomials named after Jacobi, Laguerre, and Hermite, that they arise as eigenfunctions of a (second-order) Sturm-Liouville boundary value problem. The orthogonality of the eigenfunctions belonging to different eigenvalues then follows immediately from the symmetry of the differential equation, via Lagrange's identity, and appropriate boundary conditions. But while the Jacobi, Laguerre, and Hermite polynomials are known to be the only positive definite orthogonal polynomial systems (OPS) with this property, see Bochner [2], a number of further OPS have been found which satisfy higher-order differential equations. In his pioneering work of 1938/40 [15], [16], H.L. Krall characterized all such equations of fourth-order and showed that they generate three new OPS called the Legendre-type, Jacobi-type, and Laguerre-type polynomials. Further significant contributions were made by A.M. Krall and L.L. Littlejohn (see [13], [14], [17] and the references cited there) who, among others, gave further examples of sixth and eighth order. For a general account see the survey paper [6] of Everitt and Littlejohn; in particular [6; sections 5, 6].

In these new cases it is known that the differential equation can be transformed into a formally symmetric form; also that the polynomials are orthogonal with respect to a distributional weight function which is *either* a linear combination of the Jacobi weight $(1-x)^{\alpha}(1+x)^{\beta}$ $(x \in (-1,1))$, for some $\alpha,\beta > -1$, and of delta distributions at $\pm1$, *or* a linear combination of $\delta(x)$ and the Laguerre weight $e^{-x}x^{\alpha}$, $(x \in (0,\infty))$, for some $\alpha > -1$. Moreover, the new polynomials can always be derived from the corresponding classical orthogonal polynomials by applying to the latter a certain first order differential expression depending on the degree n of the polynomials.

These observations led T.H. Koornwinder [12] to derive an explicit ${}_4F_3$-hypergeometric representation of the polynomials which are orthogonal with respect to

$$w(x) = w_{\alpha,\beta,M,N}(x) := \frac{\Gamma(\alpha+\beta+2)\,(1-x)^{\alpha}(1+x)^{\beta}}{2^{\alpha+\beta+1}\;\Gamma(\alpha+1)\;\Gamma(\beta+1)} + M\;\delta(x+1) + N\;\delta(x-1),$$

for any $\alpha,\beta > -1$ and $M,N \geq 0$.(For definition and properties of the generalized hypergeometric functions ${}_pF_q$, used throughout this paper, see [1], [3] and [4].) Denoting these polynomials by $P_n^{\alpha,\beta,N,M}(x)$, for $x \in (-1,1)$ and $n \in \mathbb{N}_0$, he then defined, by means of a confluent limit relation, the "Laguerre-type" polynomials $L_n^{\alpha,N}(x)$, $n \in \mathbb{N}_0$, $x \in (0,\infty)$, $\alpha > -1$, $N \geq 0$, which are orthogonal with respect to

$$w(x) = w_{\alpha,N}(x) := \frac{1}{\Gamma(\alpha+1)}\; e^{-x}x^{\alpha} + N\;\delta(x).$$

In detail

$$L_n^{\alpha,N}(x) := \lim_{\beta\to\infty} P_n^{\alpha,\beta,0,N}(1-\tfrac{2x}{\beta}) = \left[1 + N\begin{bmatrix}n+\alpha\\ n\end{bmatrix}\left(\frac{d}{dx}+\frac{n}{\alpha+1}\right)\right] L_n^{\alpha}(x)$$

(1.1)

$$= \left[1 + N\begin{bmatrix}n+\alpha\\ n-1\end{bmatrix}\right] L_n^{\alpha}(x) - N\begin{bmatrix}n+\alpha\\ n\end{bmatrix} L_{n-1}^{\alpha+1}(x),$$

where $L_n^{\alpha}$ denotes the classical Laguerre polynomial. Since

$$L_n^{\alpha}(x) = \begin{bmatrix}n+\alpha\\ n\end{bmatrix} {}_1F_1(-n;\alpha+1;x),$$

$L_n^{\alpha,N}$ has a ${}_2F_2$-hypergeometric representation. Moreover, the orthonormalization constant reads, see the thesis by Koekoek [11, (3.1.8)],

$$\langle L_n^{\alpha,N}, L_n^{\alpha,N}\rangle_{w_{\alpha,N}} = \begin{bmatrix}n+\alpha\\ n\end{bmatrix}\left[1 + N\begin{bmatrix}n+\alpha\\ n-1\end{bmatrix}\right]\left[1 + N\begin{bmatrix}n+\alpha+1\\ n\end{bmatrix}\right] \qquad (n \in \mathbb{N}_0).$$

The natural question whether Koornwinder's Laguerre-type polynomials are generated by a differential equation has been answered only recently by J. and R. Koekoek [10] in the affirmative. On the basis of known results for $\alpha = 0,1,2$ [14], they constructed for any $\alpha > -1$ such an equation which is of infinite order unless $\alpha \in \mathbb{N}_0$ (see (2.4) below). In the latter cases, the order reduces to $2\alpha + 4$.

Besides the classical orthogonal polynomials there are also well known non-discrete systems of special functions such as the Bessel, Whittaker, and Jacobi functions on the positive half-axis, which satisfy a (second-order) Sturm-Liouville equation. The simplest equation among them is the Bessel equation,

(1.2) $$\frac{d}{dx}\left[x^{2\alpha+1}\frac{d}{dx}y(x)\right] + \lambda^2 x^{2\alpha+1}y(x) = 0\ , \quad x \in (0,\infty) \text{ and } \alpha \ge -\tfrac{1}{2};$$

see Markett [18] and [20], and in particular the Habilitationsschrift [19]. In equivalent form this becomes

(1.2a) $$y''(x) + \frac{2\alpha+1}{x}y'(x) + \lambda^2 y(x) = 0,$$

and this equation has a regular singular point at the origin for any $a > -1/2$. Under the boundary conditions

(1.3) $y(0) = 1$ , $y'(0) = 0$, and y uniformly bounded on $[0,\infty)$,

there is a solution of equation (1.2), for any $\lambda \in (0,\infty)$, given by

$$
\begin{aligned}
y(x) = J_\lambda^a(x) &:= {}_0F_1(-;a+1;-\tfrac{1}{4}(\lambda x)^2) \\
&= 2^a\,\Gamma(a+1)\,(\lambda x)^{-a} J_a(\lambda x)
\end{aligned}
\qquad (0 \le x < \infty). \tag{1.4}
$$

Here $J_a$ denotes the Bessel function of the first kind of order $a$.

These Bessel functions generalize the trigonometric functions since

$$
J_\lambda^{-1/2}(x) = \cos \lambda x \quad , \quad J_\lambda^{1/2}(x) = \frac{\sin \lambda x}{\lambda x} . \tag{1.5}
$$

Furthermore they are related to the ultraspherical polynomials as well as to the Laguerre polynomials via the limit formulas

$$
\begin{aligned}
\lim_{n\to\infty} R_n^{a,a}(\cos \tfrac{\lambda x}{n}) &= J_\lambda^a(x), \\
\lim_{n\to\infty} R_n^{a}(\tfrac{(\lambda x)^2}{4n}) &= J_\lambda^a(x),
\end{aligned}
\qquad (0 \le \lambda, x < \infty) \tag{1.6}
$$

where $R_n^{a,a}(z) = P_n^{a,a}(z)/P_n^{a,a}(1)$, $R_n^a(z) = L_n^a(z)/L_n^a(0)$. In view of the (distributional) orthogonality relation, see Watson [22, Chapter 13],

$$
\frac{1}{2^{2a}\,\Gamma^2(a+1)}\,\lambda^{2a+1} \int_0^\infty J_\lambda^a(x)\, J_\mu^a(x)\, x^{2a+1}\, dx = \delta(\lambda-\mu) \qquad (0 < \lambda,\mu < \infty), \tag{1.7}
$$

the functions $J_\lambda^a$ give rise to the definition of the Hankel integral transform which is known to be selfdual; see Titchmarsh [21, Chapter 8] and Everitt [5, section 1]. Finally, they are closely related to the Fourier-Bessel functions which form an important (discrete) orthogonal system on the interval [0,1]; see [22, Chapter 18] and [20].

From these and other properties of the Bessel functions, and in the spirit of the extensions of the classical orthogonal polynomials mentioned above, it is natural to ask for similar generalizations of the Bessel functions which satisfy a higher-order differential equation and which give rise to a "generalized" Hankel transform and its inverse. The purpose of this paper is to introduce such "Bessel-type functions" and to discuss some of their features.

These generalizations are realized by definitions which utilise *either* linear combinations and/or linear-differential combinations of known special functions, *or* by confluent limit processes. We can best illustrate the paths we follow in this paper by presenting the following diagram:

Jacobi polynomials —— [12, (2.1)] ——> Jacobi-type polynomials

$k(\alpha,\beta)\ (1-x)^{\alpha}(1+x)^{\beta}$ $\qquad$ $k(\alpha,\beta)\ (1-x)^{\alpha}(1+x)^{\beta} + M\ \delta(x+1) + N\ (x-1)$

↓ [12, (1.6)] $\qquad$ ↓ [12, (4.8)], (1.1)

Laguerre polynomials —— [12, (4.8)] ——> Laguerre-type polynomials

$k(\alpha)\ x^{\alpha}e^{-x}$ $\qquad$ $k(\alpha)\ x^{\alpha}e^{-x} + N\ \delta(x)$

↓ (1.6) $\qquad$ ↓ (2.2)

Bessel functions —— (1.8a) ——> **Bessel-type functions**

$\kappa(\alpha)\ x^{2\alpha+1}$ $\qquad$ $\kappa(\alpha)\ x^{2\alpha+1} + M\ \delta(x)$

The function entry under each special function name indicates the non-negative weight, on the interval $(-1,1)$ or $(0,\infty)$, involved in the orthogonality property or the associated linear differential equation. It is important to note in the diagram that:

(i) a horizontal arrow $\longrightarrow$ indicates a definition process either by a linear combination of special functions of the same type but of different orders, or by linear-differential combination of special functions of the same type and order;

(ii) a vertical arrow $\downarrow$ indicates a confluent limit process of one special function to give another special function;

(iii) one seeks a Bessel-type function in the box $\boxed{\phantom{xxxx}}$ which is consistent with the two definitions obtained by both the $\longrightarrow$ and $\downarrow$ processes.

For $\alpha \geq -1/2$ and $M \geq 0$, our Bessel-type functions are defined by

$$
\text{(1.8a)} \quad
\begin{aligned}
J_\lambda^{\alpha,M}(x) &= [1 + M\,h(\alpha,\lambda) + M\,h(\alpha-1,\lambda)\,\frac{1}{2x}\frac{d}{dx}]\;J_\lambda^\alpha(x) \\
&= [1 + M\,h(\alpha,\lambda)]\;J_\lambda^\alpha(x) - M\,h(\alpha,\lambda)\;J_\lambda^{\alpha+1}(x),
\end{aligned}
$$

where $0 \leq \lambda, x < \infty$ and

$$
\text{(1.8b)} \qquad h(\alpha,\lambda) = \frac{1}{\Gamma(\alpha+2)}\left(\frac{\lambda}{2}\right)^{2\alpha+2}.
$$

Clearly one has

$$
\text{(1.8c)} \qquad J_\lambda^{\alpha,0}(x) = J_\lambda^\alpha(x) \quad , \quad J_0^{\alpha,M}(x) = J_0^\alpha(x) \equiv 1 \quad , \quad J_\lambda^{\alpha,M}(0) = 1
$$

and, in view of (1.5), the properties

$$
J_\lambda^{-1/2,M}(x) = \cos\lambda x + \frac{\lambda M}{\sqrt{\pi}}\left\{\cos\lambda x - \frac{\sin\lambda x}{\lambda x}\right\},
$$

$$
J_\lambda^{1/2,M}(x) = \frac{\sin\lambda x}{\lambda x} + \frac{\lambda^3 M}{6\sqrt{\pi}}\left\{\frac{3}{(\lambda x)^2}\cos\lambda x + \left[\,\frac{1}{\lambda x} - \frac{3}{(\lambda x)^3}\,\right]\sin\lambda x\right\}.
$$

In the subsequent section we first obtain the new functions (1.8) as a certain confluent limit of the Laguerre-type polynomials (1.1). Thus, $J_\lambda^{\alpha,M}(x)$ can be represented in terms of a ${}_1F_2$-hypergeometric function being analytic in the complex x-plane. Our first main result is to establish, for any $\alpha \in \mathbb{N}_0$, a differential equation of order $2\alpha + 4$ which has the Bessel-type functions as its solutions and in which the coefficients of all derivatives are independent of the eigenvalue parameter $\lambda$; see Theorem 2.4 below. For $M = 0$, this equation reduces to the classical Bessel equation (1.2a).

Furthermore we give a necessary and sufficient condition in terms of $2\alpha + 4$ linear equations with $\alpha + 2$ unknowns so that the new equation (2.10) can be made formally symmetric; see [6, section 3] and Theorem 2.7 below. For the first three cases $\alpha = 0,1,2$, this leads to a differential equation of the form

$$(1.9) \qquad L_x^{\alpha,M} J_\lambda^{\alpha,M}(x) + \Lambda_{\alpha,M} x^{2\alpha+1} J_\lambda^{\alpha,M}(x) = 0 \quad (0 < x < \infty,\ M \geq 0),$$

where

$$\Lambda_{\alpha,M} = M\,\lambda^{2\alpha+4} + 2^{2\alpha+2}(\alpha+2)!\,\lambda^2$$

and where the differential expression $L_x^{\alpha,M}$ is respectively given by

$$(1.10a) \qquad L_x^{0,M} y = M\{-(xy'')'' + (\tfrac{9}{x}y')'\} + 8\,(xy')' \quad (y \in C^{(4)}),$$

$$(1.10b) \qquad L_x^{1,M} y = M\{(x^3y''')''' - (33xy'')'' + (\tfrac{255}{x}y')'\} + 96(x^3y')' \quad (y \in C^{(6)}),$$

$$(1.10c) \qquad L_x^{2,M} y = M\{-(x^5y^{iv})^{iv} + (78x^3y''')''' - (1809xy'')'' + (\tfrac{11025}{x}y')'\} + 2^8 4!(x^5y')' \quad (y \in C^{(8)}).$$

In these differential equations we note that the weight function is $x^{2\alpha+1}$, for $\alpha = 0,\ 1,\ 2$ respectively.

In section 3, we provide for any $\alpha \in \mathbb{N}_0$ a pair of $\lambda$-independent differential expressions both of order $2\alpha + 2$, $A_x^{\alpha,N}$ and $B_x^{\alpha,N}$, which satisfy

$$(1.11)\qquad A_x^{\alpha,M} J_\lambda^\alpha(x) = J_\lambda^{\alpha,M}(x)$$

$$(0 \le \lambda, x < \infty)$$

$$(1.12)\qquad B_x^{\alpha,M} J_\lambda^{\alpha,M}(x) = [1 + M\,h(\alpha,\lambda)]^2\, J_\lambda^\alpha(x)$$

(with $h(\alpha,\lambda)$ as in (1.8b)); see Theorem 3.2 below. A combination of both operations then yields a differential equation of order $4\alpha + 4$ for the Bessel-type functions, which coincides with (2.10–11) only for $\alpha = 0$. It remains an open question whether for any other parameter $\alpha \in \mathbb{N}$, the order of the equation can be reduced to $2\alpha+4$ in a *canonical way* to give a new proof of Theorem 2.4.

It is of particular interest that the "inverse" expression $B_x^{\alpha,M}$ is related to the expression $A_x^{\alpha,M}$ via the identity

$$(1.13)\qquad B_x^{\alpha,M} y(x) = x^{-2\alpha-1}[\, x^{2\alpha+1} A_x^{\alpha,M}]^+ y(x) \qquad (y \in C^{(2\alpha+2)}(0,\infty)).$$

As usual, the notation + denotes the Lagrange adjoint differential expression. This is carried out in Section 4.

These results can be used to extend the orthogonality relation of the classical Bessel functions, (1.7), to one for the Bessel-type functions. In fact it may be shown that for $\alpha \in \mathbb{N}_0$,

$$(1.14)\qquad \begin{aligned} &\frac{2}{\Gamma(\alpha+1)} \int_0^b J_\lambda^{\alpha,M}(x)\, J_\mu^{\alpha,M}(x)\, x^{2\alpha+1}\, dx + M\, J_\lambda^{\alpha,M}(0)\, J_\mu^{\alpha,M}(0) \\ &= \frac{2}{\Gamma(\alpha+1)}[1+M\,h(\alpha,\lambda)][1+M\,h(\alpha,\mu)] \int_0^b J_\lambda^\alpha(x) J_\mu^\alpha(x) x^{2\alpha+1}\, dx + O(\tfrac{1}{b}) \quad (b \to \infty). \end{aligned}$$

In view of (1.7), this implies that

$$(1.15)\qquad \begin{aligned} &\frac{1}{\Gamma(\alpha+1)} (\tfrac{\lambda}{2})^{2\alpha+1} [1 + M\,h(\alpha,\lambda)]^{-2} \\ &\cdot \{\frac{2}{\Gamma(\alpha+1)} \int_0^\infty J_\lambda^{\alpha,M}(x)\, J_\mu^{\alpha,M}(x)\, x^{2\alpha+1}\, dx + M\, J_\lambda^{\alpha,M}(0)\, J_\mu^{\alpha,M}(0)\} = \delta(\lambda-\mu). \end{aligned}$$

The proof of formula (1.14) is carried out below for $\alpha = 0$. The general case $\alpha \in \mathbb{N}_0$ is left to a subsequent paper, where we also intend to present further proofs and interpretations of the orthogonality (1.15) and to discuss its impact on a "generalized" Hankel (inverse) transform.

At this place it may be remarked that in 1959, the first named author [5] investigated the differential equation

$$(1.16)\quad (-1)^k\Big[y^{(2k)}(x) + \frac{2k(\alpha+1/2)}{x}\, y^{(2k-1)}(x)\Big] = \lambda\, y(x) \quad (x \in (0,\infty)),$$

for $k \in \mathbb{N}$ and $\alpha \geq -1/2$. Obviously this equation generalizes the classical Bessel equation (1.2a) (case $k = 1$) to higher order equations in a way that is different from the case presented here. Integral representations for the solution of (1.16) were given in [5, section 4] which involve certain symmetric Fourier kernels introduced by Guinand [8]. For $k = 2$, for instance, the solution reads

$$y(x) = x^{-\alpha} J_{\alpha,2}(x) := \frac{2^{1-2\alpha}}{\Gamma(\alpha+1/2)\,\Gamma(1/2)} \cdot$$
$$\cdot \int_0^1 [\sin(xt) + \cos(xt) - e^{-xt}]\,(1 - t^4)^{\alpha-1/2}\, dt.$$

The paper [5] is mainly interested in those generalizations of the Hankel transform which preserve its symmetry as an integral operator; see [21, Chapter. 8]. The results in [5] show that unless $k = 1$, it is essentially only the case $\alpha = 0$, in which the solution of (1.16) gives rise to a symmetric orthogonal transform; in this case the differential equation (1.16) can be transformed into the symmetric form

$$(-1)^k\, (x\, y^{(k)}(x))^{(k)} = \lambda\, x\, y(x) \quad (x \in (0,\infty)).$$

This report is a shortened version of a paper to appear in the *Journal for Computational and Applied Mathematics.*

## References


1. W. N. Bailey, Generalized Hypergeometric Series (Hafner Publishing Company, New York, 1972).

2. S. Bochner, *Über Sturm-Liouvillesche Polynomsysteme*, Math. Z. 89 (1929), 730-736.

3. A. Erdelyi, W. Magnus, F. Oberhettinger and F. G. Tricomi, Higher Transcendental Functions, I (McGraw-Hill, New York, 1953).

4. A. Erdelyi, W. Magnus, F. Oberhettinger and F. G. Tricomi, Higher Transcendental Functions, II (McGraw-Hill, New York, 1953).

5. W. N. Everitt, *On a generalization of Bessel functions and a resulting class of Fourier kernels*, Quart. J. Math. Oxford (2) 10 (1959), 270-279.

6. W. N. Everitt and L. L. Littlejohn, *Orthogonal polynomials and spectral theory: a survey*, in: Orthogonal Polynomials and Their Applications (Edited by C. Brezinski, L. Gori and A. Ronveaux; J. C. Balzer AG, Scientific Publishing Company), IMACS 9 (1991), 21-55.

7. G. Gasper and M. Rahman, Basic Hypergeometric Series (Cambridge University Press, 1990).

8. A. P. Guinand, *A class of Fourier kernels*, Quart. J. Math. Oxford (2) 1 (1950), 191-193.

9. E. L. Ince, Ordinary Differential Equations (Dover Publishing Company, New York, 1956).

10. J. Koekoek and R. Koekoek, *On a differential equation for Koornwinder's generalized Laguerre polynomials*, Proc. Amer. Math. Soc. 112 (1991), 1045-1054.

11. R. Koekoek, Generalizations of the Classical Laguerre Polynomials and Some q-Analogues, Ph. D. thesis, The Technical University of Delft, Delft, The Netherlands, November 1990.

12. T. H. Koornwinder, *Orthogonal polynomials with weight function* $(1-x)^{\alpha}(1+x)^{\beta} + M\,\delta(x+1) + N\,\delta(x-1)$, Canad. Math. Bull. 27 (2) (1984), 205-214.

13. A. M. Krall, *Orthogonal polynomials satisfying fourth-order differential equations*, Proc. Roy. Soc. Edinb. (A) 87 (1981), 271-288.

14. A. M. Krall and L. L. Littlejohn, *Differential equations and Bochner-Krall orthogonal polynomials*, Technical Report, Department of Mathematics and Statistics, Utah State University, Logan, U.S.A., 1990.

15. H. L. Krall, *Certain differential equations for Tchebycheff polynomials*, Duke Math. J. 4 (1938), 705-718.

16. H. L. Krall, *On orthogonal polynomials satisfying a certain fourth-order differential equation*, The Pennsylvania State College Studies No 6, The Pennsylvania State College, State College, U.S.A., 1940.

17. L. L. Littlejohn, Nonclassical Orthogonal Polynomials and Differential Equations, Ph. D. thesis, The Pennsylvania State University, State College, U.S.A., 1981.

18. C. Markett, *Product formulas for Bessel, Whittaker, and Jacobi functions via the solution of an associated Cauchy problem*, in: Functional Analysis and Approximation (Edited by P. L. Butzer, R. L. Stens and B. Sz.-Nagy; Birkhäuser, Basel) ISNM 65 (1984), 449-462.

19. C. Markett, Produktformeln für Eigenfunktionen singulärer Sturm-Liouville Gleichungen und verallgemeinerte Translationsoperatoren, Habilitationsschrift, Rheinisch-Westfälische Technische Hochschule Aachen, Aachen, Germany, April 1985.

20. C. Markett, *Product formulas and convolution structure for Fourier-Bessel series*, Constr. Approx. 5 (1989), 383-404.

21. E. C. Titchmarsh, Fourier Integrals (Oxford University Press, 1962).

22. G. N. Watson, The Theory of Bessel Functions (Cambridge University Press, 1944).

Acknowledgement: The authors thank Professor Philip Hsieh and his colleagues in the Department of Mathematics and Statistics at Western Michigan University for the invitation to include this report in the Proceedings of the International Symposium on Ordinary Differential Equations and Applications held at the University in May 1993.

W. N. Everitt
School of Mathematics
and Statistics
University of Birmingham
BIRMINGHAM B15 2TT
England, UK.

C. Markett
Lehrstuhl A für Mathematik
RWTH Aachen
Templergraben 55
W-5100 AACHEN
Germany

# A MECHANISM FOR CONNECTION

H. GINGOLD
Department of Mathematics
West Virginia University
P.O. Box 6310
Morgantown, WV 26506-6310, USA
E-Mail UN051203@WVNVMS.WVNET.EDU

ABSTRACT

A non traditional mechanism for connecting solutions of second order linear differential equations at a turning point is presented. An application to wave propagation in inhomogeneous media is given.

**1. Introduction.** The purpose of this work is to describe a mechanism of connection which is a byproduct of studies of the second order differential equation $y''=\phi(x,\varepsilon)y$ with coalescing transition points and moving similarities, [4]. In these studies one of the asymptotic approximations developed, [1], happens to have "invariant and global properties". Since this asymptotic approximation is valid at a turning point it becomes a natural candidate to be utilized in a new and nontraditional technique of connection. This work differs considerably from the methods based on Langer type transformations, e.g. [5], and from methods of analytic continuation [6,7]. Langer transformations and special functions are not utilized. The traditional methods of connection employ the L-G (Liouville-Green) approximations which have great physical significance. The method I propose is based on a renormalization of incident and reflected waves. This renormalization contains the L-G approximation as a special case, it is valid for the wave equation with *high* and *low frequencies* as *well.* It handles turning points at infinity with ease. It has use in quantum physics [2], and in wave propagation problems in inhomogeneous media [3]. The method proposed can handle turning points of *finite* and *infinite order.*

The order of events in this article is as follows. In section 2 we describe the asymptotic decomposition which will be utilized. In section 3 the connection method is described and a connection matrix is constructed at a turning point. In section 4 it is shown that the approximations proposed could be valid for low frequencies. The derivation of a local principle is based on this. Familiarity with [1 ] is expected.

**Preliminaries.** Given a second order linear O.D.E. $y''=\phi(x)y$ on (c,b) its companion matrix system is given by

$$Y' = AY \ , \ A = \begin{bmatrix} 0 & 1 \\ \phi(x) & 0 \end{bmatrix} \tag{2.1}$$

where A and Y are two by two matrix functions. We summarize some of the transformations and results form [1] which will be needed in this paper.

ASSUMPTION 2.1. *The mapping* $\phi(x) \in C^2(c,b)$ *is not zero on* (c,b) *however,* $\phi(c^+) = 0$ *or* $\phi(b^-) = 0$ *or* $\phi(c^+) = \phi(b^-) = 0$ *is allowed.* Arg $\phi(x) \in C[c,b]$ *and* Log $\phi(x) \in C(c,b)$ *where* Arg *and* Log *denote the principal argument and logarithm respectively. Also*

$$\phi(x) + \left[\frac{1}{4}\frac{\phi'(x)}{\phi(x)}\right]^2 \neq 0 \ on \ [c,b] \tag{2.2}$$

CONVENTION 2.2. *We denote the following mappings by*

$$\sqrt[4]{\phi(x)} = \exp\frac{1}{4}\,\mathrm{Log}\,\phi(x) \quad , \quad \sqrt{\phi(x)} = \exp\frac{1}{2}\,\mathrm{Log}\,\phi(x) \tag{2.3}$$

$$\ell(x) = \frac{\frac{1}{4}\frac{\phi'(x)}{\phi(x)}}{\sqrt{\phi(x)}} \quad , \quad \lambda(x) = \sqrt{\phi(x) + \left(\frac{1}{4}\frac{\phi'(x)}{\phi(x)}\right)^2} \tag{2.4}$$

$$r(x) = -\frac{i}{2}\cdot\frac{\ell'}{1+\ell^2} \quad , \quad u(x) = \frac{[1-i\ell]^{1/4}}{[1+i\ell]^{1/4}} \tag{2.5}$$

$$[1-i\ell]^{1/4} = \exp\frac{1}{4}\,\mathrm{Log}\,(i-i\ell) \quad , \quad [1+i\ell]^{1/4} = \exp\frac{1}{4}\,\mathrm{Log}\,(1+i\ell) \tag{2.6}$$

and

$$e\,(x,\tau) = \exp\int_\tau^x \lambda(s)ds \tag{2.7}$$

$$e_+(x) := \left\{\exp\int_c^x \left[\lambda(s) - \frac{1}{4}\frac{\phi'(s)}{\phi(s)}\right]ds\right\}[\phi(x)]^{\frac{1}{4}} \ , \ x \in (c,b)$$

$$e_-(x) = \left\{\exp\int_c^x \left[\lambda(s) + \frac{1}{4}\frac{\phi'(s)}{\phi(s)}\right]\right\}[\phi(x)]^{-\frac{1}{4}} \ , \ x \in (a,c)$$

Notice that $e_+(x)$ and $e_-(x)$ are solutions of the scalar ordinary differential equation $y' = \lambda y$. Also notice that (2.2) implies that (2.6) is well defined. We will also need

ASSUMPTION 2.3 *We have* Re $\lambda(x) > 0$, $x \in [c,b]$ *and for* $c < x_1 < x_2 < b$

$$\lim_{x_1 \to c^+} \left[ \operatorname{Re} \int_{x_1}^{x_2} \lambda(s)ds \right] = +\infty \qquad \int_c^b |r(\tau)|d\tau \le \zeta < 1 \tag{2.8}$$

We are ready to formulate a theorem whose proof can be extracted from [1].

THEOREM 2.4. *Let* on each interval (c,b) and (a,c) *assumptions* 2.1, 2.3 *and convention* 2.2 *hold. Then, separately on each interval the differential system* (2.1) *possesses a fundamental solution*

$$Y = W_1 W_2 (I+P) E \tag{2.9}$$

where

$$W_1 = \begin{bmatrix} \sqrt[4]{\Phi}^{-1} & 0 \\ 0 & \sqrt[4]{\Phi} \end{bmatrix} \begin{bmatrix} 1 & 1 \\ 1 & -1 \end{bmatrix}, \quad W_2 = \left(\frac{1}{2}\right) \begin{bmatrix} u+u^{-1} & -(u-u^{-1}) \\ i(u-u^{-1}) & -i(u+u^{-1}) \end{bmatrix}$$

$$E_+ = \begin{bmatrix} e_+(x) & 0 \\ 0 & e_+^{-1}(x) \end{bmatrix}, \quad E_- = \begin{bmatrix} e_-(x) & 0 \\ 0 & e_-^{-1}(x) \end{bmatrix} \tag{2.10}$$

It is understood that on the interval (c,b), $E = E_+$ in (2.9) and on the interval (a,c), $E = E_-$. The matrix $P = (p_{jk})$, $j,k = 1,2$, is such that $p_{jk} \in C[c,b]$ or $p_{jk} \in C[a,b]$. The entries $p_{jk}$ satisfy the following decoupled integral equations;

$$p_{jk}(x) = n_{jk}(x) + L_{jk} p_{jk}, \; j,k = 1,2. \tag{2.11}$$

The linear operators $L_{jk}$ and the nonhomogeneous terms $n_{jk}(x)$ are given below.

$$L_{11}p_{11} := \int_{\alpha_{11}}^{x} \int_{\alpha_{21}}^{\tau_1} r(\tau_1) r(\tau_2) e^2(\tau_2,\tau_1) p_{11}(\tau_2)\, d\tau_2\, d\tau_1; \tag{2.12}$$

$$n_{11}(x) = \int_{\alpha_{11}}^{x} \int_{\alpha_{21}}^{\tau_1} r(\tau_1) r(\tau_2) e^2(\tau_2,\tau_1)\, d\tau_2\, d\tau_1; \tag{2.13}$$

$$L_{22}p_{22} := \int_{\alpha_{22}}^{x} \int_{\alpha_{12}}^{\tau_1} r(\tau_1) r(\tau_2) e^2(\tau_1,\tau_2) p_{22}(\tau_2)\, d\tau_2\, d\tau_1; \tag{2.14}$$

$$n_{22}(x) = \int_{\alpha_{22}}^{x} \int_{\alpha_{12}}^{\tau_1} r(\tau_1) r(\tau_2) e^2(\tau_1,\tau_2)\, d\tau_2\, d\tau_1; \tag{2.15}$$

$$L_{12}p_{12} := \int_{\alpha_{12}}^{x} \int_{\alpha_{22}}^{\tau_1} r(\tau_1) r(\tau_2) e^2(x,\tau_1) p_{12}(\tau_2)\, d\tau_2\, d\tau_1; \tag{2.16}$$

$$n_{12}(x) = \int_{\alpha_{12}}^{x} r(\tau_1)e^2(x,\tau_1)\, d\tau_1 \tag{2.17}$$

$$L_{21}p_{21} := \int_{\alpha_{21}}^{x}\int_{\alpha_{11}}^{\tau_1} r(\tau_1)r(\tau_2)e^2(\tau_1,x)p_{21}(\tau_2)\, d\tau_2\, d\tau_1; \tag{2.18}$$

$$n_{21}(x) = \int_{\alpha_{21}}^{x} r(\tau_1)e^2(\tau_1,x)\, d\tau_1. \tag{2.19}$$

On the interval (c,b) we choose

$$\alpha_{12} = b \ , \ \alpha_{11} = \alpha_{22} = \alpha_{21} = c \tag{2.20}$$

On the interval (a,c) we choose

$$\alpha_{21} = a \ , \ \alpha_{11} = \alpha_{22} = \alpha_{12} = c. \tag{2.21}$$

These choices make $L_{jk}$ contractions and they make $n_{jk}$ bounded functions. (There is a slight deviation from the notation of $e(x,\tau)$ adopted in [1].)

PROOF: This theorem is a straightforward application of theorem 3.3 in [1].

**3. A Mechanism for Connection.** In this section we assume that c is a turning point and proceed to determine a connection matrix at c. We need the following.

NOTATION 3.1 *For* $x,x_0 \in (c,b)$ *or* $x,x_0 \in (a,c)$

$$I_+(x,x_0) = \int_{x_0}^{x} \left[\lambda(s) - \frac{1}{4}\frac{\phi'(s)}{\phi(s)}\right] ds \ , \ I_-(x,x_0) = \int_{x_0}^{x} \left[\lambda(s) + \frac{1}{4}\frac{\phi'(s)}{\phi(s)}\right] ds \tag{3.1}$$

Notice that

$$I_+(x,x_0) = \int_{x_0}^{x} \frac{\phi(s)ds}{\left[\lambda(s) + \frac{1}{4}\frac{\phi'(s)}{\phi(s)}\right] ds} \ , \ I_-(x,x_0) = \int_{x_0}^{x} \frac{\phi(s)ds}{\left[\lambda(s) - \frac{1}{4}\frac{\phi'(s)}{\phi(s)}\right]} ds \tag{3.2}$$

ASSUMPTION 3.2 *Let assumption* 2.1, *convention* 2.2 *and notation* 3.1 *hold on* (a,b) *and on* (c,b). *Let*

$$\phi(c) = 0 \ , \ \text{Re}\, \ell(c^+) = +\infty \ , \ \text{Re}\, \ell(c^-) = -\infty \tag{3.3}$$

$$\text{Im}\, \ell(c) = O(1) \ \textit{as} \ x \to c^- \ \textit{and} \ x \to c^+ \tag{3.4}$$

$$r(x)\phi^{-1/2}(x) = O(1) \ \textit{as} \ x \to c^- \ \textit{and} \ x \to c^+ \tag{3.5}$$

As we proceed the reader is advised to keep in mind two generic examples $\phi(x) = \varepsilon^{-2}x^{2m}$ and $\phi(x) = \varepsilon^{-2}\exp -x^{-2m}$ $m = 1,2,\ldots$ with $a = -\infty$ , $c = 0$ and $b = +\infty$ and verify that all conditions required in the sequel for a successful connection are met. Let us focus on the evaluations of $S(c^+)$ and $S(c^-)$. To this end we will need the evaluations of $W_2(c^+)$ and $W_2(c^-)$. Put $1 - i\ell = \zeta_1 e^{i\theta_1}$ and $1 + i\ell = \zeta_2 e^{i\theta_2}$ where $\zeta_1, \zeta_2 \geq 0$ and $\theta_1, \theta_2$ are real. Let assumption 3.2 hold. Then, as $x \to c^+$ or as $x \to c^-$

$$\frac{\zeta_1}{\zeta_2} = \frac{|1 - i\ell|}{|1 + i\ell|} = \frac{|i\ell|\ |1 - (i\ell)^{-1}|}{|i\ell|\ |1 + (i\ell)^{-1}|} \quad \text{and} \quad \lim_{x\to c^+} \frac{\zeta_1(x)}{\zeta_2(x)} = 1 = \lim_{x\to c^-} \frac{\zeta_1(x)}{\zeta_2(x)} \tag{3.6}$$

Therefore,

$$u(c+) = \frac{[1 - i\ell(c^+)]^{1/4}}{[1 + i\ell(c^-)]^{1/4}} = e^{-\frac{\pi i}{4}} \quad \text{and} \quad u(c^-) = \frac{[1 - i\ell(c^-)]^{1/4}}{[1 + i\ell(c^-)]^{1/4}} = e^{i\frac{\pi}{4}} \tag{3.7}$$

Also

$$e(x,x_0) = \frac{[\phi(x)]^{1/4}}{[\phi(x_0)]^{1/4}} \exp I_+(x,x_0) = \frac{[\phi(x_0)]^{1/4}}{[\phi(x)]^{1/4}} \exp I_-(x,x_0). \tag{3.8}$$

We are ready now to proceed with the evaluation of $Y(c^+)$ and $Y(c^-)$. Notice that due to (2.9) we can write $Y = S(I + \hat{P})$ with $S = W_1 W_2 E$ and $\hat{P} = E^{-1}PE$. Of course we assume that theorem 2.4 is valid on each interval (a,c) and (c,b). Recalling the definition of $W_2$ we obtain

$$W_2(c^+) = \begin{bmatrix} \cos\frac{\pi}{4} & i\sin\frac{\pi}{4} \\ \sin\frac{\pi}{4} & -i\cos\frac{\pi}{4} \end{bmatrix} = \left(\frac{\sqrt{2}}{2}\right) \begin{bmatrix} 1 & i \\ 1 & -i \end{bmatrix} \tag{3.9}$$

Since $u(c^-) = e^{\pi i/4}$ we have

$$W_2(c^-) = \left(\frac{\sqrt{2}}{2}\right) \begin{bmatrix} 1 & -i \\ -1 & -i \end{bmatrix} \tag{3.10}$$

On the interval (c,b) we put $E_+ = DM_+$ where

$$D = \begin{bmatrix} [\phi(x)]^{1/4} & 0 \\ 0 & \phi(x)^{-1/4} \end{bmatrix}, \quad M_+ = \begin{bmatrix} \exp I_+(x,c) & 0 \\ 0 & \exp -I_+(x,c) \end{bmatrix} \tag{3.11}$$

Notice that $I_+(x,c)$ is continuous at $x=c^+$ thanks to assumptions 2.1, 2.3 and 3.2. On the interval (a,c) we put $E_- = D^{-1}M_-$ where

$$M_- = \begin{bmatrix} \exp I_-(x,c) & 0 \\ 0 & \exp -I_-(x,c) \end{bmatrix} \tag{3.12}$$

Notice that $I_-(x,c)$ is continuous at $x=c^-$. It can be verified that on (c,b)

$$S(x) = D^{-1}H\left\{W_2(c^+) + \Delta(x)\right\}DM_+ \tag{3.13}$$

with

$$H = \begin{bmatrix} 1 & 1 \\ 1 & -1 \end{bmatrix}, \quad W_2(c^+) = \begin{bmatrix} 1 & i \\ 1 & -i \end{bmatrix} \tag{3.14}$$

where $\Delta(x) = O(\ell^{-1}(x))$ and therefore

$$\lim_{x \to c^+} D^{-1}H\Delta DM_+ = O \tag{3.15}$$

Moreover, $HW_2(c^+) = \sqrt{2}\begin{bmatrix} 1 & 0 \\ 0 & i \end{bmatrix}$ (3.16)

and thus

$$S(c^+) = \sqrt{2}\begin{bmatrix} 1 & 0 \\ 0 & i \end{bmatrix} \tag{3.17}$$

The calculation of $S(c^-)$ proceeds similarly. However

$$S(x) = D^{-1}H\left\{W_2(c^-) + \Delta\right\}D^{-1}M_- \tag{3.18}$$

$$HW_2(c^-) = \sqrt{2}\begin{bmatrix} 0 & -i \\ 1 & 0 \end{bmatrix}, \; D^{-1}\begin{bmatrix} 0 & -i \\ 1 & 0 \end{bmatrix}D^{-1} = \begin{bmatrix} 0 & -i \\ 1 & 0 \end{bmatrix}, \quad S(c^-) = \sqrt{2}\begin{bmatrix} 0 & -i \\ 1 & 0 \end{bmatrix}. \tag{3.19}$$

Because of the choice of the lower limits in (2.11) and since $e^2(x,\tau) \to 0$ as $x \to c^+$ for $(\tau - c) \geq \delta > 0$ where $\delta > 0$, it can be verified that $p_{jk}(c^+) = 0$, j,k=1,2. One may then be tempted to believe that $Y(c^+) = S(c^+)$ but this will be erroneous! Indeed $Y(c^+) \neq S(c^+)$ because $\hat{P}(c^-) \neq 0$. We attempt now to calculate $\hat{P}(c^+)$. Since

$$E^{-1}PE = \hat{P} = (\hat{p}_{jk}), \; j,k = 1,2 \tag{3.20}$$

we have

$$\hat{p}_{jj}(x) = p_{jj}(x)\ ,\ j=1,2,\ \text{ and }\ p_{jj}(c^+) = 0 \tag{3.21}$$

Since

$$\hat{p}_{21}(x) = e_+^2(x)p_{21}(x) \text{ we also have } \hat{p}_{21}(c^+) = 0, \tag{3.22}$$

We are going now to calculate $\hat{p}_{12}(c^+)$. By $\hat{P} = E_+^{-1}PE_+$ we have

$$\hat{p}_{12}(x) = e_+^{-2}(x)p_{12}(x) \tag{3.23}$$

where

$$n_{12}(x) = \int_b^x r(\tau_1)e^2(x,\tau_1)d\tau_1 \tag{3.24}$$

$$L_{12}p_{12} = \int_b^x\int_c^{\tau_1} r(\tau_1)r(\tau_2)e^2(x,\tau_1)p_{12}(\tau_2)d\tau_2 d\tau_1 \tag{3.25}$$

Multiply each side of (3.26) and (3.27) by $e_+^2(x)$ to obtain

$$\hat{p}_{12} = \hat{n}_{12} + \hat{L}_{12}\hat{p}_{12} \tag{3.26}$$

where

$$\hat{n}_{12} = \int_b^x r(\tau_1)\phi^{-1/2}(\tau_1)\exp 2I_+(c,\tau_1)d\tau_1 \tag{3.27}$$

and

$$\hat{L}_{12}\hat{p}_{12} = \int_b^x\int_c^{\tau_1} r(\tau_1)r(\tau_2)e^2(\tau_2,\tau_1)\hat{p}_{12}(\tau_2)d\tau_2 d\tau_1 \tag{3.28}$$

We notice that in (3.28) we have $c \leq \tau_2 \leq \tau_1$ and therefore $|e^2(\tau_2,\tau_1)| \leq 1$ and $\|\hat{L}_{12}\hat{p}_{12}\| \leq \left[\int_c^b |r(\tau)|d\tau\right]^2 \|\hat{p}_{12}\|$. The norm $\|f\| = \sup |f(x)|$ for a function $f(x)$ defined on $[c,b]$ is obtained by taking the supremum over $x \in [c,b]$. Thus, $\hat{L}_{12}$ is a contraction mapping. Notice that in (3.24) $|e^2(x,\tau_1)| \leq 1$ for $x \leq \tau_1 \leq b$. The limit $\hat{n}_{12}(c^+)$ exists because of (3.5) in assumption 3.2 and because $I_+(c,\tau_1)$ is bounded for $c \leq \tau_1 \leq x_0$ and $x_0$ close to $c$. We conclude that

$$\hat{p}_{12}(c^+) = \sum_{v=0}^{\infty} \hat{L}_{12}\hat{n}_{12} \tag{3.29}$$

where the series on the right hand side of (3.29) is absolutely convergent and is obtained by substituting $x=c$ in the resolvent series.

A similar analysis is applicable to the interval $(a,c)$. Thus, $p_{jk}(c^-) = 0$ , $j,k=1,2$. We define $\hat{P} = (\hat{p}_{jk})$ $j,k=1,2$, in a similar manner, namely $\hat{P} = E_-^{-1}PE_-$. Consequently,

$$\hat{p}_{jj}(x) = p_{jj}(x),\ j=1,2,\ \hat{p}_{jj}(c^+) = 0. \tag{3.30}$$

Since $\hat{p}_{12}(x) = e_-^{-2}(x)p_{12}(x)$ (3.31)

we have $\hat{p}_{12}(c^+) = 0$. It remains to analyse $\hat{p}_{21}(x)$ where

$$\hat{p}_{21}(x) = e_-^2(x)p_{21}(x) \tag{3.32}$$

Multiply each side of the integral equation

$$p_{21} = n_{21} + L_{21}p_{21} \tag{3.33}$$

by $e_-^2(x)$ to obtain

$$\hat{p}_{21}(x) = \hat{n}_{21}(x) + \hat{L}_{21}\hat{p}_{21} \tag{3.34}$$

where

$$\hat{n}_{21}(x) = e_-^2(x)n_{21}(x) = \int_a^x r(\tau_1)\phi^{-1/2}(\tau_1)(\exp 2I_-(\tau_1,c))d\tau_1 \tag{3.35}$$

and

$$\hat{L}_{21}\hat{p}_{21} = \int_a^x\int_c^{\tau_1} r(\tau_1)r(\tau_2)e^2(\tau_1,\tau_2)\hat{p}_{21}(\tau_2)d\tau_2 \tag{3.36}$$

It can be easily verified that $\hat{L}_{21}$ is a contraction mapping and that $\hat{p}_{21}(c^-)$ is obtained by substituting $c$ in the resolvent series

$$\hat{p}_{21}(x) = \sum_{v=0}^{\infty}\hat{L}_{21}^v\hat{n}_{21} \tag{3.37}$$

We are ready now to calculate the connection matrix $K$ at $x=c$. It is defined by

$$S(c^-)\begin{bmatrix}1 & 0\\ \hat{p}_{21}(c^-) & 1\end{bmatrix}K = S(c^+)\begin{bmatrix}1 & \hat{p}_{12}(c^+)\\ 0 & 1\end{bmatrix} \tag{3.38}$$

Solving for $K$ we obtain

$$K = i\begin{bmatrix}0 & 1\\ 1 & [\hat{p}_{12}(c^+) - \hat{p}_{21}(c^-)]\end{bmatrix} \tag{3.39}$$

We are ready to formalize the recent discussion in

THEOREM 3.3 *Let assumptions* 2.1, 2.3 *and* 3.2 *hold on each of the intervals* (a,c) *and* (c,b). *Denote by* $Y(c^-)$ *and* $Y(c^+)$ *limiting values of certain two fundamental matrix solutions on* (a,c) *and* (c,b) *respectively. Then,* $Y^{-1}(c^-)Y(c^+) = K$ *where* $K$ *is given by* (3.39).

REMARK. The singularity introduced via $W_1$ in the L-G approximation, is at the turning point $x=c$, canceled out by $E$ in the formula $Y = W_1W_2E(I+\hat{P})$.

**4. Approximations for low frequencies and application.** Recently [3] proposed four local principles for wave propagation in inhomogeneous media. The proof was given in the setting of the one dimensional wave equation

$$y''=\phi(x)y \quad , \quad \phi(x)= -\frac{w^2}{c^2(x)} \tag{4.1}$$

where $w$ is the frequency and $c(x)$ is the speed of propagation. This was done for the range $[c'(x)]^2 < 4w^2$. It is the purpose of this section to demonstrate the validity of the proposed approximations as $w^2 \to 0$. As a byproduct we will derive a local amplitude principe. It is quite unusual in wave propagation problems to find approximations in "wavelike forms" for low frequencies, although this family of problems is of great importance. To remedy this situation we have

THEOREM 4.1 *Let* $c(x) \in C^2[a,b]$ *such that*

$$0 < m \leq c(x) \leq M \quad , \quad [c'(x)]^2 > 4w^2 \quad , \quad x \in [a,b] \tag{4.2}$$

$m,M$ *are positive constants. Then, the differential system* (2.1) *possesses a fundamental solution*

$$Y = W_1W_2(I+P)E = W_1W_2E(I+\hat{P}) \quad , \quad \hat{P} = E^{-1}PE \tag{4.3}$$

*where* $W_1,W_2$ *are given in* (2.9)

$$E = \begin{bmatrix} e(x,\alpha) & 0 \\ 0 & e^{-1}(x,\alpha) \end{bmatrix} \tag{4.4}$$

*The entries of the matrix* $P = (p_{jk})$ , $jk=1,2$, *satisfy the integral equations* (2.11) *and* $p_{jk} \in C[c,b]$ *we may choose in* (2.12) *to* (2.16), $\alpha_{jk} = \alpha \in [c,b]$, *one lower bound in all integrals. Moreover, the entries* $\hat{p}_{jk}$ *satisfy the following estimates as* $w \to 0$

$$p_{jj} = O(w^2) \quad , \quad \hat{p}_{jj} = O(w^2) \quad , \quad j=1,2 \tag{4.5}$$

$$p_{jk} = O(w^3) \quad , \quad \hat{p}_{jk} = O(w^3) \quad \text{for} \quad j,k=1,2 \quad \text{and} \quad j \neq k \tag{4.6}$$

PROOF: It can easily be verified that all conditions of theorem 3.2 in [1] are satisfied. Indeed, because of our assumptions

$$\lambda(x) = \sqrt{\left[\frac{1}{2}\frac{c'(x)}{c(x)}\right]^2 - \frac{w^2}{c^2(x)}} \in C[a,b] \ , \ r(x) = \frac{wc''(x)}{4w^2 - [c'(x)]^2} \in C[a,b] \tag{4.7}$$

Therefore, the estimates (3.57), (3.58) in [1] combined with (4.7) yield (4.5) and (4.6) for $p_{jk}$. The conclusions for $\hat{p}_{jk}$ follow because $p_{jj} = p_{jj}$, $j=1,2$, and because

$$\hat{p}_{12}(x) = e^{-2}(x,\alpha)p_{12}(x) \quad , \quad \hat{p}_{21}(x) = e^2(x,\alpha)p_{21}(x) \ , \ e^{\pm 2}(x,\alpha) \in C[a,b]. \tag{4.8}$$

We turn now to a local amplitude principle. Assume that conditions of theorem 4.1 are satisfied on [a,b]. Denote by $y_1$ and $y_2$ two linearly independent solutions related to (4.1). By a straightforward calculation we find that at $x = \alpha$ the local amplitude of the incident and reflected waves $y_1$ and $y_2$, respectively, satisfy $A := y_1\bar{y}_1 = y_2\bar{y}_2 \pm 2$. The question that we ask is how does A, the amplitude squared, vary with the measure of inhomogeneity $c'(0)$? Since $v = v(0) := \frac{c'(0)}{2w}$ it suffices to consider the variation of A with v. A calculation reveals that $v\frac{\partial A}{\partial v} < 0$ for $v^2 > 1$. This, shows that the local amplitude of the local wave functions is a monotone function of the inhomogeneity.

## 5. References

[1] H. Gingold, An invariant asymptotic formula for solutions of second-order linear ODE's. Asym. Anal. 1(1988) 317-350.

[2] H. Gingold and J. Hu, Transcendentally small reflections of waves for problems with/without turning points near infinity. A new uniform approach, J. Math. Phys. 32(1991) 3278-3284.

[3] H. Gingold, J. She & W.E. Zorumski, Local principles of wave propagation in inhomogeneous media, J. Acoust. Soc. Am. 93(1993) 599-604.

[4] H. Gingold, Approximation of solutions of $y'' = \phi(x,\varepsilon)y$ with several transition points and moving singularities. Asymp. Anal. 6(1993) 533-604.

[5] F.J. Olver, Connection formulas for second-order differential equations having an arbitrary number of turning points of arbitrary multiplicities, SIAM J. Math Anal. 9(1977) 673-700.

[6] Y. Sibuya, Global Theory of Second-Order Linear Ordinary Differential Equation with a Polynomial Coefficient, Mathematics Studies 18(North-Holland, Amsterdam, 1975).

[7] W. Wasow, Linear Turning Point Theory (Springer, Berlin, 1984).

# DYNAMICS OF THE EMIGRATION INSIDE A FUNCTIONAL ECONOMIC AREA (FEA)

JOSE M. GONZALEZ
*Department of Applied Economics, University of La Laguna*
*La Laguna, Canary Islands, Spain*

ABSTRACT

In this paper we study the growth and fluctuations of the flow Migration inside the Orotava Valley, a Functional Economic Area in the Canary Islands, Spain. With the help of econometric techniques (in particular, with the help of the "Life Cycle" Hypothesis of Ando and Modigliani) we obtain a dynamic system of equations which rules the dynamics of this flow. Then, the regression of the parameters, the stability and the conditions of sensibility of this system are analysed.

## 1. Introduction

The economic model which has taken form historically in the Canary Islands was sustained by Agriculture for Export of different cash-crops, controlled by a dominant minority (generally associated with foreign entrepreneurs). It based its high level of productivity in superb climatic and environmental conditions and on its coexistence with a subsistence agriculture or an economy of scarcity that left available plenty of cheap labour for the commercial plantations. This model gave rise to a succession of crises which cyclically impoverished the islands causing great waves of emigration as an ecological escape valve (according to Antonio Machado Carrillo's terminology [12]), whenever the demographic capacity was reached.In this historical image, the Orotava Valley on the island of Tenerife is one of the geographical areas which best fits the way in which this dependent economy acted.

In addition, the phenomenon of Mass Tourism in the 1960's substantially modified the productive structure producing a clear imbalance in the economic ecological health mechanism known up to then. Precisely the impact of this dominant economic activity on population, land occupation and standard of living conditions has converted the Orotava Valley into a Functional Economic Area, and is the subject of the present study. With the help of techniques from Systems Dynamics (Dynamical Simulation) we have elaborated three versions of a mathematical model expressed as a discrete dynamic system of equations ([7],[8],[9]). After experimental contrasting by means of a standard computer program: 'DYNAMO', this system has allowed us to draw the following conclusions:

1) Firstly a sigmoidal growth in the number of tourist places is perceived, i. e. after an accelerated growth process it follows another tempered curve until it converges into a stationary or equilibrium state around 32000 beds.

2) Together with this development of the tourist industry an overall growth in population and urbanized land which alerts us to the appearance of phenomena like overcrowding and excessive urbanization in the not too distant future.

3) This urban sprawl is due not only to the high index of natural increase on the Valley but also to the foreign residents occupying villas near the tourist complexes. In addition, this new form of settlement has given rise to a change in the net direction of migration in the Valley. Previously negative in sign, the emigrants were mainly young single males, it is now positive in sign, the loss being exceeded by the influx of outsiders.

4) Lastly, our model correctly describes the spectacular increase in per capita income among the inhabitants of the Valley area. This image is of an economic growth which is not directly corresponded by a comparable increase in the social conditions and facilities determining the citizens' Quality of Life.

Since the flow immigration-emigration could not conveniently simulated in any of our three scenarios, a revision of our model became necessary to take this element into account. For this reason we have undertaken the present new study of the migratory phenomenon.

Assisted by techniques developed in econometrics we have formulated a basic linear regression model, permitting us to accurately explain the migratory flows ocurring in the Orotava Valley in the 1960's and 1970's. For this purpose we recur to a theory similar to that proposed by Brown [5], regarding the income model in the Relative Income Hypothesis, which appears as a simplification of the Life Cycle Hypothesis formulated by Ando and Modigliani [2]. With this theory we explain in full detail the expansive growth of emigration, the structural change in 1973 produced by the oil crisis and then we study the diferent level variables of our previous models, i. e. Population, Touristic's beds and Employment Tension.

## 2. Construction of the Model

Before presenting the analytical expression of our model we must briefly explain the characteristics of the geographic framework for which it has been developed.

The Orotava Valley is an area with superb climatic and environmental conditions. Since the era of the Spanish conquest it has been the place where the major part of the colonists settled in the island, and where the earliest essential services favoring further occupation were created, i. e. tracks, houses, stores, farms, etc. For this reason, the Valley has been densely populated historically, constituting one of the most important foci attracting population in the Islands.

During this Century the Canary Islands and also this Valley are experiencing an extremely high rate of population growth (with an average annual accumulated growth rate of 1.4 %, Buriel de Orueta, [3]). Due to this, extraordinarily high levels of population density have been reached. A density has been given of 1200 inhabitants per $Km^2$, A. Alvarez, [1], p 580) on land beset with fragile support systems. With the arrival of mass tourism in the 1960's the critical population situation has become even more acute, since a movement of population has taken place from rural zones of the Islands to the tourist resorts, elevating their number of inhabitants. This has particularly occurred in Puerto de la Cruz at the center of the Valley's coastline where growth rates of 4.4 % have been known (J. F. Martín, [13]).

As can be seen, the migratory phenomenon strongly influences the demographic balance and consequently the area's economic development.

Traditionally the indices of migration in the Orotava Valley were negative, i. e. the Valley tended to lose young, single men of low academic level. This emigration was supposedly seasonal or temporary (A. Alvarez, [1], p 470) and depended linearly on population level which in turn determined the character and dimension of the "Relative Attractiveness" of the economic area. However, the tourist "boom" modified the direction of this attraction and from 1960 on, a reversal of the migratory process has been produced, the totals passing over into the positive. In this sense, the tourist phenomenon has generated a demand for well qualified labour able to attend to the demand for specialized services. At the same time there has been a demand for unskilled manpower to assist in construction or unspecialized services (J. F.

Martín, [13], p 331).

To explain the factors affecting migratory flow and its direction or sign, appears firstly the above mentioned Relative Attractiveness that the area exerts on each potential migrant, explained by means of Employment Tension or Surplus Labour Demand. A worker will emigrate to the Valley when employment there becomes attractive enough.

We additionally find that when the labour force (employment supply) is greater than employment demand, Employment Tension is negative and emigration is the result. Nevertheless this does not happen automatically, since it is logical to think that a worker might well be averse to moving somewhere else without a job to go to. In any case there is bound to be a certain delay between the moment of finding oneself out of a job and deciding to move.

This last circumstance also alerts us to the fact that the relative attraction of the zone and its nature will be appreciated somewhat later by a potential emigrant. That is, the decision to move on was taken after a certain lapse of time.

As a consequence, emigrant mobility, i.e. freedom to move and willingness to change their place of residence is influenced by a certain habit which must be noted in earlier periods. That is, we are speaking of the same circumstance recognized by Brown [5] in the propensity to consume, that in his words:

*The habits, customs, standards and levels associated with real consumption previously enjoyed become "impressed" on the human physiological and psychological systems and this produces an inertia or "hysteresis" in consumer behaviour. Because of this inertia, consumer demand reacts to changes in consumer income with a certain slowness, and thus past real consumption exerts a stabilizing effect on current consumption...*

Therefore taking this idea up again in our discussion, the emigration-immigration must be a function of the present Employment Tension and of the migration flow found in the past. So, parting from the hypothesis that the effect of the persistence of habits is, besides being continuous, an inverse function of time, the migration flow in the immediately previous period is that which must exert a greater influence on the present. In consequence we may use the econometric techniques to find the equation (1) as a basic model of linear regression.

$$(1) \quad E(t) = A + B\times ET(t) + C\times E(t-1) + u(t)$$

where Employment Tension: ET acts as exogenous variable, u represents a stochastic disturbance term and the endogenous variable Emigration: E also depends on itself as a lagged variable. The parameters A, B and C are calculated by the method of least squares and it is supposed that (1) fulfills all the regularity conditions of any linear econometric model: homoscedasticity, absence of linear dependece on ET, etc.

At a first estimation of the parameters we find that the multiple regression carried out does not respond adequately to reproduction of the actual data known for E and ET in the period 1960–79. Thus the factor R-squared is very small, the t-Student test gives variables wchich are not statistically significant and the test of the significance of the regression as a whole offers us a critical value of Snedecor's F, rejecting the validation of the hypothesis. All this is a consequence of the migration flow changing sign around the year 1973. The positive migrations become negative in sign because of the crisis in the tourist sector due to rising fuel prices. This circumstance, appreciable in the real data we have about emigration in the Valley leads us to establish the hypothesis that at that time a structural change in the model arises. So in order to verify the existence of this change and with an eye to improving our equations we introduce a dummy variable $D_1$, leaving (1) in the form:

$$(2) \quad E(t) = A + B\times ET(t) + C\times E(t-1) + \alpha\times D_1(t) + u(t)$$

Let us now proceed to the estimation of the parameters by least sqares:

A = 821,3541 B = -0.090609 C = 0.028228

the new ordinate associated with $D_1$ being: -980.3681

With this, new estimation of the actual data for E, those fitted into (2) and their respective residual errors are presented in Table 1..

**Table 1.**

| Observation | Actual | Fitted | Residual |
|---|---|---|---|
| 1961 | 323.0000 | 257.5388 | 65.4612 |
| 1962 | 331.0000 | 315.0902 | 15.9098 |
| 1963 | 345.0000 | 366.1474 | -21.1474 |
| 1964 | 360.0000 | 409.9441 | -49.9441 |
| 1965 | 380.0000 | 446.1579 | -66.1579 |
| 1966 | 453.0000 | 475.5360 | -22.5360 |
| 1967 | 476.0000 | 499.2520 | -23.2520 |
| 1968 | 490.0000 | 512.4052 | -22.4052 |
| 1969 | 506.0000 | 520.2303 | -14.2303 |
| 1970 | 536.0000 | 519.9571 | 16.0429 |
| 1971 | 564.0000 | 517.9045 | 46.0955 |
| 1972 | 566.0000 | 521.4131 | 44.5869 |
| 1973 | 567.0000 | 535.4233 | 31.5767 |
| 1974 | -432.0000 | -417.1904 | -14.8096 |
| 1975 | -435.0000 | -418.7508 | -16.2492 |
| 1976 | -332.0000 | -374.7997 | 42.7997 |
| 1977 | -331.0000 | -320.6984 | -10.3016 |
| 1978 | -327.0000 | -260.7780 | -66.2220 |
| 1979 | -133.0000 | -197.7827 | 64.7827 |

A clear intercorrelation is seen in the table between the actual and fitted data stands out in Graph 1., where the curves described by each series are plotted.

**Graph 1.**

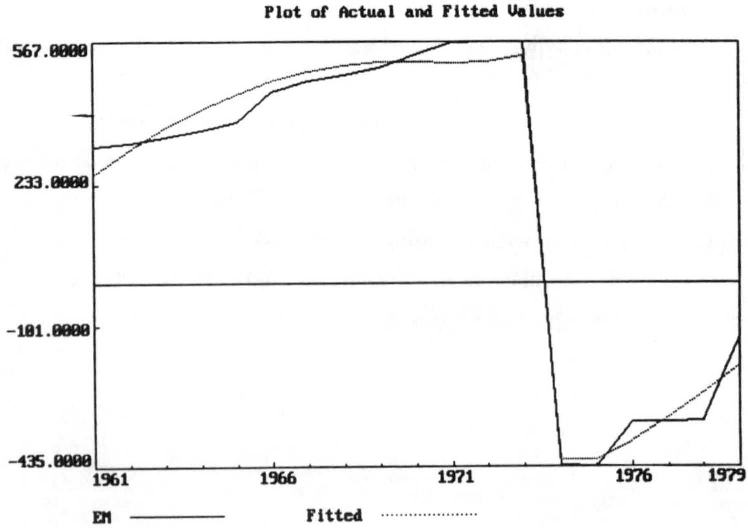


The hypothesis of a structural change produced around 1973 is perfectly confirmed by simply evaluating the high degree of significance shown by Student's t-test for the variable $D_1$ which gives us a quantity of -21.1256, also confirmable with the aid of Chow's contrast. This furnishes us with a critical value of 7.7192 that when is confronted with the values of a Snedecor's F test with 5 and 11 degrees of freedom, leads us to accept the improvement in the estimation reached after considering the structural change for both 5% and 1% levels of significance.

Furthermore, Student's t contrast which most clarifies the degree of actual significance of each variable in the regression supplies the following critical values:

15.7133 for A     -8.2605 for ET, and     0.58516 for E(t-1)

Thus, we can see that only the critical value obtained for E(t-1) is not significant enough in the contrast with a Student's t test of 17 degrees of freedom. This result alerts us to the fact that the net migration produced in the period preceding the one we wish to evaluate has little effect on the latter. On the other hand, the magnitude attained by Employment Tension ET does powerfully determine the population movement phenomenon.

We must include now the previous equations into our demographic model, which will be showed as a Dynamical System of differential equations which will contain the following variables: Total Population: P, Places of touristic resorts: T and Employment Tension ET.

In this new model the equation ruling the growth in tourist places available coincides with that utilized in the first of the models developed ([8]). We are dealing with a logistic equation of the type:

$$(3)\ \frac{dT}{dt} = a_1 T\,[1 - T/a_2]$$

where we have asumed $a_2$ = 32000 in the real carrying capacity of the hotels beds.

The equation ruling population development is linear and depends on the natural growth $b_1P$ and on emigration EM:

$$(4)\ \frac{dP}{dt} = b_1 P + EM$$

with $b_1 = 13.4 \times 10^{3}$.

In order to obtain the values for variable ET we base ourselves on the fundamental hypothesis of all the previous simulations, i.e. the Employment generated as a direct funtion of tourist development -the number of hotel places available T-. The precise expression of this function has been taken from a study applying Systems Dynamics to an explanation of the general economy of the Canary Islans as a whole [3], and is as follows:

$$(5)\quad ET(t) = d_1 \, [\, P(t) + T(t) \,]^{d_2} / TR(t) \; - \; d_3 \, P(t)$$

where $d_1$ and $d_2$ are known parameters and TR(t) indicates the degree of terciarization of the economy. This progresses with the passing of time, modifying the factor $d_1[P + T]^{d_2}$ which only reflects the employment generated by all the services. In addition, $d_3$ will be another parameter, coinciding in this case with the activity rate of population.

Finally, the equation govering the Terciarization Rate TR is again of a logistic type, where the carrying capacity, i.e. maximun possible terciarization has been situated at 90%, following data collected from countries or regions strongly affected by tourism, e.g. Puerto Vallarta, in Mexico, Bermuda, Caribbean islands, etc.

$$(6)\quad \frac{dTR}{dt} = e_1 \, TR \, [\, 1 - TR/0.9 \,]$$

### 3. Discussion of the Model

The equations (2)-(6) have been solved numerically with the aid of the computer language DYNAMO and the results obtained are presented in figures **2**, **3**, **4**, and **5** for the period 1960-1990.

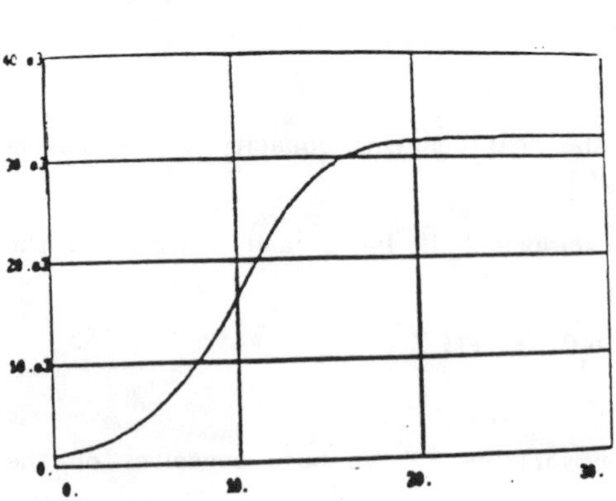


**Figure 2.** Tourist Resorts

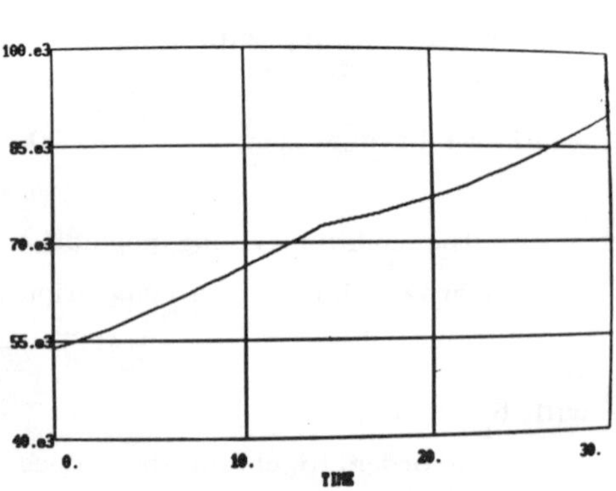


**Figure 3.** Population

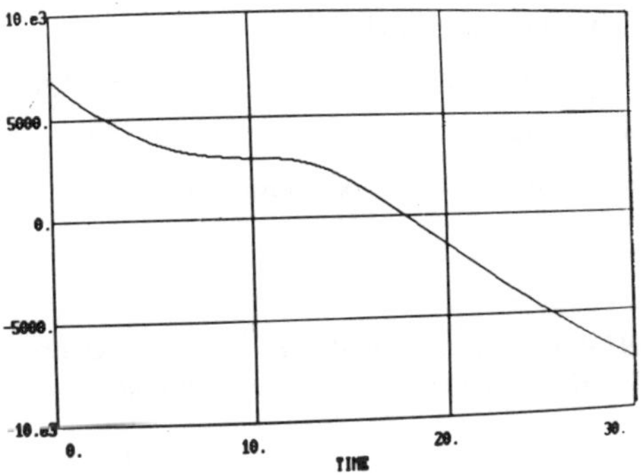


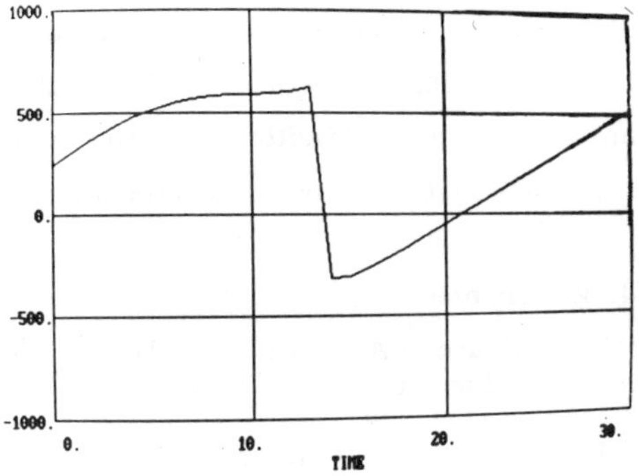


**Figure 4.** Employment Tension

**Figure 5.** Emigration-Immigration

Graph **2.** reproduces the same variation through time of tourist places available as presented in our previous studies. The equation is in fact the same.

The level variable Population (Graph **3.**) and the auxiliary variable of Employment Tension now appear simulated with a closer approximation to the actual data. Their differences are 0.66% and 1.2% respectively, confirming the goodnes of fit of our model in the reality simulation process under study. Only the Emigration variable EM (Graph **5.**) does not adequately fit ist actual development. In any case, let us then accept that our new model simulates the economic-demographic reality of the Orotava Valley with a high level of precission and, then, we can obtain the following conclusions:

$1^{st}$ The variation in tourist places available reproduces faithfully their actual development through the time period investigated. A process of accelerated growth is clearly visible in the early years, in which the boom in tourism was produced, then followed by a levelling of the curve, product of the setting down of development in the tourist industry.

$2^{nd}$ The curve that defines the changes in the variable T corresponds well to other similar curves for the variable P, with less pronounced slopes peculiar to the real evolution of the said variables.

$3^{th}$ Finally, the curve described defining Emigration indicates difficulty in quantifying that factor, as it varies continuously in sign and in magnitude. Nevertheless these characteristics correspond well to reality, as the external factors of the day (oil-crisis, political change in Spain, recession in the early 80's) have all had a notable influence on emigration.

**4. References**

[1] A. Alvárez, *Agricultura y Turismo en el Valle de La Orotava: un modelo de Articulación*, Doctoral Thesis, facultad de Geografía e Historia, Universidad de La Laguna, 1983

[2] A. Ando, F. Modigliani, *The life-cycle hypothesis of savings: Aggregate implications and tests*, American Economic Review, **53**, 55-84

[3] A. Ardaiz, A. Iñíguez, S. Martínez, A. Valentín, *Un modelo de planificación del agua: CANAGUA*, S/C de Tenerife, 1978

[4] E. Beltrami, *Mathematics for Dynamic Modeling*, Academic Press, Inc, San Diego, 1987

[5] T. M. Brown, *Habit persistence and lags in consumer behaviour*, Econometrica, **20**, 355-371

[6] E. L. Burriel de Orueta, *Canarias: Población y Agricultura en una sociedad dependiente*, OIKOS-TAU S. A. Ediciones, Barcelona, 1981

[7] J. M. González, *Turismo, Demografía y Ocupación del suelo en el Valle de La Orotava. La Dinámica de Sistemas en la Simulación de un Modelo de Dinámica Regional*, Documento de Trabajo N. 37, Facultad de Ciencias Económicas y Empresariales, Universidad de La Laguna, Junio, 1992

[8] J. M. González, *Tourism, Demography and Occupation of the Territory. Simulation of the impact of a dominant economic activity on the development of a FEA (Functional Economic Area)*, Acts of Twenty-third Annual Pittsburgh Conference on Modelling and Simulation, May, 1992, Volume 23, 845-857

[9] J. M. González, *Tourism, Demography and Occupation of the Territory. Simulation of the impact of a dominant economic activity on the development of a FEA (Functional Economic Area) II*, Acts of Imacs International Symposium on Mathematical Modelling and Scientific Computing, Bangalore, India, December, 1992

[10] J. Guckenheimer, Ph. Holmes, *Nonlinear Oscillations, Dynamical Systems, and Bifurcations of Vector Fields*, Springer Verlag, New York, 1983

[11] J. J. Johnston, *Econometric Methods*, $Ma_c$Graw Hill, Inc, 1984

[12] A. Machado Carrillo, *Ecología, Medio Ambiente y Desarrollo Turístico en Canarias*, Consejería de la Presidencia, Gobierno de Canarias, S/C de Tenerife, 1990

[13] J. F. Martín Ruiz, *Desarrollo Demográfico y Crecimiento espacial de las áreas turísticas de la isla de Tenerife*, Anuario de Estudios Atlánticos, **30**, 1984, 317-340

[14] A. Pulido, *Modelos Econométricos*, Editorial Pirámide, S. A., Madrid, 1987

[15] J. L. Raymond, E. Uriel, *Investigación econométrica aplicada: un caso de estudio*, Editorial AC, Madrid, 1987

[16] J. T. Sandefur, *Discret Dynamical Systems. Theory and Applications*, Clarendon Press, Oxford, 1990

# REGULAR VARIATION AND SOLUTIONS OF SECOND ORDER DIFFERENTIAL EQUATIONS

L. J. GRIMM AND L. M. HALL

University of Missouri-Rolla

ABSTRACT. The indices of variation of principal and nonprincipal solutions at $x = \infty$ of $y'' = f(x)y$, where $f(x) \geq 0$, depend on growth and integrability properties of $f$. In this note, we show the relations between these indices. Our results are illustrated by several examples.

## 1. Introduction.

The subject of regular variation of functions has been developed extensively since its introduction by J. Karamata in 1930 in connection with his study of Tauberian theorems. The Karamata theory has been effective in the study of a variety of problems in analysis, see, for example, the monographs [2], [3], [9].

A natural problem in the asymptotic theory of differential systems is to determine the rate of variation of solutions from the coefficients. V. Marić and M. Tomić [6] [7] and E. Omey [8] have characterized solutions of

$$y'' = f(x)y,$$

where f is positive and measurable. In a previous paper [4], we extended some of the Marić and Tomić results to more general linear equations and certain functional-differential equations. Here we consider the regularity of variation of principal and nonprincipal solutions of (1).

We begin with definitions of Karamata's functions.

**Definition.** *Let $g(x)$ be a positive measurable function on $[x_0, \infty)$. If, for each $\lambda > 1$, there exists $\alpha$, $-\infty \leq \alpha \leq \infty$, such that*

$$\lim_{x \to \infty} \frac{g(\lambda x)}{g(x)} = \lambda^\alpha$$

*then, at infinity, $g(x)$ has index $\alpha$ and is said to be*

*(i) slowly varying if $\alpha = 0$,*

*(ii) regularly varying if $\alpha \neq 0$, $-\infty < \alpha < \infty$,*

*(iii) rapidly varying if $\alpha = -\infty$ or $\alpha = \infty$.*

## 2. Application to Differential Equations.

We consider the equation

$$y'' = f(x)y \tag{1}$$

where f is positive and continuous on $[x_0, \infty)$. This equation has real-valued solutions $y(x)$ for which $\int^\infty \frac{dx}{y^2(x)}$ is convergent and other solutions for which this integral is divergent.

**Definition.** *A solution of (1) for which $\int^\infty \frac{dx}{y^2(x)}$ is divergent is called a principal solution (at $x = \infty$); a solution for which the integral converges is called a nonprincipal solution.*

The principal solution is uniquely determined up to a constant factor, and if $u(x)$ and $v(x)$ are principal and nonprincipal solutions of (1), then $\frac{u(x)}{v(x)} \to 0$ as $x \to \infty$. If $u(x)$ is a principal solution of (1), then $v(x) = u(x) \int_\gamma^x \frac{dt}{u^2(t)}$ is a nonprincipal solution on $[\gamma, \infty)$, and if $v(x)$ is a nonprincipal solution on $[\gamma, \infty)$ then $u(x) = v(x) \int_x^\infty \frac{dt}{v^2(t)}$ is a principal solution on $[\gamma, \infty)$. These and other results concerning principal and nonprincipal solutions can be found in Chapter XI of [5].

Marić and Tomić [7] have proved the following result.

**Theorem MT.** *Let $y$ be a positive decreasing (principal) solution of (1) on $[x_0, \infty)$. Then $y$ is*

*(i) slowly varying if and only if*

$$\lim_{x\to\infty} x\int_x^\infty f(t)\,dt = 0;$$

*(ii) regularly varying with index* $\alpha = \frac{1-\sqrt{1-4c}}{2}$ *if and only if*

$$\lim_{x\to\infty} x\int_x^\infty f(t)\,dt = c > 0;$$

*(iii) rapidly varying if and only if*

$$\lim_{x\to\infty} x\int_x^{kx} f(t)\,dt = \infty$$

*for each constant* $k > 1$.

*Remarks.* Note that the integral condition in case (iii) has a form different from those in cases (i) and (ii). Also note that a solution of (1) will be rapidly varying at infinity if and only if (1) has an irregular singular point at infinity.

Omey [8] has obtained the following theorem regarding nonprincipal solutions.

**Theorem O.** *Let* $y(x)$ *be a positive solution on* $[x_0,\infty)$ *of (1) tending to* $\infty$ *as* $x\to\infty$. *If* $f(x)$ *satisfies*

$$\lim_{x\to\infty} x^2 f(x) = c, 0 < c < \infty,$$

*then* $y(x)$ *is regularly varying with index* $\alpha = \frac{1+\sqrt{1+4c}}{2}$; *if* $\lim_{x\to\infty} x^2 f(x) = 0$ *and* $\int^\infty x f(x) = \infty$, *then* $y(x)$ *is regularly varying with index 1.*

**3. Examples.**

*Example 1.* Consider the equation $y'' = \frac{4}{(x^2-1)(x-1)}y$. Here $\lim_{x\to\infty}\int_x^\infty \frac{4dt}{(t^2-1)(t-1)} = 0$. The function $y_1 = \frac{x+1}{x-1}$ is a positive decreasing solution with $\lim_{x\to\infty}\frac{y(\lambda x)}{y(x)} = 1 = \lambda^0$. Note that $\lim_{x\to\infty}\frac{4}{(x^2-1)(x-1)} = 0$, but $\int^\infty \frac{4x}{(x^2-1)(x-1)}dx < \infty$. Hence Omey's theorem does not apply, but a second (nonprincipal) solution is given by $y_2 = \frac{x^2+x-4(x-1)\log(x+1)}{x-1}$, which has index one.

*Example 2.* Consider the Cauchy-Euler equation $y'' = \frac{2}{x^2}y$. Here $\lim_{x\to\infty} \int_x^\infty \frac{2}{t^2}dt = 2$. Thus, the principal solution, $y_1 = \frac{1}{x}$, has index $\alpha = -1$. By Omey's theorem, nonprincipal solutions have index 2, and for this equation it is easy to see that nonprincipal solutions have the form $y = c_1\frac{1}{x} + c_2x^2$ where $c_2 \neq 0$.

*Example 3.* For $y'' = y$, Theorem MT shows that the principal solution $y_1 = e^{-x}$ is rapidly varying. The nonprincipal solution $y_2 = e^x$ is also rapidly varying.

*Example 4.* For Airy's equation, $y'' = xy$, Theorem MT shows that the principal solution $Ai(x)$ is rapidly varying; the nonprincipal solution $Bi(x)$ is also rapidly varying.

**4. Relation Between Indices.**

In Examples 1 and 2, the sum of the indices of the principal and nonprincipal solutions is 1. This relationship holds in general when the solutions are slowly or regularly varying. The following theorem is proved in [6].

**Theorem.** *Let $y_1(x)$ be a principal solution of $y'' = f(x)y$ where $f(x)$ is measurable and nonnegative on $[x_0, \infty)$. Then if $y_1$ is slowly varying (with index 0) or regularly varying with index $\alpha$, each nonprincipal solution $y_2$ will be regularly varying with index $1 - \alpha$. (Note that $\alpha \leq 0$ since $y_1$ is decreasing.)*

This theorem is equivalent to:

**Corollary.** *Let $y_1(x)$ be a slowly or regularly varying principal solution and $y_2(x)$ a nonprincipal solution of $y'' = f(x)y$, where $f(x)$ is measurable and nonnegative on $[x_0, \infty)$. Then $g(x) = y_1(x)y_2(x)$ is regularly varying with index one.*

*Remark.* The condition $\lim_{x\to\infty} x\int_x^\infty f(t)dt = c$ holds if the hypothesis of Theorem O holds, so Theorem O is a corollary of this result and Theorem MT.

In Examples 3 and 4, since the solutions are rapidly varying, the indices are infinite, but the product $y_1(x)y_2(x)$ still has finite index. However, Examples 3 and 4 show that this index is not always the same. In this case, the index of the product depends on f, as described in the following theorem.

**Theorem.** *In (1), let f be positive and continuous on* $[x_0, \infty)$ *with* $\lim_{x\to\infty} x^2 f(x) = \infty$. *If f is slowly or regularly varying with index* $\beta$ *then the index of the product of the principal solution* $y_1$ *and a nonprincipal solution* $y_2$ *is* $\frac{-\beta}{2}$.

*Proof.* Since $\lim_{x\to\infty} x^2 f(x) = \infty$, the solutions $y_1$ and $y_2$ are given asymptotically (see[1]) by

$$y_1(x) \backsim (f(x))^{-1/4} e^{-\int_{x_0}^{x} \sqrt{f(t)}dt}$$

and

$$y_2(x) \backsim (f(x))^{-1/4} e^{\int_{x_0}^{x} \sqrt{f(t)}dt}$$

Hence the product $y_1 y_2$ satisfies

$$\lim_{x\to\infty} \frac{y_1(\lambda x) y_2(\lambda x)}{y_1(x) y_2(x)} = \lim_{x\to\infty} \sqrt{\frac{f(x)}{f(\lambda x)}} = \lambda^{\frac{-\beta}{2}}.$$

## References


1. C. Bender and S. Orszag, *Advanced Mathematical Methods for Scientists and Engineers*, McGraw-Hill, New York, 1978.
2. N. H. Bingham, C. M. Goldie, and J. L. Teugels, *Regular variation*, Encyclopedia of Mathematics and its Applications, vol. 27, Cambridge University Press, Cambridge, 1987.
3. J. L. Geluk and L. de Haan, *Regular Variation, Extensions and Tauberian Theorems*, CWI Tract 40, Stichting Mathematisch Centrum, Centrum voor Wiskunde en Informatica, Amsterdam, 1987.
4. L. J. Grimm and L. M. Hall, *Self adjoint differential equations and Karamata functions*, Proceedings of the Third International Colloquium on Differential Equations, to appear, VSP International Science Publishers, 1993.
5. P. Hartman, *Ordinary Differential Equations*, Wiley, New York, 1964.
6. V. Marić and M. Tomić, *A trichotomy of solutions of second order linear differential equations*, Rev. Res. Fac. Sci. Univ. Novi Sad Math. Ser. **14** (1984), 1–11.
7. V. Marić and M. Tomić, *A classification of solutions of second order linear differential equations by means of regularly varying functions*, Publ. Inst. Math. (Beograd) (N. S.) **48 (62)** (1990)), 199–207.
8. E. Omey, *Regular variation and its applications to second order linear differential equations*, Bull. Soc. Math. Belg. , Ser. B **33** (1981), 207–229.
9. E. Seneta, *Regularly Varying Functions*, Lecture Notes in Mathematics vol. 508, Springer-Verlag, Berlin, 1976.


Rolla, MO 65401, USA

# The Vibrated Pendulum With a Fixed Mass and a Sliding Mass

Bertram A. Schmidt and George W. Grossman
Central Michigan University
Mt. Pleasant, MI 48859

**Abstract:**

A pendulum with a fixed mass and a sliding mass is subjected to a horizontal vibration. The vibration causes new stable equilibrium positions.

## I Introduction

Researchers have found that pendulums can have surprising equilibrium positions if oscillatory motions are applied to the pivot. Stephenson[1] (1907) analyzed the plane pendulum with a vertical oscillation imposed on the pivot. He noted the inverted position. Lowenstern[2] (1932) investigated some pendulums including the spherical pendulum with a vertical imposed oscillation. His analysis showed the inverted position. Phelps and Hunter[3] (1965) analyzed the plane pendulum subjected to a vertical oscillation at an unrestricted frequency. Their analysis displayed the inverted position. Ryland and Meirovitch[4] (1977) studied the plane pendulum with radial flexibility subjected to a vertical vibration at an unrestricted frequency. Their work showed conditions for stability in the upward and downward vertical positions. Schmidt[5,6] (1980) and (1990) investigated several complex pendulums with imposed oscillations. Interesting equilibrium positions were shown.

In this paper we consider a pendulum made up of two concentrated and equal masses and a weightless rod. One mass is attached rigidly to the rod and the other is allowed to slide freely along the rod (see Figure 1). A small amplitude high frequency harmonic oscillation is applied to the pivot in the horizontal direction. The position of the sliding mass is described by the radius r measured from the pivot in direction of the fixed mass.

The pendulum is constructed so that r can be positive, negative or zero.

In this investigation, we find approximate stable equilibrium values for the angle of the pendulum and the position of the sliding mass. In section II, we express the equations of motion in dimensionless variables and parameters including a small parameter which expresses both the smallness of the amplitude of the oscillation and the largeness of the frequency of the oscillation. A transformation is used to arrange the problem in the standard form for the method of averaging which is described in Bogoliuboff and Mitropolskiy[7], (1962)). The averaging process is applied. We then examine the averaged equations for equilibrium and stability.

## II The Analysis

The equations of motion are given by

$$m(R^2 + r^2)\ddot{\theta} + 2mr\dot{r}\dot{\theta} + mg(R + r)\sin\theta - mA\nu^2(R + r)\cos\theta\cos\nu t + C_1\dot{\theta} = 0$$

$$\text{and} \quad m\ddot{r} - mr(\dot{\theta})^2 - mA\nu^2\sin\theta\cos\nu t - mg\cos\theta + C_2\dot{r} = 0 \quad ,$$

where m is the mass, R is the radius of the fixed mass measured from the pivot, r is the radius of the sliding mass, θ is the angle between the pendulum and downward vertical, g is the gravitational constant, A and ν are the amplitude and frequency of the imposed oscillation, $C_1$ and $C_2$ are damping coefficients, t is time and the dot indicates differentiation with respect to time.

It can be observed that the governing equations are second order, nonlinear in two unknowns θ and r. A direct approach is a numerical solution, however it is more interesting from the mathematical and physical point of view to find qualitative solutions. We examine stability characteristics of transformed equations to determine equilibrium positions of the pendulum. The geometric theory of ordinary differential equations is well summarized in such research books as by Lefschetz[8].

We restrict the problem so that $\nu$ is large compared to the natural frequency of the pendulum without the sliding mass ($\sqrt{g/R}$). We introduce a small dimensionless parameter $\varepsilon$ such that $\varepsilon$ is of the order of $\frac{\sqrt{g/R}}{\nu}$. We establish other dimensionless parameters and variables. Let $\tau = \nu t$, $x = \frac{r}{R}$, $\varepsilon^2\delta = \frac{g/r}{\nu^2}$, $\varepsilon\gamma = A/R$, $\varepsilon a_1 = C_1/mR^2\nu$, $\varepsilon a_2 = C_2/m\nu$, and let the prime indicate the derivative with respect to $\tau$. The parameters $\delta$, $\gamma$, $a_1$ and $a_2$ are positive and independent of $\varepsilon$. In terms of the new variables and parameters, the equations of motion become

$$(1 + x^2)\,\theta'' + 2xx'\theta' + \varepsilon^2\delta\,(1 + x)\sin\theta - \varepsilon\gamma\,(1 + x)\cos\theta\cos\tau + \varepsilon a_1\theta' = 0$$

and

$$x'' - x(\theta')^2 - \varepsilon\gamma\sin\theta\cos\tau - \varepsilon^2\delta\cos\theta + \varepsilon a_2 x' = 0. \qquad (1a, b)$$

To express the problem in the standard form for the method of averaging, we introduce the transformation

$$\theta' = \varepsilon\phi + \varepsilon\gamma\,(1 + x^2)^{-1}(1 + x)\cos\theta\sin\tau \quad \text{and} \quad x' = \varepsilon y + \varepsilon\gamma\sin\theta\sin\tau \qquad (2a, b)$$

Combining Eqs (1) and Eqs (2) gives

$$\theta' = \varepsilon\,[\,\phi + \gamma\,(1 + x^2)^{-1}(1 + x)\cos\theta\sin\tau\,] \quad ,$$

$$\begin{aligned}\phi' = \varepsilon\,(1 + x^2)^{-1}\,[&-2x\,(y + \gamma\sin\theta\sin\tau)(\phi + \gamma\,(1 + x^2)^{-1}(1 + x)\cos\theta\sin\tau)\\ &- \delta\,(1 + x)\sin\theta - \gamma\,(y + \gamma\sin\theta\sin\tau)\cos\theta\sin\tau\\ &+ 2\gamma\,(1 + x^2)^{-1}x\,(x + 1)\,(y + \gamma\sin\theta\sin\tau)\cos\theta\sin\tau\\ &+ \gamma\,(1 + x)\,(\phi + \gamma\,(1 + x^2)^{-1}(1 + x)\cos\theta\sin\tau)\sin\theta\sin\tau\\ &- a_1\,(\phi + \gamma\,(1 + x^2)^{-1}(1 + x)\cos\theta\sin\tau\,] \quad ,\end{aligned}$$

$$x' = \varepsilon\,[\,y + \gamma \sin\theta \sin\tau]\ ,$$

$$y' = \varepsilon\,[\,-\gamma \cos\theta\,(\,\phi + \gamma(\,1 + x^2\,)^{-1}\,(\,1 + x\,) \cos\theta \sin\tau\,) \sin\tau + x\,(\,\phi + \gamma(\,1 + x^2\,)^{-1}\,(\,1 + x\,) \cos\theta \sin\tau\,)^2 + \delta \cos\theta - a_2\,(\,y + \gamma \sin\theta \sin\tau\,)\,]. \qquad (3a, b, c, d)$$

Eqs (3) are in the standard form for the method of averaging. We apply the method of averaging, i.e., the averaging operator is applied to the right hand side of Eqs (3) and the variables $\theta$, $\phi$, $x$ and $y$ are replaced by their approximations $\Theta$, $\Phi$, X and Y. This gives

$$\Theta' = \varepsilon\Phi\ ,$$

$$\Phi' = \varepsilon\,(\,1 + X^2\,)^{-1}\,[\,-2XY\Phi + \gamma^2 X\,(\,1 + X^2\,)^{-1} \sin\Theta \cos\Theta - \delta\,(\,1 + X\,) \sin\Theta - a_1\Phi\,],$$

$$X' = \varepsilon Y\ ,$$

$$Y' = \varepsilon\,[\,\tfrac{1}{2}\gamma^2\,(\,X + 1\,)(\,X - 1\,)(\,1 + X^2\,)^{-2} \cos^2\Theta + X\Phi^2 + \delta \cos\Theta - a_2 Y\,]\ . \qquad (4a, b\ c, d)$$

We seek equilibrium solutions of Eqs (4) and stability of these equilibrium solutions. The equilibrium solutions are described by

$$\gamma^2 X_e(\,1 + X_e^2\,)^{-1} \sin\Theta_e \cos\Theta_e - \delta\,(\,1 + X_e) \sin\Theta_e = 0 \quad \text{and}$$

$$\tfrac{1}{2}\gamma^2\,(\,X_e^2 - 1\,)(\,1 + X_e^2\,)^{-2} \cos^2\Theta_e + \delta \cos\Theta_e = 0\ , \qquad (5a, b)$$

where $X_e$ and $\Theta_e$ are the equilibrium values of X and $\Theta$. Eqs (5) describe equilibrium. We present eight cases. In this we introduce $\beta = \dfrac{\gamma^2}{2\delta}$ .

A) $\cos\Theta_e = -\dfrac{(1+X_e^2)^{-2}}{\beta(X_e^2-1)}$ , $3X_e^3 + X_e^2 + X_e - 1 = 0$

For this case $-\frac{\pi}{2} < \Theta_e < \frac{\pi}{2}$, $0 < X_e < 1$ .

B) $\Theta_e = \frac{\pi}{2}$ or $\frac{3}{2}\pi$ , $X_e = -1$

C) $\Theta_e = 0$, $0 \le X_e < 1$, $\beta \ge 1$, $X_e = \frac{1}{\sqrt{2}}\left(-2-\beta+\sqrt{\beta^2+8\beta}\right)^{1/2}$

D) $\Theta_e = 0$, $-1 < X_e < 0$, $\beta > 1$, $X_e = -\frac{1}{\sqrt{2}}\left(-2-\beta+\sqrt{\beta^2+8\beta}\right)^{1/2}$

E) $\Theta_e = \pi$, $X_e \ge \sqrt{3}$, $\beta \ge 8$, $X_e = \frac{1}{\sqrt{2}}\left(-2+\beta+\sqrt{\beta^2-8\beta}\right)^{1/2}$

F) $\Theta_e = \pi$, $1 < X_e < \sqrt{3}$, $\beta > 8$, $X_e = \frac{1}{\sqrt{2}}\left(-2+\beta-\sqrt{\beta^2-8\beta}\right)^{1/2}$

G) $\Theta_e = \pi$, $-\sqrt{3} < X_e < -1$, $\beta > 8$, $X_e = -\frac{1}{\sqrt{2}}\left(-2+\beta+\sqrt{\beta^2-8\beta}\right)^{1/2}$

H) $\Theta_e = \pi$, $X_e \le -\sqrt{3}$, $\beta \ge 8$, $X_e = -\frac{1}{\sqrt{2}}\left(-2+\beta-\sqrt{\beta^2-8\beta}\right)^{1/2}$

We examine stability of these equilibrium positions. For this we let $a_1 = a_2 = 0$. The linear approximations of Eqs (4) near the equilibrium values $\Theta_e$ and $X_e$ are

$$\Theta'_1 = \varepsilon\Phi_1$$

$$\Phi'_1 = \varepsilon[a\Theta_1 + bX_1]$$

$$X'_1 = \varepsilon Y_1$$

$$Y'_1 = \varepsilon\,[\, c\Theta_1 + dX_1 \,]\,, \qquad (6a, b, c, d)$$

where $\Theta_1 = \Theta - \Theta_e$, $\Phi_1 = \Phi$, $X_1 = X - X_e$, and $Y_1 = Y$ and

$$a = \gamma^2 X_e (1 + X_e^2)^{-2} \cos 2\Theta_e - \delta\,(1 + X_e^2)^{-1}(1 + X_e) \cos\Theta_e$$

$$b = (1 + X_e^2)^{-3}\left[\frac{1}{2}\gamma^2 (1 - 3X_e^2) \sin 2\Theta_e + \delta\,(1 + X_e^2)(-1 + 2X_e + X_e^2) \sin\Theta_e\right]$$

$$c = -\gamma^2 (X_e^2 - 1)(1 + X_e^2)^{-2} \sin\Theta_e \cos\Theta_e - \delta \sin\Theta_e$$

$$d = \gamma^2 (1 + X_e^2)^{-3} (3X_e - X_e^3) \cos^2\Theta_e\,. \qquad (7a, b, c, d)$$

The eigenvalues of Eqs (6), to a factor of $\varepsilon^2$, are described by

$$\lambda^2 = \frac{1}{2}\left( a + d \pm \sqrt{(a - d)^2 + 4bc}\right). \qquad (8)$$

We consider stability for case A. Eqs. (5) give

$$\cos\Theta_e = \frac{-(1 + X_e^2)^2}{\beta\,(X_e^2 - 1)} \quad \text{and} \quad 3X_e^3 + X_e^2 + X_e - 1 = 0\,.$$

These give $X_e = .4694\ \ldots$ and $\cos\Theta_e = \dfrac{1.91\ \ldots}{\beta}$. Recognizing these in Eqs (7) yields $a > 0$, or $a < 0$ or $a = 0$ and $\dfrac{b}{\sin\Theta_e} > 0$, $\dfrac{c}{\sin\Theta_e} > 0$, and $d > 0$, which along with Eq (8) indicates instability.

For case B, Eqs (7) indicate $a < 0$, $\dfrac{b}{\sin\Theta_e} < 0$, $\dfrac{c}{\sin\Theta_e} < 0$, and $d = 0$. Combining these with Eq (8) shows instability.

For case C, Eqs (7) show $a < 0$, or $a = 0$, or $a > 0$ and $b = 0$, $c = 0$, and $d \geq 0$. These along with Eq (8) indicate instability.

For case D, Eqs (7) show $a < 0$, $b = 0$, $c = 0$, $d < 0$. Eq (8) indicates a stable equilibrium position.

For case E, Eqs (7) show $a > 0$, $b = 0$, $c = 0$, $d \geq 0$. Eq (8) indicates instability.

For case F, Eqs (7) give $a > 0$, $b = 0$, $c = 0$, $d > 0$. Eq (8) indicates instability.

For case G, Eqs (7) give $a < 0$, $b = 0$, $c = 0$, $d < 0$. Eq (8) indicates a stable equilibrium position.

For case H, Eqs (7) show $a < 0$, $b = 0$, $c = 0$, $d \geq 0$. Eq (8) indicates instability.

Therefore, the stable equilibrium positions are described in cases D and G.

## III. Conclusion

We have found two stable equilibrium positions for the vibrated pendulum with a fixed mass and a sliding mass. They are

$$1)\quad \Theta_e = 0,\ X_e = -\frac{1}{\sqrt{2}}\left(-2-\beta+\sqrt{\beta^2+8\beta}\right)^{1/2},\quad -1 < X_e < 0$$

for $\beta = \dfrac{\gamma^2}{2\delta} > 1$, i.e., the pendulum hangs vertically downward and the sliding mass has a negative radial position.

$$2)\quad \Theta_e = \pi,\ X_e = -\frac{1}{\sqrt{2}}\left(-2+\beta-\sqrt{\beta^2-8\beta}\right)^{1/2},\quad -\sqrt{3} < X_e < -1$$

for $\beta = \dfrac{\gamma^2}{2\delta} > 8$, i.e., the pendulum is inverted and the sliding mass has a negative radial position. These equilibrium positions are shown in Fig. 2.

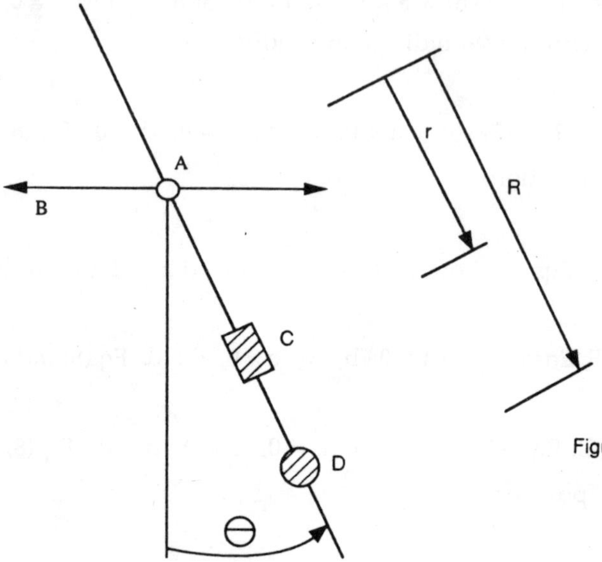


Figure 1

A - pivot
B - vibration
C - sliding mass
D - rigid mass

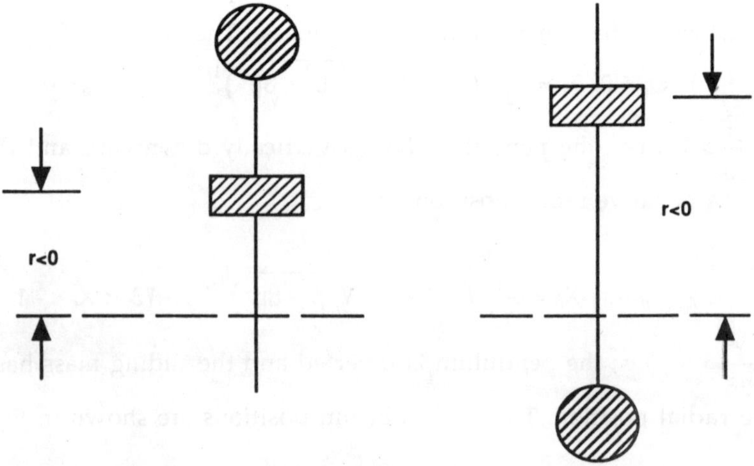


**Figure 2.**

**References**


1. Stephenson, A., 1907, "On a New Type of Dynamical Stability," Memoirs and Proceedings of the Manchester Literary and Philosophical Society, Vol. **52**, pp. 1 - 10.
2. Lowenstern, E.R., 1932, "The Stabilizing Effect of Imposed Oscillations of a High Frequency on a Dynamical System," Philosophical Magazine, Vol. **13**, pp. 458 - 486.
3. Phelps, F.M., III, and Hunter, J.H., Jr., 1965, "An Analytical Solution of the Inverted Pendulum," American Journal of Physics, Vol. **33**, pp. 285 - 295.
4. Ryland, G., II, and Meirovitch, L., 1977, "Stability Boundaries of a Swinging Spring with Oscillating Support," Journal of Sound and Vibration, Vol. **51**, pp. 547 - 560.
5. Schmidt, B.A., 1980, "Vibrated Pendulum with a Mass Free to Move Radially," ASME Journal of Applied Mechanics, Vol. **47**, pp. 428 - 430.
6. Schmidt, B.A., 1990, "The Rotationally Flexible Pendulum Subjected to a High-Frequency Excitation," ASME Journal of Applied Mechanics, Vol. **57**, pp. 725 - 730.
7. Bogoliuboff, N.M., and Mitropolskiy, Yu.A., 1962, Asymptotic Methods in the Theory of Nonlinear Oscillations, Gordon and Breach.
8. Lefschetz, S., 1977, " Differential Equations: Geometric Theory", Dover Books.

# DISTRIBUTION OF REAL EIGENVALUES IN STURM-LIOUVILLE PROBLEMS WITH INFINITELY MANY TURNING POINTS

HARRY GINGOLD
*Department of Mathematics, West Virginia University*
*Morgantown, West Virginia 26506*

and

T J HEMPLEMAN
*Department of Mathematics, West Virginia University*
*Morgantown, West Virginia 26506*

ABSTRACT

The order of growth of eigenvalues in Sturm-Liouville problems with infinitely many turning points is studied. The distribution of eigenvalues is obtained for differential equations with piecewise continuous and piecewise smooth coefficients. Prüfer's method is used to obtain upper and lower bounds for all real eigenvalues. The derivation of Sturm's oscillation theorems from Prüfer's method is noted.

## 1. Introduction.

Sturm-Liouville eigenvalue problems have attracted substantial attention in the literature of mathematics and mathematical physics.

Given a differential equation of the form

$$(P(t)y')' + [\lambda r(t) + q(t)]y = 0 \tag{1}$$

and boundary conditions on an interval $[a, b]$

$$A_1 y(a) + A_2 P(a) y'(a) = 0 \tag{2(i)}$$

$$B_1 y(b) + B_2 P(b) y'(b) = 0. \tag{2(ii)}$$

Eq.s (1), (2(i)), & (2(ii)) make up the general Sturm-Liouville boundary value problem.

The classical theory of existence and distribution of eigenvalues is well developed, especially for the case where $r(t)$ is bounded away from zero.

Several methods of attack on the Sturm-Liouville eigenvalue problem exist, and we were most concerned with those that considered piecewise continuity in the coefficient functions and an infinite number of sign changes in $r(t)$.

As far as the mere existence of real eigenvalues when $r(t)$ changes sign is concerned, several general results exist (e.g. [6] & [7] and their references). A result which goes back to Hilbert guarentees the existence of real eigenvalues when $r(t)$ changes sign and is piecewise continuous. However, there are restrictions imposed on the coefficient function $q(t)$ in Eq. (1) as well as on the signs of the constant coefficients in Eq.s (2(i)) & (2(ii)). Various existence theorems of real eigenvalues are given in [7], for general Sturm-Liouville problems in which the approach is based on an integral equation reformulation of the Sturm-Liouville problems with relaxed conditions on the coefficients. A substantial survey of literature can also be found in [7]. In [1], Jörgen's conjecture is attacked using Prüfer's method. The asymptotic distribution of eigenvalues of Sturm-Liouville problems with an indefinite weight $r(t)$ is then studied.

There is also a family of powerful asymptotic approaches, based on the L-G approximation and its generalization, which can provide very accurate asymptotic estimates for large (real or complex) eigenvalues, as well as for their corresponding eigenfunctions. These require that the coefficients $r(t)$ and $q(t)$ be analytic, e.g. [9] & [11], or at least possessed of a certain degree of smoothness, e.g. [8].

In the case where $r(t)$ is smooth with a finite number of turning points, a sophisticated connection theory has evolved, e.g. [8], [9], & [11] and their references. A new mechanism of connection, which would yield numerical approximations to small modulus eigenvalues and to their eigenfunctions is proposed in [4].

Functional theoretic approaches can obtain order of growth and distribution of eigenvalues, but the case for $r(t)$ changing signs many times has not yet been resolved (compare with [10]).

The purpose of this note is to obtain growth estimates on all real eigenvalues of Sturm-Liouville problems, demonstrating the power of Prüfer's method and its derivative, an angle comparison theorem. We also point out that Sturm's two fundamental oscillation theorems can be obtained as a corollary to the angle comparison theorem.

## 2. Prüfer's Method.

*2.1 Prüfer's Construction*

Given a solution $y(t,\lambda)$ of Eq. (1), Prüfer let two new functions $\rho(t,\lambda)$ and $\theta(t,\lambda)$ be defined as

$$\begin{aligned} y(t,\lambda) &= \rho(t,\lambda)\sin\theta(t,\lambda) \\ P(t)y'(t,\lambda) &= \rho(t,\lambda)\cos\theta(t,\lambda). \end{aligned}$$

By taking derivatives (note that all derivatives are taken with respect to the variable $t$ and not the parameter $\lambda$) we arrive at

$$\theta' = \frac{1}{P}\cos^2\theta + g\sin^2\theta \tag{3}$$

$$\rho' = (\frac{1}{P} - g)\rho\sin\theta\cos\theta. \tag{4}$$

For any functions $\theta$ and $\rho$ which satisfy Eq.s (3) & (4), $y = \rho\sin\theta$ is a solution to Eq. (1) (see [5] for complete derivation).

Thus Prüfer's method has reduced the second order ordinary differential equation in Eq. (1) to a system of two first order ordinary differential equations.

*2.2 A Useful Theorem*

**Angle Comparison Theorem.** *Let*

$$(P_i y_i') + g_i y_i = 0$$

*(for i = 1,2 with $g_i = \lambda r_i + q_i$) where $P_i$, $r_i$, and $q_i$ are piecewise continuous, and $P_i(t) > 0$ for $t \in [a,b]$. If*

$$g_2(t) \geq g_1(t)$$

$$\&$$

$$\frac{1}{P_2(t)} \geq \frac{1}{P_1(t)}$$

*$\forall\, t \in [a,b]$, and if $y_i(t,\lambda) = \rho_i(t,\lambda)\sin\theta_i(t,\lambda)$, then*

$$\theta_2(t,\lambda) \geq \theta_1(t,\lambda) \quad iff \quad \theta_2(a,\lambda) \geq \theta_1(a,\lambda) \tag{5a}$$

*and, in fact,*

$$\theta_2(t,\lambda) > \theta_1(t,\lambda) \quad if \quad \theta_2(a,\lambda) > \theta_1(a,\lambda). \tag{5b}$$

*Moreover, let $t_1, t_2$ be two consecutive zeros of a nontrivial solution $y_1(t,\lambda)$, then any nontrivial solution $y_2(t,\lambda)$ vanishes (at least) once in the interval $[t_1,t_2]$.*

*Proof.* In Eq. (5a), the second is immediately implied by the first, so it remains to show the reverse, which will also prove Eq. (5b). Recall from Prüfer's construction

$$\theta_i' = \frac{\cos^2\theta_i}{P_i} + g_i\sin^2\theta_i.$$

Then we have

$$(\theta_2 - \theta_1)' = C(t)(\theta_2 - \theta_1) + h(t) \tag{6}$$

where $h = (\frac{1}{P_2} - \frac{1}{P_1})\cos^2\theta_1 + (g_2 - g_1)\sin^2\theta_1 \geq 0$, and $C(t) = \frac{1}{P_2}\frac{\cos^2\theta_2 - \cos^2\theta_1}{\theta_2 - \theta_1} + g_2\frac{\sin^2\theta_2 - \sin^2\theta_1}{\theta_2 - \theta_1}$, a piecewise continuous function.

Eq. (6) has the solution

$$\theta_2 - \theta_1 = [\exp\int_a^t C(s)\,ds][\theta_2(a,\lambda) - \theta_1(a,\lambda)] + \int_a^t [\exp\int_u^t C(s)\,ds]h(u)\,du.$$

If $(\theta_2(a,\lambda) - \theta_1(a,\lambda)) \geq 0$, then the first term is always non-negative, and the second, as an integral of non-negative factors is non-negative, so $(\theta_2(t,\lambda) - \theta_1(t,\lambda)) \geq 0\ \forall\, t \in [a,b]$. Likewise, if $(\theta_2(a,\lambda) - \theta_1(a,\lambda)) > 0$ then $(\theta_2(t,\lambda) - \theta_1(t,\lambda)) > 0\ \forall\, t \in [a,b]$, and thus we have both (5a) & (5b).

Sturm's fundamental comparison theorems are derived by noticing that if $\theta = \theta(t,\lambda)$ is a solution of Eq. (3), then $\theta(t,\lambda) + k\pi$, with k a fixed integer, is also a solution of Eq. (3). Hence, given a nonnegative solution $y(t,\lambda)$ of Eq. (1), without loss of generality we may assume that $0 \leq \theta(a,\lambda) < \pi$. Given two consecutive zeros $t_1$ and $t_2$ of a nontrivial solution $y_1(t,\lambda)$ of $(P_1 y_1')' + g_1 y_1 = 0$ we can assume without loss of generality that $\theta_1(t_1,\lambda) = 0$, $\theta_1(t_2,\lambda) = \pi$ and $0 < \theta_1(t,\lambda) < \pi$ for $t_1 < t < t_2$. Moreover, let $y_2(t,\lambda)$ be a nontrivial solution of $(P_2 y_2')' + g_2 y_2 = 0$. Without loss of generality assume that $0 < \theta_2(t_1,\lambda) < \pi$. Then, $\theta_2(t_1,\lambda) > \theta_1(t_1,\lambda) = 0$. By the angle comparison theorem, $\theta_2(t_2,\lambda) > \theta_1(t_2,\lambda) = \pi$. Since $\theta_2(t,\lambda)$ is continuous on $[t_1,t_2]$, $\exists\ \hat{t} \in (t_1,t_2)$ such that $\theta_2(\hat{t},\lambda) = \pi$ and the result follows.

## 3. Distribution of Eigenvalues..

*3.1 Statement of Theorem*

**A Theorem.** *Suppose the coefficient functions of a Sturm-Liouville type differential equation*

$$(Py')' + (\lambda r - q)y = 0$$

*are such that*

*a) $P(t)$ is piecewise continuous and $\exists$ positive numbers $m_1$ and $m_2$ for which $0 < m_1 < P(t) < m_2\ \forall t \in [a,b]$.*

*b) $r(t)$ & $q(t)$ are piecewise continuous on $[a,b]$, possessed of perhaps an infinite number of sign changes on the interval, but both bounded on $[a,b]$, and $q(t) \leq 0$.*

*Let $\alpha^2 > max\{r(t) - q(t)\}$ on $[a,b]$, and $\alpha_2$ be a constant such that on some non-degenerate interval $[a_1,b_1] \subset [a,b]$, $\lambda\alpha_2^2 < \lambda r(t) - q(t)$ and $\lambda \geq 0$.*

*Then the eigenvalue problem of Eqs. (1), (2(i)), & (2(ii)) possesses an infinite sequence of positive eigenvalues $\lambda_n$ such that*

$$\frac{\sqrt{m}_1(n-2)\pi}{\alpha(b-a)}^2 < \lambda_n < \frac{\sqrt{m}_2(n+5)\pi}{\alpha_2(b_1-a_1)}^2,$$

*for all n = 2, 3, ....*

*3.2 Lemmata*

**Lemma 1.** *Suppose that there is some constant $\alpha$ such that Eq. (1) can be expressed as*

$$y'' - \lambda\alpha^2 y = 0. \tag{7}$$

*Let $y(t,\lambda) = \rho(t,\lambda)\sin\theta(t,\lambda)$ & $y'(t,\lambda) = \rho(t,\lambda)\cos\theta(t,\lambda)$. Then on any interval $[t_1,t_2]$ and for any $\lambda > 0$, we have $|\theta(t_2,\lambda) - \theta(t_1,\lambda)| < \pi$.*

*Proof.* Eq. (7) has the solution

$$y = c_1 e^{\sqrt{\lambda}\alpha t} + c_2 e^{-\sqrt{\lambda}\alpha t}.$$

If we consider the ratio

$$\begin{aligned}\frac{y'}{y} = \frac{\rho\sin\theta}{\rho\cos\theta} &= \tan\theta \\ &= \frac{1}{\sqrt{\lambda}\alpha}\frac{c_1 e^{\sqrt{\lambda}\alpha t} + c_2 e^{-\sqrt{\lambda}\alpha t}}{c_1 e^{\sqrt{\lambda}\alpha t} - c_2 e^{-\sqrt{\lambda}\alpha t}}.\end{aligned}$$

Since this quotient has no more than one zero or pole, the function $\tan\theta$ cannot cross more than one zero or pole, and it's argument is bounded by at most one period. Then $|\theta(t_2,\lambda) - \theta(t_1,\lambda)| < \pi$.

**Lemma 2.** *Suppose that there is some constant $\alpha$ such that Eq. (1) can be expressed as*

$$y'' + \lambda\alpha^2 y = 0. \tag{8}$$

*Let $y(t,\lambda) = \rho(t,\lambda)\sin\theta(t,\lambda)$ & $y'(t,\lambda) = \rho(t,\lambda)\cos\theta(t,\lambda)$. Then on any interval $[t_1,t_2]$ and for any $\lambda > 0$, the difference*

$$\sqrt{\lambda}\alpha(t_2 - t_1) - \pi < \theta(t_2,\lambda) - \theta(t_1,\lambda) < \sqrt{\lambda}\alpha(t_2 - t_1) + \pi.$$

*Proof.*

From Eq. (8) we see that $P(t) \equiv 1$ and $g(t,\lambda) = \lambda\alpha^2$. Then Eq. (3) becomes

$$\begin{aligned}\theta' &= \cos^2\theta + \lambda\alpha^2\sin^2\theta \\ &= \frac{1}{2}(1+\lambda\alpha^2) + \frac{1}{2}(1-\lambda\alpha^2)\cos 2\theta.\end{aligned}$$

This differential equation is separable, and the right hand side is always positive, so there is no singularity to the following definite integral.

$$\int_{\theta(t_1,\lambda)}^{\theta(t,\lambda)} \frac{2}{a + b\cos 2\theta}\, d\theta = \int_{t_1}^{t} ds$$

where $a = 1 + \lambda\alpha^2$ and $b = 1 - \lambda\alpha^2$. Evaluating the integral yields

$$\frac{2}{\sqrt{a^2 - b^2}}\left[\tan^{-1}\frac{(a-b)\tan\theta(t,\lambda)}{\sqrt{a^2-b^2}} - \tan^{-1}\frac{(a-b)\tan\theta(t_1,\lambda)}{\sqrt{a^2-b^2}}\right] = t - t_1.$$

Defining $\tilde{\theta}(t_1,\lambda) = \tan^{-1}\sqrt{\lambda}\alpha\tan\theta(t_1,\lambda)$, this is equivalent to

$$\sqrt{\lambda}\alpha\tan\theta(t,\lambda) = \tan[\sqrt{\lambda}\alpha(t-t_1) + \tilde{\theta}(t_1,\lambda)].$$

For $\lambda$ fixed, this relation is well defined for $t = t_1$, and as t varies from $t_1$ to $t_2$ the argument of the right hand side is a monotone increasing function. Further, the two arguments must agree at each zero and pole, and both sides must agree as to sign. Without loss of generality, we may assume that $0 \le \theta(t_1,\lambda) < \pi$. Then

$$|\theta(t_2,\lambda) - \sqrt{\lambda}\alpha(t_2-t_1) - \tilde{\theta}(t_1,\lambda)| < \frac{\pi}{2}|\tilde{\theta}(t_1,\lambda) - \theta(t_1,\lambda)| < \frac{\pi}{2}.$$

Finally, we see

$$\theta(t_1,\lambda) + \sqrt{\lambda}\alpha(t_2-t_1) - \pi < \theta(t_2,\lambda) < \theta(t_1,\lambda) + \sqrt{\lambda}\alpha(t_2-t_1) + \pi$$

which establishes the bounds on $\theta(t_2,\lambda) - \theta(t_1,\lambda)$.

*3.3 Proof of Theorem.* Let

$$m_1 y_1'' + g_1(t,\lambda)y_1 = 0$$
$$m_2 y_2'' + g_2(t,\lambda)y_2 = 0$$

where

$$g_1(t,\lambda) = \lambda\alpha^2$$

$$g_2(t,\lambda) = \left\{\begin{array}{ll} -\lambda{\alpha_1}^2 & t \in [a,a_1] \\ \lambda{\alpha_2}^2 & t \in (a_1,b_1) \\ -\lambda{\alpha_3}^2 & t \in [b_1,b] \end{array}\right\}.$$

If $y_i = \rho_i(t,\lambda)\sin\theta_i(t,\lambda)$ for $i = 1,2$ and $\theta_2(a,\lambda) \le \theta(a,\lambda) \le \theta_1(a,\lambda)$, then by the angle comparison theorem $\theta_2(b,\lambda) \le \theta(b,\lambda) \le \theta_1(b,\lambda)$, independently of $\lambda$.

*case 1* $(i = 1)$.

$m_1 y_1'' + g_1(t,\lambda)y_1 = 0$ or $y_1'' + \frac{\lambda\alpha^2}{m_1}y_1 = 0$. By Lemma 2,

$$\theta_1(b,\lambda) - \theta_1(a,\lambda) < \sqrt{\frac{\lambda\alpha^2}{m_1}}(b-a) + \pi$$

$$\theta_1(b,\lambda) < \theta_1(a,\lambda) + \sqrt{\frac{\lambda\alpha^2}{m_1}}(b-a) + \pi$$

$$\theta_1(b,\lambda) < \sqrt{\frac{\lambda\alpha^2}{m_1}}(b-a) + 2\pi. \tag{9}$$

*case 2*$(i = 2)$.

$m_2 y_2'' + g_2(t,\lambda) y_2 = 0$, but as $g_2$ is piecewise defined this case must be considered by intervals.

for $t \in [a, a_1]$

By Lemma 1,

$$|\theta_2(a_1,\lambda) - \theta_2(a,\lambda)| < \pi$$
$$\theta_2(a,\lambda) - \pi < \theta_2(a_1,\lambda). \tag{10}$$

for $t \in (a_1, b_1)$

By Lemma 2 and Eq. (10),

$$\sqrt{\frac{\lambda\alpha_2^2}{m_2}}(b_1 - a_1) - \pi < \theta_2(b_1,\lambda) - \theta_2(a_1,\lambda)$$
$$\theta_2(a_1,\lambda) + \sqrt{\frac{\lambda\alpha_2^2}{m_2}}(b_1 - a_1) - \pi < \theta_2(b_1,\lambda)$$
$$\theta_2(a,\lambda) + \sqrt{\frac{\lambda\alpha_2^2}{m_2}}(b_1 - a_1) - 2\pi < \theta_2(b_1,\lambda). \tag{11}$$

for $t \in [b_1, b]$

By Lemma 1 and Eq. (11),

$$\theta_2(b_1,\lambda) - \pi < \theta_2(b,\lambda)$$
$$\theta_2(a,\lambda) + \sqrt{\frac{\lambda\alpha_2^2}{m_2}}(b_1 - a_1) - 3\pi < \theta_2(b,\lambda)$$
$$\sqrt{\frac{\lambda\alpha_2^2}{m_2}}(b_1 - a_1) - 4\pi < \theta_2(b,\lambda). \tag{12}$$

By Eq.s (9) & (12) and the angle comparison theorem,

$$\sqrt{\frac{\lambda\alpha_2^2}{m_2}}(b_1 - a_1) - 4\pi < \theta(b,\lambda) < \sqrt{\frac{\lambda\alpha^2}{m_1}}(b - a) + 2\pi.$$

If $n\pi \le \theta(b,\lambda_n) < (n+1)\pi$, then

$$\sqrt{\frac{\lambda_n\alpha_2^2}{m_2}}(b_1 - a_1) - 4\pi < (n+1)\pi$$
$$\sqrt{\lambda_n} < \frac{\sqrt{m_2}(n+5)\pi}{\alpha_2(b_1 - a_1)}$$
$$\lambda_n < \left[\frac{\sqrt{m_2}(n+5)\pi}{\alpha_2(b_1 - a_1)}\right]^2.$$

and

$$n\pi < \sqrt{\frac{\lambda_n \alpha^2}{m_1}}(b-a) + 2\pi$$

$$\frac{\sqrt{m}_1(n-2)\pi}{\alpha(b-a)} < \sqrt{\lambda_n}$$

$$\left[\frac{\sqrt{m}_1(n-2)\pi}{\alpha(b-a)}\right]^2 < \lambda_n.$$

Finally we find the following boundaries on $\lambda_n$:

$$\left[\frac{\sqrt{m}_1(n-2)\pi}{\alpha(b-a)}\right]^2 < \lambda_n < \left[\frac{\sqrt{m}_2(n+5)\pi}{\alpha_2(b_1-a_1)}\right]^2,$$

for n = 2, 3, ....

To consider $q(t)$ without sign restrictions requires only a simple change in the bounding function $g_2(t, \lambda)$.

**4. Concluding Remarks.**

If $r(t)$ is negative on some non-degenerate interval $[a_1, b_1] \subset [a, b]$, then similar arguments apply and we obtain an infinite sequence of real but negative eigenvalues $\lambda_n$.

It turns out that each maximal subinterval of $[a, b]$ on which $r(t)$ does not change sign and is not identically zero, contributes an infinite sub-sequence of real eigenvalues to the discrete spectrum.

The condition of piecewise continuity can be relaxed, of course, but we will not be concerned with this in the current note. Also, one can improve both the upper and lower bounds using a more stringent method, but this has been omitted here in the interest of brevity.

Our analysis applies to most general Sturm-Liouville problems. If the Sturm-Liouville problem is not self-adjoint, we would have complex eigenvalues in addition to the real eigenvalues (compare with e.g. [7] and its references).

## References

1. F. V. Atkinson and A. B. Mingarelli, *Asymptotics of the number of zeros and of the eigenvalues of general weighted Sturm-Liouville problems*, Journal für Mathematik **375/376** (1987), 380-393.
2. E. A. Coddington and N. Levinson, *Theory of Ordinary Differential Equations*, McGraw-Hill, New York, 1955.
3. N. Dunford and J. T. Schwartz, *Linear Operators, Part II*, Interscience Publishers, New York, 1963.
4. H. Gingold, *Approximations of solutions of $y'' = \phi(x, \epsilon)y$ with several transition points and moving singularities*, Asymptotic Analysis **6** (1993), 335-364.

5. E. Hille, *Lectures on Ordinary Differential Equations*, Addison-Wesley, Reading, MA, 1964.
6. E. L. Ince, *Ordinary Differential Equations*, Dover Publications, Inc., New York, New York, 1956.
7. A. B. Mingarelli, *Volterra - Stieltjes Integral Equations and Generalized Ordinary Differential Expressions*, Springer-Verlag, Berlin, 1983.
8. F. J. Olver, *Connection formulas for second-order differential equations having an arbitrary number of turning points of arbitrary multiplicities*, SIAM J. Math Anal. **9** (1977), 673-700.
9. Y. Sibuya, *Global Theory of Second-Order Linear Ordinary Differential Equation with a Polynomial Coefficient*, Mathematics Studies, North-Holland, Amsterdam, 1975.
10. Sh. Strelitz and S. Abramovich, *On the existence of eigenvalues of differential operators dependant on a parameter*, Transactions of the AMS **258** (1980), 407-429.
11. W. Wasow, *Linear Turning Point Theory*, Springer-Verlag, Berlin, 1984.

# ON RADIAL SOLUTIONS TO THE EULER-LAGRANGE EQUATIONS OF INVARIANT ENERGY FUNCTIONALS

**SALIM M. HAIDAR**
*Department of Mathematics and Statistics*
*Grand Valley State University*
*Allendale, MI 49401 U.S.A.*

**Abstract**

There are very few known results on existence and regularity of weak equilibrium solutions to the Euler-Lagrange equations corresponding to nonlinear energy functionals. In this note, we present a new result on the question of existence of equilibria in radial elasticity. For invariant energy functionals, a seemingly promising approach to the question of regularity that replaces the delicate phase plane analysis method will be discussed.

## 1 Introduction

Consider a nonhomogeneous isotropic hyperelastic material body which, in its reference configuration, occupies the open bounded subset $\Omega$ of $\mathbb{R}^3$. The material particle at $\boldsymbol{x} \in \Omega$ is displaced by the deformation $\boldsymbol{g} : \Omega \to \mathbb{R}^3$ to $\boldsymbol{g}(\boldsymbol{x})$. In the absence of external forces, the total energy associated with a deformation $\boldsymbol{g}(\cdot)$ is given by the functional:

$$\boldsymbol{g} \to J(\boldsymbol{g}) := \int_\Omega W(\boldsymbol{x}, \nabla \boldsymbol{g}(\boldsymbol{x}))d\boldsymbol{x} \tag{1.1}$$

where $W : \Omega \times M_+^{3\times3} \to [0, \infty]$ denotes the corresponding stored energy function and $M_+^{3\times3}$ denotes the space of all $3 \times 3$ real matrices with positive determinants. The Euler-Lagrange equations associated with Eq. (1.1) are given by:

$$\operatorname{div}\left[\frac{\partial W}{\partial F}(\boldsymbol{x}, F)\right] = 0, \quad F := \nabla \boldsymbol{g}(\boldsymbol{x}). \tag{1.2}$$

For the case of homogeneous material, Ball and Murat [1984] discuss the questions of existence and regularity of solutions to the problem:

$$\operatorname{Inf}\{J(\boldsymbol{g}) : \boldsymbol{g} \in C_\lambda\}, \quad \lambda > 0 \tag{1.3}$$

where

$$C_\lambda := \{\boldsymbol{g} \in W^{1,1}(\Omega;\mathbb{R}^3) : \boldsymbol{g}(\boldsymbol{x}) = \lambda\boldsymbol{x} \text{ on } \partial\Omega, \det\nabla\boldsymbol{g}(\cdot) > 0 \text{ a.e, } J(\boldsymbol{g}) < \infty\}. \quad (1.4)$$

Like minimization problems in the finite dimensional case, the framework in which this typical problem Eq. (1.3), Eq. (1.4) is studied consists of two vital components: 1) Compactness, and, 2) Lower semi-continuity (in some topological sense). The nonlinear constraint of $\det F > 0$ makes the convexity property of $W(\boldsymbol{x},\cdot)$ totally unrealistic in more than one space dimension as, among many other reasons, it contradicts with the principle of frame indifference. Due to the lack of this property, lower semi-continuity of $J(\cdot)$ becomes a serious problem and a big obstacle. Since 1950, several attempts to treat this issue have been made, but it is yet to be universally settled (refer to Morrey [1952], Truesdell and Noll [1965], Ball [1977b]).

In Section 2 we present an existence result which extends to the nonhomogeneous case that of Ball [1982] and, like the results of Ball, is partly motivated by the work of Ogden [1972]. In addition, our discussion of the regularity of weak solutions of Eq. (1.3) represents a novel approach that does not require the consideration of the delicate phase plane analysis. In Section 3 we describe the physical significance of the homogeneity property used for the discussion of regularity in Section 2.

## 2 Existence and Regularity of Solutions

Let $\Omega = B(0,1) := \{\boldsymbol{x} \in \mathbb{R}^3 : |\boldsymbol{x}| < 1\}$. The class of admissible deformations for the problem Eq. (1.3), Eq. (1.4) which we consider corresponds to radial displacements of the form: $\boldsymbol{g}(\boldsymbol{x}) = \frac{r(R)}{R}\boldsymbol{x}$, where $R := |\boldsymbol{x}|$.

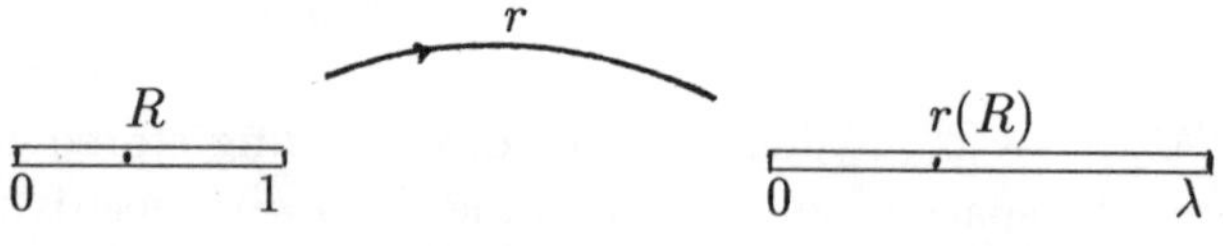


In this case, the stored energy function $W$ can be expressed as a symmetric function $\phi(R,.,.,.)$ of the principal stretches of $\nabla\boldsymbol{g}(\boldsymbol{x})$. In particular, $W(\boldsymbol{x},\nabla\boldsymbol{g}(\boldsymbol{x})) = \phi(R,\nu_1,\nu_2,\nu_3)$ where the singular values of $\nabla\boldsymbol{g}(\boldsymbol{x})$ are $\nu_1 = r'$, $\nu_2 = \nu_3 = r/R$ (refer to Rivlin and Ericksen [1955]). The problem Eq. (1.3), Eq. (1.4) becomes:

$$\operatorname{Inf}\{I(r) : r \in \mathcal{U}_{ad}\} \quad (2.1)$$

$$\mathcal{U}_{ad} := \{r \in W^{1,1}(0,1) : \dot{r}(R) > 0 \text{ a.e., } r(0) = 0,\ r(1) = \lambda,\ I(r) < \infty\} \quad (2.2)$$

where $J(\boldsymbol{g}) = 4\pi\, I(r)$ with $I(r) := \int_0^1 R^2\phi(R; r', r/R, r/R)dR$. The Euler-Lagrange equation corresponding to the functional $I$ is:

$$\frac{d}{dR} R^2\phi_{,1} = 2R\phi_{,2}. \tag{2.3}$$

Following Ball [1982, Theorem 4.2], we say that $\boldsymbol{g} \in W^{1,1}(B; \mathbf{R}^3)$ is a weak equilibrium solution of Eq. (1.2) if and only if: $r'(R) > 0$ a.e. $R \in (0,1)$, $R^2\phi_{,1}$ and $R^2\phi_{,2} \in L^1(0,1)$, and $R^2\phi_{,1} = 2\int_1^R \rho\phi_{,2}(\rho)d\rho + \text{const.}$, where

$$\phi_{,i}(R) := \frac{\partial\phi}{\partial\nu_i}\left(R; r'(R), \frac{r(R)}{R}, \frac{r(R)}{R}\right).$$

In order to obtain a good correlation between theory and experiment, Ogden and J. Ericksen et al. found that expansions of $W(\cdot)$ in symmetric forms such as $\sum \alpha_n\psi(\nu_n)$ do a good job for homogeneous incompressible material. To account for volume change effects one may then consider the form $\sum \alpha_n\psi(\nu_n) + h(\nu_1\nu_2\nu_3)$. For the nonhomogeneous case, we present the following result in which $W(\boldsymbol{x}, \cdot)$ admits to such expansions.

**Theorem 2.1.** Let the functions $\phi, \psi$, and $h$ be given such that:

(H0) $\phi(R; \nu_1, \nu_2, \nu_3) \geq \sum_{i=1}^3 \psi(R, \nu_i) + h(R, \nu_1\nu_2\nu_3)$ for $R \in (0,1)$, $\nu_i \in (0, +\infty)$.

(H1) $\phi_{,11}(R; \nu_1, \nu_2, \nu_3) > 0$.

(H2) $h : (0,1) \times (0, +\infty) \to \mathbf{R}$ is $C^1$ and strictly convex in its second argument.

(H3) $h(\cdot, \sigma)$ is nondecreasing.

(H4) $\lim_{\sigma\to 0+} h(R, \sigma) = \lim_{\sigma\to\infty} h(R, \sigma)/\sigma^{1+\varepsilon} = +\infty$, $\varepsilon > 0$.

(H5) $\psi : (0,1) \times (0, +\infty) \to [0, +\infty)$ is $C^1$ and convex in its second argument with $\lim_{\nu\to\infty} \psi(R, \nu) = +\infty$.

Then $I$ attains an absolute minimum on $\mathcal{U}_{ad}$.

**Proof:** Let $\rho = R^3$ and $u(\rho) = r^3(R)$. Then $\dot{u} = du/d\rho = r'(r/R)^2 = \det\nabla\boldsymbol{g}(x)$. In terms of this change of variables, the energy functional takes the form:

$$I(u) = \int_0^1 f(\rho, u, \dot{u})d\rho$$

where

$$f(\rho, u, p) = \phi\left(\rho^{\frac{1}{3}}, (\frac{\rho}{u})^{\frac{2}{3}}p, (\frac{u}{\rho})^{\frac{1}{3}}, (\frac{u}{\rho})^{\frac{1}{3}}\right) \geq \sum_{i=1}^3 \psi(\rho^{\frac{1}{3}}, \nu_i) + h(\rho^{\frac{1}{3}}, \dot{u}). \tag{2.4}$$

Put $L := \inf\{I(u) : u \in \mathcal{U}_{ad}\}$ and let $\{u_n\}_{n\in N}$ be a minimizing sequence for $I$ in $\mathcal{U}_{ad}$.

For each integer $m > 1$, Eq. (1.8) and (H3) imply:

$$\int_{\frac{1}{m}}^{1} h\left((\frac{1}{m})^{\frac{1}{3}}, \dot{u}_n(\rho)\right) d\rho \le \int_0^1 h(\rho^{\frac{1}{3}}, \dot{u}_n) \le I(u_n) \le D,$$

where $D$ is a constant independent of $n$. Hence, by (H4), the family $\{\dot{u}_n\}_{n\in N}$ is sequentially weakly relatively compact in $L^1(\frac{1}{m}, 1)$. Define $\{u_{1,n}\}$ by $u_{1,n} := u_n$ for all $n$. By De la Valleé Poussin's criterion we can choose the following sequences inductively: $\{u_{m,n}\}_{n=1}^{\infty}$ to be a subsequence of $\{u_{m-1,n}\}_{n=1}^{\infty}$ with the property that $\dot{u}_{m,n}$ converging weakly in $L^1(\frac{1}{m}, 1)$ to, say, $Z_m$.

Let $Z$ be the function: $Z : (\frac{1}{k}, 1) \to [0, \lambda]$ given by $Z(\rho) := Z_k(\rho)$ where $k$ is chosen so that $\rho \in (\frac{1}{k}, 1)$. Following Sivaloganathan [1985], put

$$u(\rho) := \lambda - \int_{\rho}^{1} Z(t)dt \text{ and } u_m := u_{m,m} \tag{2.5}$$

for all $m > 1$.

For each $\delta \in (0, 1)$, choose $m$ large enough so that $\delta > \frac{1}{m}$. It is not difficult to show that $u_m \rightharpoonup u$ in $L^1(\delta, 1)$ and thereby conclude that

$$u_m \rightharpoonup u \text{ in } W^{1,1}(\delta, 1). \tag{2.6}$$

To establish the (sequentially) weak lower semi-continuity of $I(\cdot)$, we extend the definition of $f$ to $(0, 1) \times \mathbf{R} \times \mathbf{R}$ by setting

$$f(\rho, u, p) = +\infty \text{ if } u \text{ or } p \le 0. \tag{2.7}$$

This makes $f(\rho, \cdot, \cdot)$ a continuous function from $\mathbf{R} \times \mathbf{R}$ to $\bar{\mathbf{R}}$ for each $\rho \in (0, 1)$. Since $f(\rho, u, \cdot)$ is also convex we may apply a result of Ball, Currie and Olver [1981] to conclude that

$$\int_{\delta}^{1} f(\rho, u, \dot{u})d\rho \le \liminf_{m\to+\infty} \int_{\delta}^{1} f(\rho, u_m, \dot{u}_m)d\rho.$$

From this we see that the nonnegative monotone functions $\delta \mapsto \int_{\delta}^{1} f(\rho, u, \dot{u})d\rho$ have a least upper bound $L$. The monotone convergence theorem and the definition of $L$ now imply: $\int_0^1 f(\rho, u, \dot{u})d\rho = L$. Hypothesis (H4), Eq. (2.5), and Eq. (2.7) imply that $\dot{u}(\rho) > 0$ for a.e. $\rho \in (0, 1)$ and $\dot{u} \in L^1(\delta, 1)$ for each $\delta \in (0, 1)$. Moreover $\int_{\delta}^{1} \dot{u}(t)dt \le \lambda$. The monotone convergence theorem implies that $u \in W^{1,1}(0, 1)$ and Eq. (2.6) gives $u_m \to u$ uniformly. Thus $u(0) \ge 0$. By (H4), $u \in W^{1,1+\varepsilon}(0, 1)$ and the Sobolev Imbedding Theorem yields that $u \in C_b(0, 1)$, the space of bounded continuous functions on $(0,1)$ and consequently $u(0) = 0$. It now follows that $u \in \mathcal{U}_{ad}$. $\square$

It is natural to ask how regular can this minimizer be? Could it be in a smaller space? Is it Lipschitz or does it satisfy Eq. (2.3)? Within this context, very little is known about this very complex problem. Our new approach to this problem is based on concepts from the Hamilton-Jacobi theory, field theory of the Calculus of Variations, and the theory of groups of continuous transformations as established in Haidar [1989].

*Artificial example*:

Consider $A(r) := \int_0^1 (r^2 - R)^2 (\dot{r})^6 dR$ with $r(0) = 0$, $r(1) = 1$. Clearly, $r^*(R) := R^{1/2}$ is the absolute minimizer of $A(\cdot)$ in $W^{1,t}(0,1)$ for $t < 2$. If we restrict the class of competing curves to $W^{1,t}(0,1)$ for $t \geq 2$, then $r^*$ is no longer admissible. Let $V(R_0, r_0)$ denote the value of $\mathrm{Inf}\{A(r)\}$ over all $r \in W^{1,1}(0,1)$ satisfying $r(0) = 0$, $r(R_0) = r_0$. The value function $V(\cdot,\cdot)$ provides a lower bound for $A(r)$ at each point $(R_0, r_0)$ and is given by the Hamilton-Jacobi equation

$$\begin{cases} V_R(R,r) + f^*(R,r,V_r(R,r)) = 0 \\ V(0,0) = 0 \end{cases} \tag{2.8}$$

where $f^*(R,r,z) = \sup_p [pZ - f(R,r,p)]$ with $f(R,r,p) =$ integrand of $A(r)$. Moreover, the solution of the o.d.e. $\dot{r} = \hat{p}(R,r)$ is optimal, where $\hat{p}(R,r)$ is the value of the function $p(\cdot,\cdot)$ at which $f^*(R,r,z)$ is attained. (Refer to Rund [1966, page 18].) For the functional $A(r)$, it is easy to verify that $\hat{p}(R,r) = \frac{1}{6^{1/5}} \cdot (r^2 - R)^{-2/5} \cdot V_r^{1/5}$ and that Eq. (2.8) becomes:

$$V_R = \alpha R^{-2/5} \left[ \left( \frac{r}{\sqrt{R}} \right)^2 - 1 \right]^{-2/5} V_r^{6/5} = 0, \; V(0,0) = 0, \tag{2.9}$$

where $\alpha = \frac{5}{6^{6/5}}$.

The functional $A(r)$ is invariant under the action of the transformations: $\bar{R} = \varepsilon R$, $\bar{r} = \varepsilon^{1/2} r$, $\varepsilon > 0$. Accordingly, we construct (similarity) solutions of Eq. (2.9) of the form $V(R,r) = h \circ \xi(R,r)$ where $\xi(R,r) = rR^{-1/2}$. Using these solutions, one may easily verify that $\hat{p}(R,r) = \frac{3}{5} r R^{-1}$ and that $\hat{r}(R) = R^{3/5}$ is optimal. Moreover, the solution $\hat{r} \in W^{1,t}(0,1)$ for $t \in [2, \frac{5}{2})$.

Of course, the integrand of $A(r)$ does not serve as an example for the stored energy function $\phi$. However, the work shown above can be generalized (see Haidar [1989]) to lead in a very natural way to energy functionals which are invariant under the action of the group of continuous transformations: $\bar{R} = \varepsilon R$, $\bar{r} = \varepsilon^\gamma r$

with

$$f(\varepsilon R, \varepsilon^{\gamma} r, \varepsilon^{\gamma-1} r') = \varepsilon^{-1} f(R, r, r'), \ \varepsilon > 0, \ \gamma \in (0,1). \tag{2.10}$$

Setting $\varepsilon = \frac{1}{R}$ in Eq. (2.10) yields

$$f(R, r, r') = R^{-1} f(1, rR^{-\gamma}, r'R^{1-\gamma}) := R^{-1} e(P, X) \tag{2.11}$$

where $X(R,r) = rR^{-\gamma}$ and $P(R,r') = r'R^{1-\gamma}$. We call the attention of the reader to the fact that $e(P,X)$ of Eq. (2.11) is the restriction to the plane $X_1 = X_2 = X$ of the symmetric quantity $E(P, X_1, X_2)$ associated with $\phi(R; \nu_1, \nu_2, \nu_3)$. For some $\lambda > 0$, observe that an $r(\cdot)$ of the form $r^*(R) = \lambda R^{\gamma}$ must be an absolute minimizer for $I(r)$ in $W^{1,t}(0,t)$ for $t < (1-\gamma)^{-1} =: t_0$ because: along curves of the form $r^*(R) = \lambda R^{\gamma}$ and in light of Eq. (2.11) one has $I(r) = \int_0^1 R^{-1} e(\lambda\gamma, \lambda) dR$ which yields the value zero only if there is a zero of $e(\cdot,\cdot)$ of the form $(\lambda\gamma, \lambda)$ in the $PX$-plane. By restricting the competing curves to $\mathcal{U}_{ad} \cap W^{1,t}(0,1)$ for $t \geq t_0$, the curve $r^*(\cdot)$ is no longer admissible. In this case, it is possible to produce a more regular solution curve of the form $\hat{r}(R) = \lambda R^{\theta}$ for $\theta < \lambda$ but we shall not address it here. In Haidar [1992], singular equilibria of this form corresponding to strongly elliptic isotropic energies are established.

Uniqueness results for this case cannot be expected even under very favorable conditions on $W$. For example, the uniqueness results of Knops and Stuart [1984, Proposition 4.1 and Corollary 4.3] in the homogeneous case do not carry over to the nonhomogeneous case. We mention the following result (Haidar [1992]): There exist positive numbers $\lambda$ and $c$ such that the problem Eq. (2.1), Eq. (2.2) corresponding to

$$e(P,X) = (PX^2)^{5/3} - 2(PX^2)^{2/3} + (PX^2)^{-1/3} + (P-c)^2(X-c)^4$$

admits nontrivial (i.e., $r(R) \neq \lambda R$) solutions of the form $r^*(R) = \lambda R^{\gamma}, \ R \in [0,1]$.

# 3 Physical Interpretation of the Homogeneity Property

The constitutive property Eq. (2.10) provides for a plausible model of the onset of fracture in nonhomogeneous elastic materials. Let us take $e(P,X) = (P-X)^2(PX)^{-1}$ and assume that the material body $B(0,1)$ undergoes a rigid deformation. The stored energy function $\phi$ now behaves like $R^{-2-\gamma}$ and, consequently, $\phi \to \infty$ as $R \to 0^+$. Hence, the material under consideration must be denser (coarser) near the center of the ball than it is elsewhere (e.g., near $R = 1$). In general, when the principal stretches $\nu_1 = \nu_2 = \nu_3 = \nu =$ constant, $\phi$ as a function of $R$ behaves like $R^{-3}q(R)$, where $q(R) := e(\nu R^{1-\gamma}, \nu R^{1-\gamma})$. By the natural growth condition (H4) of Theorem 2.1, $\phi \to +\infty$ as $R \to 0^+$. To account for the possibility of the material to fracture we propose the following argument: Let us expose the unit bar to expansion or compression and think of the intensity of this action as propagating through material layers of different conductivities.

**Case 1:** Expansion (i.e., $\varepsilon > 1$). In this situation, one may arrange the layers as follows:

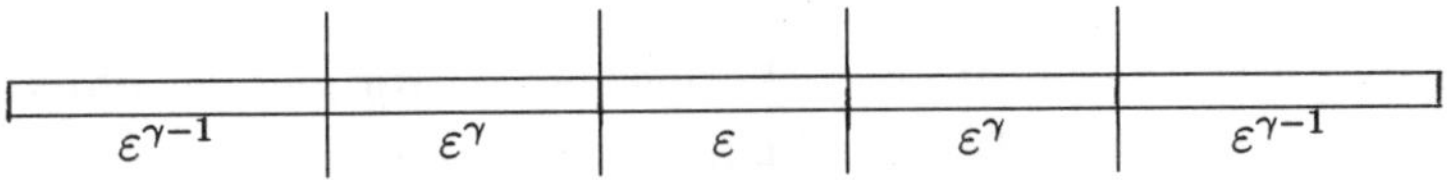


**(Conductivity Distribution)**

Since $0 < \gamma < 1$ the conductivity near the center of the bar is higher than it is elsewhere. This reflects weaker resistance near the center than elsewhere in the bar.

**Case 2:** Compression (i.e., $\varepsilon < 1$). In this situation, one may arrange the layers as follows:

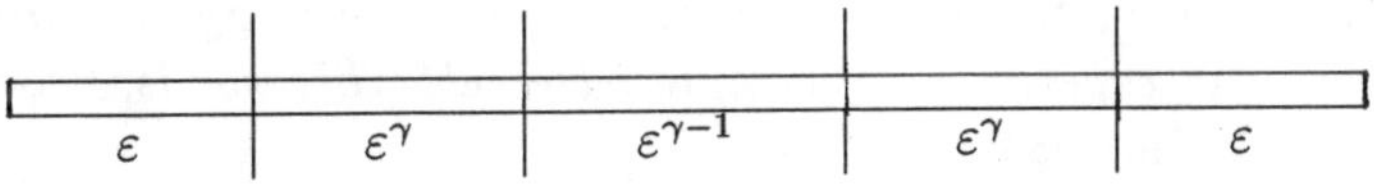


**(Conductivity Distribution)**

This also reflects weaker resistance near the center of the bar leading to the same outcome as in Case 1.

The above arguments may easily be adapted for the homogeneous case as well as for the incompressible case.

*Acknowledgement.* The author thanks the R & D office of GVSU for its support in the preparation of this note.

# 4 References

J. Ball [1977b], *Constitutive inequalities and existence theorems in nonlinear elasticity*, in *Nonlinear Analysis and Mechanics: Heriot-Watt Symposium* (ed. R.J. Knops) Vol. I, 187-241, London: Pitman.

J. Ball [1982], *Discontinuous equilibrium solutions and cavitation in nonlinear elasticity*, Phil. Trans. Roy. Soc. Lond., A306, 557-611.

J. Ball and F. Murat [1984], *$W^{1,P}$-quasiconvexity and variational problems for multiple integrals*, J. Fcnl. Anal., 58, 225-253.

J. Ball, J. Currie, and P. Olver [1981], *Null Lagrangians, weak continuity, and variational problems of arbitrary order*, J. Fcnl. Anal., 41, 135-174.

S. Haidar [1989], *The Lavrentiev Phenomenon in Nonlinear Elasticity*, Ph.D. Dissertation, Department of Mathematics, Carnegie-Mellon University, Pittsburgh, PA.

S. Haidar [1992], *An example of J. Ball revisited*, submitted for publication.

C. Morrey [1952], *Quasiconvexity and the lower semi-continuity of multiple integrals*, Pacific J. Math., 2, 25-53.

R. Ogden [1972], *Large deformation isotropic elasticity: On the correlation of theory and experiment of compressible rubberlike solids*, Proc. Roy. Soc. Lond., A326, 567-583.

R. Rivlin and J. Ericksen [1955], *Stress-deformation relations for isotropic materials*, J. Rat. Mech. Anal., 4, 323-425.

H. Rund [1966], *The Hamilton-Jacobi Theory in the Calculus of Variations*, D. Van Nostrand Company, Inc., London.

J. Sivaloganathan [1985], *Uniqueness of Regular and Singular Equilibria in Radial Elasticity*, manuscript, Heriot-Watt University, Scotland.

C. Truesdell and W. Noll [1965], *The Nonlinear Field Theories of Mechanics*, in Handbuch der Physik, Vol. III/3, ed. S. Flügge, Springer, Berlin.

# A GENERALIZED SINGULARITY OF THE FIRST KIND

William A Harris, Jr.
Department of Mathematics, University of Southern California
Los Angeles, California 90089-1113, U.S.A.

Yasutaka Sibuya
School of Mathematics, University of Minnesota
Minneapolis, Minnesota 55455, U.S.A.

ABSTRACT

We are concerned with the linear differential system $t\frac{dy}{dt} = A(t^q, t)y$, $q > 1$, irrational, where the $n \times n$ matrix $A(\tau, t)$ is holomorphic for $|\tau| < a$, $|t| < a$. We show that this system has the same general properties as a linear differential system with an isolated singularity of the first kind: formal-log solutions are convergent; a fundamental matrix can be computed in algorithmic fashion; and first kind implies regular singularity.

## 1. Introduction:

We are concerned with the linear differential system:

$$t\,\frac{dy}{dt} = A(t^q, t)\,y\,, \tag{1.1}$$

where $q$ is irrational, $q > 1$, and the $n \times n$ matrix $A(\tau, t)$ is holomorphic for $|\tau| < a$, $|t| < a$.

Such problems occur in the study of the asymptotic behavior of solutions of the differential equation:

$$y'' + 2\,b\,t^{\alpha-1}\,y' + c\,t^{2\beta-2}\,y = 0\,,$$

$\alpha \leq 0$, $\beta \leq 0$, $\frac{\alpha}{\beta}$ irrational, arising in the study of radial ground states for linear elliptic differential equations [2,4].

We prove

**Theorem 1.** *Let $A(\tau,t)$ be holomorphic for $|\tau| < a$, $|t| < a$, with $A(\tau,t) = \sum_{k=0}^{+\infty} t^k \ A_k(\tau)$ , and let*

$$y(\tau,t) = \sum_{k=0}^{+\infty} t^k \ y_k(\tau) \qquad ( \tau = t^q ) \tag{1.2}$$

*be a "formal" solution of (1.1) in the sense that*

$$q\tau \frac{dy_0(\tau)}{d\tau} = A_0(\tau) \ y_0(\tau) \ , \tag{1.3$_0$}$$

$$q\tau \frac{dy_k(\tau)}{d\tau} = ( A_0(\tau) - k \ I ) \ y_k(\tau) + \sum_{j=0}^{k-1} A_{k-j}(\tau) \ y_j(\tau) \tag{1.3$_k$}$$

*for $k \geq 1$, then $y(\tau,t)$ is convergent for $|\tau| < a$, $|t| < a$, i.e. $y(t^q,t)$ is a true solution of (1.1).*

**Theorem 2.** *Let $\lambda_1, \cdots, \lambda_n$ be the eigenvalues of $A_{00} = A(0,0)$ and let $\lambda_i = \lambda_j$ or $\lambda_i - \lambda_j \neq k + mq$ for $k$, $m$ nonnegative integers, not both zero, for $i$, $j = 1, \cdots, n$. Then system (1.1) has a fundamental matrix of the form:*

$$Y(t^q,t) = P(t^q,t) \ t^{A_{00}} \ , \tag{1.4}$$

*where $P(\tau,t)$ is holomorphic for $|\tau| < a$, $|t| < a$, and $P(0,0) = I$.*

**Theorem 3.** *There exists a transformation*

$$y(\tau,t) = P(\tau,t) \ u(\tau,t) \qquad ( \tau = t^q ) \tag{1.5}$$

*such that the $n \times n$ matrix $P(\tau,t)$ is holomorphic for $|\tau| < a$, $|t| < a$, the matrix $P(0,0)$ is nonsingular, and (1.5) transforms (1.1) into*

$$t \frac{du}{dt} = B(\tau,t) \ u \qquad ( \tau = t^q ) \ ,$$

*in such a way that the $n \times n$ matrix $B(\tau,t)$ is a polynomial in $(\tau,t)$ of the following form:*

$$t^{-D_1} \ \tau^{-D_2} \ B(\tau,t) \ \tau^{D_2} \ t^{D_1} = C + D_1 + q \ D_2 \ ,$$

*where $D_1$ and $D_2$ are diagonal matrices with integer entries, and $C$ is a constant matrix.*

Thus, system (1.1) has a fundamental matrix with the structure:

$$Y(\tau,t) = P(\tau,t)\,\tau^{D_2}\,t^{D_1}\,t^{C} \qquad (\tau = t^q),$$

or first kind implies regular singularity.

## 2. Proof of Theorem 1:

If $y(\tau,t) = \sum_{k=0}^{+\infty} t^k\, y_k(\tau)$ is a formal solution of system (1.1) as indicated in equations $(1.3_0)$ and $(1.3_k)$, then

$$\mathcal{L}_k[y_k](\tau) = H_k(\tau)\,, \tag{2.1}$$

where

$$\mathcal{L}_k[v](\tau) = q\,\tau\,\frac{dv(\tau)}{d\tau} - (\,A_0(\tau) - k\,I\,)\,v(\tau)$$

and

$$H_k(\tau) = \sum_{j=0}^{k-1} A_{k-j}(\tau)\,y_j(\tau)\,, \qquad H_0 \equiv O\,.$$

Since formal solutions of linear homogeneous and nonhomogeneous differential systems with a singularity of the first kind are convergent [3], all coefficients $y_k(\tau)$ are holomorphic for $|\tau| < a$.

Let $\rho_0$ be any positive number such that $0 < \rho_0 < a$. For a power series $v = \sum_{\ell=0}^{+\infty} \tau^\ell\, v_\ell$ set

$$\|\,v\,\| = \max_{\ell=0}^{+\infty} \rho_0^\ell\,|v_\ell|\,.$$

Then, it is not difficult to show that $\mathcal{L}_k[v] = h$ implies that $\|v\| \leq \dfrac{K}{k}\,\|h\|$ for $k \gg 1$, with $K$ independent of $k$. Hence

$$\|y_k\| \leq \frac{K}{k}\sum_{j=0}^{k-1} \|A_{k-j}\|\,\|y_j\| \qquad \text{for} \quad k \gg 1\,.$$

Set $M_k = \max_{\ell=0}^{k} \rho^\ell\,\|y_\ell\|$, where $\rho$ is any positive number such that $0 < \rho < \rho_0$. Since the power series $\sum_{k=0}^{+\infty} t^k\, A_k(\tau)$ converges for $|\tau| < a$, $|t| < a$, there exists a positive number $L$ such that $\rho_0^k \|A_k\| \leq L$ for $k \geq 0$. This implies that, for $k \gg 1$, we have

$$\begin{aligned}
\rho^k \|y_k\| &\leq \left( \frac{K}{k} \sum_{j=0}^{k-1} \rho^{k-j} \|A_{k-j}\| \right) M_{k-1} \\
&\leq \left( \frac{L\,K}{k} \sum_{j=0}^{k-1} \left(\frac{\rho}{\rho_0}\right)^{k-j} \right) M_{k-1} \\
&\leq \left( \frac{L\,K}{k\left(1 - \frac{\rho}{\rho_0}\right)} \right) M_{k-1} .
\end{aligned}$$

Therefore

$$M_k = M_{k-1} \qquad \text{for} \qquad k \gg 1$$

and the sequence $\{ \rho^k \|y_k\| \ ; \ k = 0, 1, 2 \cdots \}$ is bounded. This completes the proof of Theorem 1.

## 3. Proofs of Theorems 2 and 3:

From the construction of the matrices $B$ and $P$ of Theorem 3 it will be clear that the hypotheses of Theorem 2 allow the choice $B = C = A_{00}$ and $P_0 = I$. We shall prove Theorem 3 in this section.

The transformation

$$y(\tau, t) = P(\tau, t)\, u(\tau, t) \qquad (\tau = t^q)$$

transforms the system

$$t \frac{du}{dt} = A(\tau, t)\, u \qquad (\tau = t^q)$$

into a system

$$t \frac{du}{dt} = B(\tau, t)\, u \qquad (\tau = t^q),$$

if and only if

$$q\tau \frac{\partial P(\tau, t)}{\partial \tau} + t \frac{\partial P(\tau, t)}{\partial t} = A(\tau, t)\, P(\tau, t) - P(\tau, t)\, B(\tau, t) . \tag{3.1}$$

With

$$P(\tau,t) = \sum_{k=0}^{+\infty} t^k P_k(\tau) , \qquad A(\tau,t) = \sum_{k=0}^{+\infty} t^k A_k(\tau) ,$$

$$B(\tau,t) = \sum_{k=0}^{+\infty} t^k B_k(\tau) ,$$

we have

$$q\tau P_0'(\tau) = A_0(\tau) P_0(\tau) - P_0(\tau) B_0(\tau) , \tag{3.2$_0$}$$

and

$$\begin{aligned} q\tau P_k'(\tau) &= A_0(\tau) P_k(\tau) - P_k(\tau) ( B_0(\tau) + k I ) \\ &+ \sum_{j=1}^{k-1} ( A_{k-j}(\tau) P_j(\tau) - P_j(\tau) B_{k-j}(\tau) ) \\ &+ A_k(\tau) P_0(\tau) - P_0(\tau) B_k(\tau) \end{aligned} \tag{3.2$_k$}$$

for $k \geq 1$ .

For $k = 0$, as before, with

$$P_0(\tau) = I + \sum_{j=0}^{+\infty} \tau^j P_{0j} , \qquad A_0(\tau) = \sum_{j=0}^{+\infty} \tau^j A_{0j} ,$$

$$B_0(\tau) = \sum_{j=0}^{+\infty} \tau^j B_{0j} ,$$

we obtain from equation $(3.2_0)$,

$$\begin{cases} O = A_{00} - B_{00} , \\ q\ell P_{0\ell} = A_{00} P_{0\ell} - P_{0\ell} A_{00} + A_{0\ell} - B_{0\ell} \\ \qquad + \sum_{j=1}^{\ell-1} ( A_{0\ell-j} P_{0j} - P_{0j} B_{0\ell-j} ) \end{cases}$$

or

$$\begin{cases} B_{00} = A_{00} , \\ A_{00} P_{0\ell} - P_{0\ell} ( A_{00} + q\ell I ) \\ \qquad = B_{0\ell} - A_{0\ell} - \sum_{j=1}^{\ell-1} ( A_{0\ell-j} P_{0j} - P_{0j} B_{0\ell-j} ) \qquad \text{for} \quad \ell > 1 . \end{cases}$$

Thus, if $\lambda_1, \cdots, \lambda_n$ are the eigenvalues of $A_{00}$ and $\lambda_i - \lambda_j = 0$ or $\dfrac{\lambda_i - \lambda_j}{q}$ is not an integer for $i, j = 1, \cdots, n$, then we may choose $B_{0\ell} = O$ and solve for the $P_{0\ell}$ $(\ell = 1, \cdots)$.

Otherwise, assume that $A_{00}$ has the Jordan canonical form:

$$\begin{bmatrix} \lambda_1 & 0 & 0 & \cdots & 0 & 0 \\ \delta_2 & \lambda_2 & 0 & \cdots & 0 & 0 \\ \cdots & \cdots & \cdots & \cdots & \cdots & \cdots \\ \cdots & \cdots & \cdots & \cdots & \cdots & \cdots \\ 0 & 0 & 0 & \cdots & \delta_n & \lambda_n \end{bmatrix},$$

with

$$\Re[\lambda_1] \leq \Re[\lambda_2] \leq \cdots \leq \Re[\lambda_n].$$

Then, $B_0(\tau) - A_{00}$ can be chosen as a lower triangular matrix with entries

$$(\, B_0(\tau) - A_{00}\,)_{ij} = \begin{cases} 0 & \text{if } (\lambda_i - \lambda_j)/q \text{ is not a positive integer}, \\ c_{ij}\ \tau^{(\lambda_i - \lambda_j)/q} & \text{if } (\lambda_i - \lambda_j)/q \text{ is a positive integer}. \end{cases}$$

We may continue in this fashion for $k \geq 1$ (see, equation $(3.2_k)$) by replacing conditions on $B_0(\tau)$ by conditions on $B_0(\tau) + kI$, i.e.

$$\begin{aligned} A_{00}\, P_{k\ell} \;-\; P_{k\ell}\,(\, A_{00} + (k + q\ell)\, I\,) \\ = B_{k\ell} - A_{k\ell} - \sum_{j=1}^{\ell-1} (\, A_{k\ell-j}\, P_{kj} - P_{kj} B_{k\ell-j}\,) \qquad \text{for } \ell > 1, \end{aligned}$$

yielding appropriate triangular matrices as above when $\dfrac{\lambda_i - \lambda_j - k}{q}$ is a positive integer.

Since $B(\tau, t)$ is a polynomial by construction, $A(\tau, t)$ is holomorphic by assumption, and $P(\tau, t)$ is a formal solution of the linear (matrix) differential equation $t\dfrac{dP}{dt} = AP - PB$, Theorem 1 shows that $P(\tau, t)$ is holomorphic.

**4. An Example:**

Let

$$A_{00} = \begin{bmatrix} \lambda_1 & 0 \\ 0 & \lambda_2 \end{bmatrix}$$

with $\Re[\lambda_1] < \Re[\lambda_2]$. There can only be one $k$ and $\ell$ such that $\lambda_2 = \lambda_1 + k + q\ell$, $q > 1$, $q$ irrational. Thus

$$B(\tau,t) = \begin{bmatrix} \lambda_1 & 0 \\ 0 & \lambda_2 \end{bmatrix} + t^k \tau^\ell \begin{bmatrix} 0 & 0 \\ c & 0 \end{bmatrix} .$$

Let

$$D_1 = \begin{bmatrix} 0 & 0 \\ 0 & k \end{bmatrix}, \quad D_2 = \begin{bmatrix} 0 & 0 \\ 0 & \ell \end{bmatrix}, \quad C = \begin{bmatrix} \lambda_1 & 0 \\ c & \lambda_1 \end{bmatrix},$$

then

$$Y(\tau,t) = P(\tau,t)\, \tau^{D_2}\, t^{D_1}\, t^C \qquad (\tau = t^q)$$

is a fundamental matrix for

$$t\,\frac{dy}{dt} = A(\tau,t)\, y \qquad (\tau = t^q).$$

## 5. Algebraic lemma:

The simplification of Theorem 3 is based on the matrix equation:

$$A\,X - X\,(A + \alpha I) = F - G, \tag{5.1}$$

where $A$ and $G$ are given and we wish to determine an $F$ with the maximal number of zero elements so that the equation may be solved. Clearly, if the eigenvalues of $A$ do not differ by $\alpha$, we may choose $F \equiv O$ and determine $X$ as the unique solution of

$$A\,X - X\,(A + \alpha I) = -G.$$

We give the details for the choice of $F$ in the $2 \times 2$ case.

Let $A = \begin{bmatrix} \lambda_1 & 0 \\ \delta & \lambda_2 \end{bmatrix}$, $\delta = 0$ or $= 1$, and if $\delta = 1$, then $\lambda_1 = \lambda_2$. Write $X = (x_1, x_2)$, $F = (f_1, f_2)$, $G = (g_1, g_2)$, then equation (5.1) with $\alpha > 0$ can be written

$$\begin{cases} ((\alpha + \lambda_1)\, I - A)\, x_1 = -\delta\, x_2 - f_1 + g_1, \\ ((\alpha + \lambda_2)\, I - A)\, x_2 = -f_2 + g_2. \end{cases} \tag{5.2}$$

Assume that $Re[\lambda_1] \leq \Re[\lambda_2]$, then

$$(\alpha + \lambda_2)\, I - A = \begin{bmatrix} \alpha + (\lambda_2 - \lambda_1) & 0 \\ -\delta & \alpha \end{bmatrix}$$

is nonsingular and we may choose $f_2 = 0$ and solve the second equation in (5.2) for $x_2$. If $\lambda_2 - \lambda_1 \neq \alpha$, we may also choose $f_1 = 0$ and determine $x_1$. Therefore assume that $\lambda_2 - \lambda_1 = \alpha$. Set

$$x_1 = \begin{bmatrix} x_{11} \\ 0 \end{bmatrix} \quad \text{and} \quad f_1 = \begin{bmatrix} 0 \\ c \end{bmatrix},$$

then the first equation in (5.2) becomes

$$\begin{bmatrix} \alpha\, x_{11} \\ -\delta x_{11} \end{bmatrix} + \begin{bmatrix} 0 \\ c \end{bmatrix} = -\delta\, x_2 + g_1$$

which determines $x_{11}$ and $c$ by the first and second components respectively.

## 6. Formal log-series solutions:

We show that the convergence of the more general logarithmic–formal solution

$$y(\tau, t) = t^{\lambda} \sum_{\ell=1}^{p} \frac{(\ln t)^{\ell-1}}{(\ell-1)!}\, u_{\ell}(\tau, t)$$

with

$$u_{\ell}(\tau, t) = \sum_{k=0}^{+\infty} t^{k}\, u_{\ell k}(\tau) \qquad (\tau = t^{q})$$

can be put in the present frame work.

Let $y(\tau, t)$ be a formal solution of $t\dfrac{dy}{dt} = A(\tau, t)\, y$, then

$$u(\tau, t) = (\, u_1(\tau, t)\,,\, u_2(\tau, t)\,, \cdots,\, u_p(\tau, t)\,)^T$$

is a formal solution of the linear system $t\dfrac{du}{dt} = \mathcal{A}(\tau, t)\, u$ , where

$$\mathcal{A}(\tau, t) = \begin{bmatrix} A(\tau,t) - \lambda I & -I & O & O & \cdots & O \\ O & A(\tau,t) - \lambda I & -I & O & \cdots & O \\ \cdots & \cdots & \cdots & \cdots & \cdots & \cdots \\ \cdots & \cdots & \cdots & \cdots & \cdots & \cdots \\ O & O & O & \cdots & \cdots & -I \\ O & O & O & \cdots & \cdots & A(\tau,t) - \lambda I \end{bmatrix}$$

and convergence follows as before.

**Acknowledgements:** This paper was completed while William A. Harris, Jr. was on sabbatical leave from the University of Southern California as a Senior Fellow, AHPCRC, University of Minnesota, supported in part by the U.S.Army under Grant DAAL 03-89-c-0038. Yasutaka Sibuya was partially supported by a grant from the National Science Foundation.

## REFERENCES

[1] W.A.Harris, Jr., Logarithmic solutions of linear differential systems with a singularity of the first kind, Applicable Analysis, 8 (1978) 171–174.

[2] W.A.Harris, Jr., P.Pucci and J.Serrin, Asymptotic behavior of solutions of a nonstandard second order differential equation, Differential and Integral Equations, to appear.

[3] W.A.Harris, Jr., Y.Sibuya and L.Weinberg, Holomorphic solutions of linear differential systems at singular points, Arch. Rational Mech. Analy., 35(1969) 245–248.

[4] P.Pucci and J.Serrin, Asymptotic properties for solutions of strongly nonlinear second order differential equations, Rend. Sem. Mat., Univ. Pol. di Torino, Fasc. Speciale (1989) 121–129.

# Persistence of Singular Perturbation Solutions in Noisy Environments

by

F.C.Hoppensteadt
Department of Mathematics and Department of Statistics and Probability,
Michigan State University
East Lansing, MI 48824

## Abstract

This presentation reviewed recent work on aspects of singular perturbation solutions that persist in the presence of noise. First, a small deviation theory was described where small amplitude but highly irregular perturbations are applied to quasistatic state problems [1]. Conditions are given under which the quasistatic state approximation remains valid.

Next, two large deviation theories were described: In the first case, Gaussian approximations derived in [2] were described for random advection problems. In the second case, an averaging principle derived in [3] was described for a certain general class of dynamical systems in Hilbert spaces, and some applications of it to initial-boundary value problems for partial differential equations were presented.

**References**:

1. F.C.Hoppensteadt, Analysis and Simulation of Chaotic Systems, Springer-Verlag, February, 1993.
2. R.Z. Hashminskii, F.C. Hoppensteadt, H. Salehi, Asymptotic solutions of linear partial differential equations of first order having random coefficients, Random Operators and Stochastic Equations, in press.
3. A.V.Skorokhod, F.C.Hoppensteadt, H. Salehi, An averaging principle for dynamical systems in Hilbert space with Markov random perturbations, in preparation.

Acknowledgments: This work was done in part with the support of NSF Grant No. 92 006677 and in part in collaboration with R.Z.Khashminskii, H. Salehi, and A.V. Skorokhod.

# Globally Analytic Decompositions of a Matrix Function and a System of Differential Equations

HARRY GINGOLD
DEPARTMENT OF MATHEMATICS
WEST VIRGINIA UNIVERSITY
MORGANTOWN, WV 26506 USA
AND
PO-FANG HSIEH
DEPARTMENT OF MATHEMATICS AND STATISTICS
WESTERN MICHIGAN UNIVERSITY
KALAMAZOO, MI 49008 USA

## Abstract

The analytic decomposition is an important process in the study of a system of differential equations. In this paper we will study the globally analytic singular value decomposition (SVD) of a rectangular matrix function, which is relevant to the study of an over- or under-determined singular system of equations. For an $m$ by $n$ complex-valued matrix $M(t) \not\equiv 0$ and analytic on $[a, b]$, $(-\infty \leq a < b \leq \infty)$, we introduce two kinds of *singular values*, one *analytic* and the other *nonnegative*, which are shown to be continuous and piecewise analytic. Based on these, two kinds of singular value decompositions are developed. If $M(t)$ and all of its singular values are periodic in $(-\infty, \infty)$, then both kinds of decompositions are also proved to be periodic with the same period. The corresponding decompositions of the pseudoinverse are also obtained.

**1. Introduction.** The analytic decomposition process is an important step in the study of a system of differential equations. Many results for simplifying a square matrix function were obtained in the process of studying a system of differential equations. (e.g., Braaksma [2], Gingold [6], Gingold and Hsieh [7], Hsieh and Sibuya [13], Sibuya [16] and Wasow [18]). The method of singular value decomposition (SVD) is an efficient tool to study a rectangular constant matrix and is well developed in the last couple of decades for both theoretical and numerical treatment of such matrices. Various applications of SVD of a constant matrix is given in Appendix A of Strang [17], while many numerical algorithms of a constant matrix by means of SVD can be found in Golub and Van Loan [10]. (See also Horn and Johnson [12, Chapter 3]). For a matrix funtion with analytic entries, the SVD in analytic factors has been studied in the last few years by numerical methods (e.g. see Bunse-Gerstner, Byers, Mehrmann and Nichols [3], Dieci, Russell and Van Vleck [5] and Wright [19]). Recently, the authors [8] developed a constructive method to triangularize a square matrix function that is analytic, and whose eigenvalues are all analytic, on a real interval to an upper triangular matrix by an analytic unitary similarity on the original interval. This method also yields a diagonalization for a Hermitian matrix. If the matrix and all of its eigenvalues are analytic and periodic on the entire real line, then such a triangularization process preserves the periodicity as well as the analyticity.

In this paper we will apply these results to study the SVD of an $m$ by $n$ matrix function analytic on a real interval. We will introduce two kinds of singular values, one is *analytic singular value* and the other *nonnegative singular value* for such a rectangular matrix. It will be seen in Section 3 that we do not need any additional condition to have the SVD preserving analyticity on the original interval if we do not require that the singular values are nonnegative, as in the case of SVD for a constant matrix. If we require that the singular values are nonnegative, the worst we could have for this kind of SVD is the presence of algebraic branch points of the singular values at their zero points. In this case the nonnegative singular values are continuous and piecewise analytic on the original interval. Furthermore, if the matrix function under consideration and all of its singular values have the same periodicity, then both kinds of SVD preserve its periodicity. As the magnitude of singular values are important in application of the SVD method to a matrix, such as in numerical algorithm and image processing (e.g., see Golub and Van Loan [11] and Strang [17]), the fact that the singular values of a matrix function are analytic, or continuous, is useful in determining their magnitudes on the interval in consideration. These kinds of decompositions are also important for the study of over- and under-determined singular systems of differential equations (cf. Campbell [4]). In Section 2, recent results on globally analytic simplification of matrix functions will be summarized. The theorems of this paper, involving two kinds of singular values of an analytic matrix function as well as their corresponding SVD of such matrix, in both non-periodic and periodic cases, will be given in Section 3. The decompositions of the pseudoinverse of such matrix in terms of these two kinds of singular values are also to be given in this section. The proof of the main theorem will be given in Section 4. Two examples to illustrate different features of the theorems of this paper can be found in Gingold and Hsieh [9]. The algorithm developed in Gingold and Hsieh [8] plays an important role in the proof as well as in the examples.

**2. Globally Analytic Triangularization of Matrix Functions.** We will summarize in this section recent results of the authors [8] on globally analytic triangularization of a square matrix function $A(t)$ which is analytic in a finite or infinite real interval, when all of its eigenvalues are analytic, or just simply real-valued, in this real interval, to an upper–triangular matrix by a *unitary analytic* matrix. The main result of this section is the following

**Theorem 2.1.** *Let $A(t)$ be an $n$ by $n$ matrix function with analytic entries on $[a,b]$, where $-\infty \leq a < b \leq \infty$. Assume that every eigenvalue of $A(t)$ is analytic on $[a,b]$. Then, there exists a unitary matrix $U(t)$ analytic on $[a,b]$ such that*

$$B(t) = U^{-1}(t)\, A(t)\, U(t), \tag{2.1}$$

*where $B(t)$ is an upper–triangular matrix whose entries are analytic functions of $t$ on $[a,b]$.*

Here the analyticity of a function on $[a,b]$ means the analyticity in an open interval containing $[a,b]$.

The analyticity of the eigenvalues of $A(t)$ plays an important role in the construction of $U(t)$. The following lemma gives a condition which assures the analyticity of all eigenvalues of $A(t)$ on $[a,b]$.

**Lemma 2.2.** *Let $A(t)$ be an $n$ by $n$ matrix function with analytic entries on $[a,b]$. If every eigenvalue of $A(t)$ is real valued on $[a,b]$, then all the eigenvalues are analytic on $[a,b]$.*

This lemma is proved in Gingold and Hsieh [8] and Rellich [14, 15]. The labeling of the eigenvalues is important to assure the analyticity of these eigenvalues. For instance, for

$$F(t) = \begin{bmatrix} 0 & t \\ t & 0 \end{bmatrix}, \qquad t \in (-\infty, \infty), \tag{2.2}$$

we take $\pm t$ as eigenvalues, rather than $\pm|t|$. In this paper, we will assume that *the eigenvalues of a square matrix are properly labeled such that each eigenvalue is an analytic function of $t$ on $[a,b]$.*

As a special case of Theorem 2.1, we have the following theorem for a Hermitian matrix.

**Theorem 2.3.** *Let $A(t)$ be an $n$ by $n$ matrix function with analytic entries on $[a,b]$. If $A(t)$ is Hermitian on $[a,b]$, then there exists a unitary matrix $U(t)$ analytic on $[a,b]$ such that*

$$B(t) = U^{-1}(t)A(t)U(t) \tag{2.3}$$

*is a diagonal matrix whose elements are analytic on $[a,b]$.*

This is the global version of the Rellich Theorem [14, 15] (Baumgärtel [1] mentioned this global version but no proof is furnished). In fact, since $A(t)$ is Hermitian, all of its eigenvalues are real. It is easy to see that $B(t)$ given by (2.1) satisfies the relation $B^*(t) = B(t)$ in addition to being upper–triangular. Thus $B(t)$ is diagonal. It is noteworthy that Theorem 2.3 is treated in [1, 14, 15] as the perturbation of a result for a constant matrix, while in [8] it is obtained as an immediate consequence of Theorem 2.1 by a different algorithm.

**Remark 1.** Thereom 2.3 is used to study the gloabally analytic diagonalization of a system of equations by Gingold, Hsieh and Sibuya [10].

**Remark 2.** Thereoms 2.1 and 2.3 are true also when the interval $[a,b]$ is replaced by an open or half open interval.

**Remark 3.** If the matrix $A(t)$ in Thereoms 2.1 and 2.3 and all of its eigenvalues are periodic with period $\omega$, then the matrices $B(t)$, and $U(t)$ are also periodic with period $\omega$.

**3. The Theorems.** In this section, Theorem 2.3 will be used to develop the singular value decompositions (SVD's) of an analytic rectangular matrix. Two kinds of singular values are to be introduced: one is analytic and the other is nonnegative but continuous. These two kinds of SVD's are established in the following

**Theorem 3.1.** *Let $M(t)$ be an $m$ by $n$ matrix function, not identically zero, with analytic entries on $[a,b]$, where $-\infty \le a < b \le \infty$. Then, $M(t)$ can be factored as*

$$M(t) = V(t)\Sigma(t)U(t)^*, \tag{3.1}$$

*where $V(t)$ is an $m$ by $m$ unitary matrix analytic on $[a,b]$ consisting of normalized eigenvectors of $M(t)M(t)^*$, $U(t)$ is an $n$ by $n$ unitary matrix analytic on $[a,b]$ consisting of normalized eigenvectors of $M(t)^*M(t)$ and $\Sigma(t)$ is an $m$ by $n$ matrix of the form*

$$\Sigma(t) = D(t) \oplus 0, \tag{3.2}$$

*with*

$$D(t) = diag\{\sigma_1(t), \sigma_2(t), \cdots, \sigma_k(t)\}, \qquad (1 \le k \le \min(m, n)). \tag{3.3}$$

*Here, $\sigma_j(t)$, $(1 \le j \le k)$, are analytic on $[a, b]$, where $\lambda_j(t)$, $(1 \le j \le n)$, are eigenvalues of $M(t)^* M(t)$ analytic on $[a, b]$ such that*

$$\lambda_j(t) = \sigma_j(t)^2 \not\equiv 0, \quad (1 \le j \le k), \quad \lambda_j(t) \equiv 0, \quad (k+1 \le j \le n), \quad t \in [a, b]. \tag{3.4}$$

*Moreover, let $S(t)$ be the $m$ by $m$ nonsingular matrix*

$$S(t) = diag\{s_1(t), s_2(t), \cdots, s_k(t), 1, \cdots, 1\}, \tag{3.5}$$

*where*

$$s_j(t) = \frac{|\sigma_j(t)|}{\sigma_j(t)}, \qquad j = 1, 2, \cdots, k. \tag{3.6}$$

*Then, $M(t)$ can also be factored as*

$$M(t) = W(t)\hat{\Sigma}(t)U(t)^*, \tag{3.7}$$

*where*

$$W(t) = V(t)S(t), \qquad \hat{\Sigma}(t) = S(t)\Sigma(t) = \hat{D}(t) \oplus 0 \tag{3.8}$$

*with*

$$\hat{D}(t) = diag\{\hat{\sigma}_1(t), \hat{\sigma}_2(t), \cdots, \hat{\sigma}_k(t)\}, \quad \hat{\sigma}_j(t) = |\sigma_j(t)|, \quad (1 \le j \le k). \tag{3.9}$$

*Similar to (3.5) and (3.7), let $\hat{S}(t)$ be the $n$ by $n$ nonsingular matrix*

$$\hat{S}(t) = diag\{s_1(t), s_2(t), \cdots, s_k(t), 1, \cdots, 1\}. \tag{3.10}$$

*Then, $M(t)$ can further be factored as*

$$M(t) = V(t)\hat{\Sigma}(t)X(t)^*, \tag{3.11}$$

*where*

$$X(t) = U(t)\hat{S}(t). \tag{3.12}$$

The functions $\sigma_1(t)$, $\sigma_2(t)$, $\cdots$, $\sigma_k(t)$ are called the *analytic singular values* of $M(t)$. On the other hand, the functions $\hat{\sigma}_1(t)$, $\hat{\sigma}_2(t)$, $\cdots$, $\hat{\sigma}_k(t)$ are called the *nonnegative singular values* of $M(t)$. As will be shown, $\lambda_j(t) \ge 0$ for $t \in [a, b]$,

$(1 \leq j \leq k \leq n)$. Thus $\sigma_j(t)$, $(1 \leq j \leq k)$, given in (3.4) are meaningful. Moreover, every eigenvalue of $M(t)^* M(t)$, which is not identically zero, is also an eigenvalue of $M(t)M(t)^*$, and vice versa. As we have seen in matrix (2.2), the labeling of eigenvalues is important to assure the analyticity of all the eigenvalues.

The positive integer $k$ given in (3.4) is called the *analytic rank* of the matrix $M(t)$. As some of $\lambda_j(t)$, $(1 \leq j \leq k)$, may vanish at some points of $[a, b]$, this is a distinct difference between the singular values of a constant matrix and those of a matrix function.

As in the case of a constant matrix, it is desirable to have the singular values nonnegative. The nonnegative singular values $\hat{\sigma}_j(t) = |\sigma_j(t)|$, $(1 \leq j \leq k)$, meet this requirement and the SVD's in the form of (3.7) and (3.11) are in terms of these nonnegative singular values. It is noteworthy that the points where the nonnegative singular values $\hat{\sigma}_j(t)$, $(1 \leq j \leq k)$, and the matrix $W(t)$ (or $X(t)$) fail to be analytic are the zeros of $\lambda_j(t)$ of orders $\not\equiv 0 \pmod 4$, $(1 \leq j \leq k)$.

**Remark 4.** Note that both analytic singular values and nonnegative singular values are not identically zero.

For some special cases of the analytic singular values, we have

**Corollary 3.2.** *Let $M(t)$ be an $m$ by $n$ matrix function, not identically zero, with analytic entries on $[a, b]$, where $-\infty \leq a < b \leq \infty$. If all the analytic singular values do not have zeros on $[a, b]$, or, when some of them vanish but not identically, all of their zeros are of even order, then all the analytic singular values of $M(t)$ given in* Theorem 3.1 *are nonnegative and $\hat{\Sigma}(t) \equiv \Sigma(t)$.*

Utilizing Theorem 3.1, we have the following decompositions $M(t)^+$ and $\hat{M}(t)^+$ of the pseudoinverse (Moore-Penrose inverse) of a matrix function $M(t)$ in terms of analytic and nonnegative singular values, respectively.

**Theorem 3.3.** *Let $M(t)$ be an $m$ by $n$ matrix function, not identically zero, with analytic entries on $[a, b]$, where $-\infty \leq a < b \leq \infty$. Then, there is a decomposition $M(t)^+$, an $n$ by $m$ matrix, of the pseudoinverse of $M(t)$ given by*

$$M(t)^+ = U(t)\Sigma(t)^+ V(t)^*, \tag{3.13}$$

*where $V(t)$ and $U(t)$ are given in* Theorem 3.1 *and, in particular,*

$$\Sigma(t)^+ = D(t)^{-1} \oplus 0, \tag{3.14}$$

*is an $n$ by $m$ matrix meromorphic on $[a, b]$ with $D(t)$ given by (3.3).*

*Moreover, there are decompositions* $\hat{M}(t)^+$, *an* $n$ *by* $m$ *matrix, of the pseudoinverse of* $M(t)$ *given by*

$$\hat{M}(t)^+ = U(t)\hat{\Sigma}(t)^+ W(t)^*, \tag{3.15}$$

*and*

$$\hat{M}(t)^+ = X(t)\hat{\Sigma}(t)^+ V(t)^*, \tag{3.16}$$

*where* $W(t)$ *and* $X(t)$ *are given in* Theorem 3.1 *and, in particular,*

$$\hat{\Sigma}(t)^+ = \hat{D}(t)^{-1} \oplus 0, \tag{3.17}$$

*is an* $n$ *by* $m$ *matrix analytic on* $[a, b]$, *except possibly at countably many points, with* $\hat{D}(t)$ *given by (3.9).*

The matrix $M(t)^+$ given by (3.13) is called the *meromorphic decomposition of the pseudoinverse* and $\hat{M}(t)^+$ given by (3.15) (or (3.16)) is called the *nonnegative decomposition of the pseudoinverse* of $M(t)$, respectively. Since $S(t)^2 = I_m$ and $\hat{S}(t)^2 = I_n$, $M(t)^+ = \hat{M}(t)^+$ with (3.13) and (3.15) (or (3.16)) giving different decompositions in terms of analytic and nonnegative singular values, respectively. It is noteworthy that $M(t)^+$ and $\hat{M}(t)^+$ are not necessarily continuous for an analytic matrix $M(t)$ as can be seen in an example in Gingold and Hsieh [9].

The proof of Theorem 3.1 will be given in Section 4. As $M(t)^*M(t)$ is Hermitian and analytic on $[a, b]$, the results of Gingold and Hsieh [8] will be used to prove this theorem. Theorem 3.3 follows immediately from Theorem 3.1, by the results of Gingold and Hsieh [8].

**Remark 5.** For the special cases given in Cororalley 3.2, $S(t) = I_m$, $\hat{S}(t) = I_n$ and $\Sigma(t) = \hat{\Sigma}(t)$. Thus, two kinds of SVD of $M(t)$ given by (3.1) and (3.7) coincide, and consequently, the corresponding decompositions $M(t)^+$ and $\hat{M}(t)^+$ of the pseudoinverse given by (3.13), (3.15) and (3.16) also coincide.

**Remark 6.** Similar to Remark 2, Theorems 3.1, 3.3 and Corollary 3.2 are true also if the interval $[a, b]$ is replaced by an open or a half open interval.

**Remark 7.** Similar to Remark 3, if all the singualar values of $M(t)$ in Theorems 3.1 and 3.3 are periodic with period $\omega$ then the SVD of $M(t)$ and that of its pseudoinverses also preserve the same periodicity (see [9]).

**4. Proof of Theorem 3.1.** To prove Theorem 3.1, note first that if $\lambda_j(t)$ is an eigenvalue of $M(t)^*M(t)$, which is not identically zero, with corresponding eigenvector $u_j(t)$, then $\lambda_j(t)$ is also an eigenvalue of $M(t)M(t)^*$ with corresponding eigenvector $M(t)u_j(t)$ at the points where $\lambda_j(t) \neq 0$ (see Remark 8 below). Furthermore, $M(t)^*M(t)$ is Hermitian and analytic on $[a,b]$. Thus, its eigenvalues $\{\lambda_1(t), \lambda_2(t), \cdots, \lambda_n(t)\}$ are real valued on $[a,b]$. By Lemma 2.2, they are analytic on $[a,b]$. Moreover, $\lambda_j(t) \geq 0$ for $t \in [a,b]$, $(j = 1, 2, \cdots, n)$, since for an eigenvalue $\lambda_j(t)$ with associated eigenvector $u_j(t)$,

$$0 \leq [M(t)u_j(t)]^* \cdot M(t)u_j(t) = u_j(t)^* M(t)^* M(t) \cdot u_j(t) = \lambda_j(t) u_j(t)^* \cdot u_j(t). \tag{4.1}$$

Let $k$ be the index such that

$$\lambda_j(t) \not\equiv 0, \quad (1 \leq j \leq k \leq \min(m,n)); \qquad \lambda_j(t) \equiv 0, \quad (k+1 \leq j \leq n), \tag{4.2}$$

for $t \in [a,b]$. By Theorem 2.3, there are $n$ normalized analytic eigenvectors $u_j(t)$, $(j = 1, 2, \cdots, n)$, associated with $\lambda_j(t)$, respectively, such that the matrix

$$U(t) = [u_1(t) \;\; u_2(t) \;\; \cdots \;\; u_n(t)]. \tag{4.3}$$

is analytic and unitary on $[a,b]$. As $\lambda_j(t) \geq 0$ for $t \in [a,b]$, the zeros of $\lambda_j(t)$ are all of even order. Let $t = t_{jp}$ be the zeros of $\lambda_j(t)$, $(p = 1, 2, \cdots N_j)$, with order $2\alpha_{jp}$, ($\alpha_{jp}$ : positive integers), respectively. Then, for each $\lambda_j(t)$, there exists a function $h_j(t)$, positive and analytic on $[a,b]$ such that

$$\lambda_j(t) = \prod_{p=1}^{N_j} (t - t_{jp})^{2\alpha_{jp}} h_j(t), \quad (j = 1, 2, \cdots, k). \tag{4.4}$$

Let

$$\sigma_j(t) = [\lambda_j(t)]^{\frac{1}{2}} = \prod_{p=1}^{N_j} (t - t_{jp})^{\alpha_{jp}} [h_j(t)]^{\frac{1}{2}}, \qquad (j = 1, 2, \cdots, k). \tag{4.5}$$

Then each function $\sigma_j(t)$ is analytic on $[a,b]$. Let

$$v_j(t) = \frac{1}{\sigma_j(t)} M(t) u_j(t), \qquad (j = 1, 2, \cdots, k). \tag{4.6}$$

Then, $v_j(t)$, $(j = 1, 2, \cdots, k)$, are eigenvectors of $M(t)M(t)^*$ corresponding to $\lambda_j(t)$, $(j = 1, 2, \cdots, k)$, respectively, and meromorphic in $t$ for $t \in [a,b]$. Furthermore,

$$\begin{aligned} v_j(t)^* \cdot v_\ell(t) &= \frac{1}{\sigma_j(t)\sigma_\ell(t)} u_j(t)^* M(t)^* \cdot M(t) u_\ell(t) \\ &= \frac{\sigma_j(t)}{\sigma_\ell(t)} u_j(t)^* \cdot u_\ell(t) = \begin{cases} 1, & \text{for } j = \ell, \\ 0, & \text{for } j \neq \ell. \end{cases} \end{aligned} \tag{4.7}$$

Thus, $v_j(t)$, $(j = 1, 2, \cdots, k)$, are orthonormal and analytic on $[a, b]$. By the algorithm given in Gingold and Hsieh [8], there exist $m - k$ analytic vectors $v_{k+1}(t), \cdots, v_m(t)$ such that the set $\{v_1(t), v_2(t), \cdots, v_m(t)\}$ is analytic and orthonormal on $[a, b]$. Let

$$V(t) = [v_1(t) \;\; v_2(t) \;\; \ldots \;\; v_m(t)] \tag{4.8}$$

and

$$\Sigma(t) = V(t)^* M(t) U(t). \tag{4.9}$$

Then $\Sigma(t)$ is in the form of (3.2) with (3.3).

The expression (3.7) follows immediately from $S(t)^2 = I_m$, (3.1) and (3.8). Similarly, the expression (3.11) follows immediately from $\hat{S}(t)^2 = I_n$, $\Sigma(t)\hat{S}(t) = \hat{\Sigma}(t)$, (3.1) and (3.12). **q.e.d.**

**Remark 8.** By (4.6), the eigenvector $M(t)u_j(t) = \sigma_j(t)v_j(t)$ of $M(t)M(t)^*$ is a zero vector at the zero points of $\sigma_j(t)$, namely those of $\lambda_j(t)$. Moreover, as $\lambda_j(t) \not\equiv 0$ $(j = 1, 2, \cdots, k)$, the zero points of $\sigma_j(t)$ are at most countably many and have no finite accumulation point.

**Remark 9.** Let $W(t) = [w_1(t) \;\; w_2(t) \;\; \cdots \;\; w_m(t)]$, then, by (3.5), (3.6), (3.8) and (4.6), we have

$$\begin{aligned} w_j(t) &= \frac{1}{\hat{\sigma}_j(t)} M(t) u_j(t), \qquad (j = 1, 2, \cdots, k), \\ w_j(t) &= v_j(t), \qquad (j = k+1, k+2, \cdots, m). \end{aligned} \tag{4.10}$$

## References


[1] H. Baumgärtel, *Analytic Perturbation Theory for Matrices and Operators*, Birkhäuser Verlag, Basel, 1985.

[2] B. L. J. Braaksma, Global reduction of linear differential systems involving a small singular parameter, *SIAM J. Math. Anal.*, 2(1971), 149–165.

[3] A. Bunse-Gerstner, R. Byers, V. Mehrmann and N. K. Nichols, Numerical computation of an analytic singular value decomposition of a matrix valued function, *Num. Math.*, 60(1991), 1–39.

[4] S. L. Campbell, *Singular Systems of Differential Equations*, Pitman Advanced Publ. Program, San Francisco, 1980.

[5] L. Dieci, R. D. Russell and E. S. Van Vleck, Unitary integrators and applications to continuous orthonormalizatiopn techniques, to appear in *SIAM J. Num. Anal.*

[6] H. Gingold, On continuous triangularization of matrix functions, *SIAM J. Math. Anal.*, 10(1979), 709–720.

[7] H. Gingold and P. F. Hsieh, Global approximation of perturbed Hamiltonian differential equations with several turning points, *SIAM J. Math. Anal.*, 18(1987), 1275–1293.

[8] H. Gingold and P. F. Hsieh, Globally analytic triangularization of a matrix function, *Lin. Alg. and Its Appl.*, 169(1992), 75–101.

[9] H. Gingold and P. F. Hsieh, Analytic singular value decompositions of a matrix function, preprint.

[10] H. Gingold, P. F. Hsieh and Y. Sibuya, Globally analytic simplification and the Levinson theorem, to appear in *J. Math. Anal. and Appl.*

[11] G. H. Golub and C. F. Van Loan, *Matrix Computations*, The Johns Hopkins Univ. Press, Baltimore, Maryland, 1983.

[12] R. A. Horn and C. R. Johnson, *Topics in Matrix Analysis,* Cambridge Univ. Press, Cambridge, 1991.

[13] P. F. Hsieh and Y. Sibuya, A global analysis of matrices of functions of several variables, *J. Math. Anal. and Appl.* 14(1966), 332–340.

[14] F. Rellich, Störungstheorie der Spektralzerlegung, *I, Math. Ann.*, 113(1936), 600 – 619, ; II, ibid., 113(1937), 677 – 685; III, ibid., 116(1939), 555 – 570; IV, ibid., 117(1940), 356 – 382, V, ibid., 118(1942), 462 – 484.

[15] F. Rellich, *Perturbation Theory of Eigenvlaue Problems,* Gordon and Breach, New York, 1969

[16] Y. Sibuya, Some global properties of matrices of function of one variables, *Math. Ann.*, 161(1965), 67–77.

[17] G. Strang, *Linear Algebra and Its Aplications,* Third ed., Harcourt Brace Jovanovich Publ., San Diego, 1988.

[18] W. Wasow, On holomorphically similar matrices, *J. Math. Anal. Appl.*, 4(1962), 202 – 206.

[19] K. Wright, Differential equations for the analytic singular value decomposition of a matrix, *Num. Math.*, 62(1992), 283–295.

# On summability of formal solutions and the Stokes phenomenon in a class of difference equations of level $1^+$

G.K. Immink
Institute of Econometrics, University of Groningen
P.O.Box 800, 9700 AV Groningen
The Netherlands

**Abstract**

The difference equation

$$y(z+1) - a(z)y(z) = b(z)$$

where $a, b \in z^{-1}\mathbf{C}\{z^{-1}\}$ and $\lim_{z\to\infty} za(z) \neq 0$, possesses a unique formal solution $\hat{f} = \sum_{n=1}^{\infty} a_n z^{-n}$. The coefficients $a_n$ satisfy a growth condition which is slightly stronger than the Gevrey condition of order 1, but weaker than the Gevrey condition of order $k$ for any $k > 1$. Accordingly, we say that the equation is of level $1^+$. It is the simplest nontrivial example of a difference equation in which level $1^+$ occurs. We compare two equivalent ways of summing the formal solution, using Laplace integrals of the ordinary formal Borel transform of $\hat{f}$ and Laplace integrals of the formal Borel transform of $\hat{f}$ with respect to the variable $z(\log z + i\theta)$, for some real value of $\theta$. The latter Borel transform turns out to be a quasi-analytic function on the positive real axis, except for a countable number of "singular" values of $\theta$. This is in agreement with the results on weak accelerations and quasi-analytic functions obtained by J. Ecalle. For singular values of $\theta$, the Borel transform has a quasi-analytic singularity at the point 1. We discuss the nature of this singularity and its relation to the Stokes phenomenon.

## 1 First summation method

We consider the difference equation

$$y(z+1) - \left(\frac{a}{z} + \frac{\alpha(z)}{z^2}\right)y(z) = \frac{b}{z} + \frac{\beta(z)}{z^2} \tag{1}$$

where $a, b \in \mathbb{C}_*$, and $\alpha, \beta \in \mathbb{C}\{z^{-1}\}$. This equation has a formal solution $\hat{f} = \sum_{n=1}^{\infty} a_n z^{-n}$. It is our purpose to sum this formal solution so as to obtain actual (analytic) solutions, represented by $\hat{f}$ in different regions of the complex plane, and to study the Stokes phenomenon.

$\hat{f}$ is Gevrey of order 1, i.e. there exist positive constants $A$ and $C$, such that, for all $n \in \mathbb{N}$,

$$|a_n| \leq CA^n n!$$

In fact, we have a better estimate : for $n \geq 2$,

$$|a_n| \leq CA^n (\frac{n}{\log n})^n$$

(cf. [3, 4]. Consequently, the formal Borel transform $\hat{\mathcal{B}}\hat{f}$ of $\hat{f}$, defined by

$$\hat{\mathcal{B}}\hat{f} = \sum_{n=1}^{\infty} a_n \frac{t^{n-1}}{(n-1)!}$$

defines an entire function of $t$ to be denoted by $\phi$. This function satisfies the convolution equation (1*) obtained from (1) by formally inserting a Laplace integral $\int \phi(t)e^{-tz}dt$ into (1). For example, suppose that $\alpha \equiv \beta \equiv 0$. In that case the convolution equation (1*) reads

$$e^{-t}\phi(t) - a\int_0^t \phi(\tau)d\tau = b$$

and its solution is given by

$$\phi(t) = be^{-a}\exp(t + ae^t)$$

$\phi$ is bounded in the half plane $\mathrm{Re}\, t \leq 0$ and grows at most exponentially as $\mathrm{Re}\, t \to \infty$ in the strips where $\cos(\mathrm{Im}\, t + \arg a) \leq 0$. It has supra-exponential growth as $\mathrm{Re}\, t \to \infty$ and $\cos(\mathrm{Im}\, t + \arg a) > 0$. These properties remain valid in the general case.

Returning to the general case, we can now define the following sums of $\hat{f}$. To begin with, the Laplace integral $\int_0^{-\infty} \phi(t)e^{-tz}dt$ is an ordinary Borel-sum of $\hat{f}$. It defines a solution $f$ of (1) which is analytic in the complex plane with the exception of a neighbourhood of $O$ and a strip around the positive real axis, and is represented asymptotically by $\hat{f}$ as $z \to \infty$, $0 < \arg z < 2\pi$. Furthermore, let

$$\theta_n := \arg a + 2n\pi, \qquad n \in \mathbb{Z}$$

where we take the value of $\arg a$ in $[0, 2\pi)$, and let $\Gamma_n$ be a path from $O$ to $\infty$ such that $\mathrm{Re}\, t \to \infty$ on $\Gamma_n$ and, for sufficiently large $|t|$,

$$\frac{\pi}{2} < \mathrm{Im}\, t + \theta_{-n} < \frac{3\pi}{2}$$

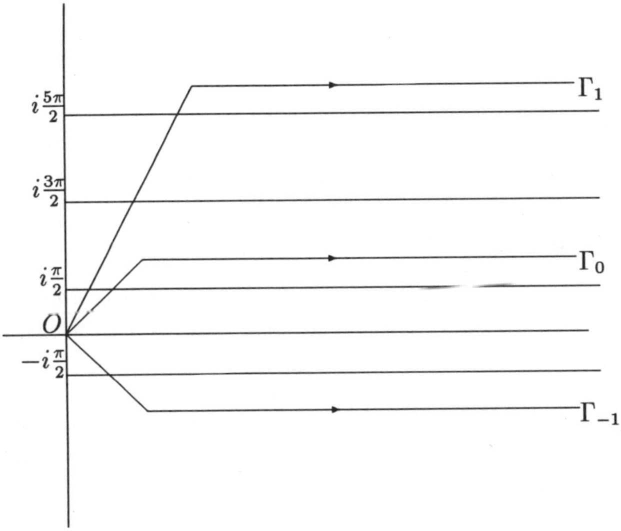


Figure 1 ($\arg a = 0$)

For each $n \in \mathbb{Z}$, the integral

$$\int_{\Gamma_n} \phi(t) e^{-tz} dt$$

defines a solution $f_n$ of (1), analytic in the complex plane, with the exception of a neighbourhood of $O$ and a strip around the negative real axis. Similar Laplace integrals are used in [1]. It can be shown that $f_n$ is represented asymptotically by $\hat{f}$ as $z \to \infty$ such that $\operatorname{Re} z(\log z + i\theta)$ remains bounded from below for some $\theta \in (\theta_{n-1}, \theta_n)$. The solutions $f_n$ are characterized by their asymptotic behaviour. The Stokes phenomenon is described by the following relation, valid for all $n \in \mathbb{Z}$,

$$f_{n+1}(z) - f_n(z) = P_n e^{-2n\pi i z} y_0(z)$$

where $y_0$ is a (fixed) solution of the homogeneous equation corresponding to (1). The numbers $P_n$ are Stokes multipliers of the equation (1). Together with the Stokes multipliers of the homogeneous equation they form a complete system of analytic invariants of equation (1).

# 2 Second summation method

There is another method of summation, by which to obtain the solutions $f_n$, which is essentially due to J. Ecalle, and which is closer to the summation methods used in the theory of differential equations. This summation method works as follows. First, we note that the solutions of the homogeneous difference equation have a dominant factor of the form

$$\exp\{-z\log z + (\log(ae) + 2n\pi i)z\} \qquad n \in \mathbb{Z}$$

A change of variable

$$z \mapsto \psi_\theta(z) := z(\log z + i\theta), \qquad \theta \in \mathbb{R}$$

takes the formal solution $\hat{f}(z)$ into a formal series $\hat{g}_\theta(\psi_\theta(z))$ (which is no longer an ordinary power series), whose formal Borel transform defines a function $\phi_\theta$, analytic for $|t| < 1$ on the Riemann surface of $\log t$. $\phi_\theta$ is a so-called 'weak' accelerate of $\phi$ (cf. [2]). We can prove the following results (cf. [5]).

1). If $\theta \in (\theta_{n-1}, \theta_n)$, $\phi_\theta$ is quasi-analytic on $(0, \infty)$. More precisely, it is analytic on $(0,1) \cup (1,\infty)$, infinitely many times differentiable at $t = 1$ and its derivatives at $t = 1$ satisfy an inequality of the form

$$|\phi_\theta^{(m)}(1)| \leq CA^m (m \log m)^m \qquad m \in \mathbb{N}$$

where $A$ and $C$ are positive constants. Moreover, $\phi_\theta$ grows at most exponentially as $t \to \infty$, and we have

$$f_n(z) = \int_0^\infty \phi_\theta(t) e^{-t\psi_\theta(z)} dt \qquad \text{Re}\, \psi_\theta(z) > c_\theta$$

2).If $\theta = \theta_n$ for some $n \in \mathbb{Z}$, $\phi_\theta$ can be written as a sum

$$\phi_\theta = \phi_n^1 + \phi_n^2$$

where $\phi_n^1$ is quasi-analytic on $(0, \infty)$, while $\phi_n^2$ is analytic in the sector $-\frac{\pi}{2} < \arg(t-1) < \frac{5\pi}{2}$. For $0 < t < 1$ and $\arg(t-1) = \pi$ it can be represented by the integral

$$\frac{P_n}{2\pi i} \int_R^\infty y_0(z) e^{-2n\pi i z + t\psi_\theta(z)} d\psi_\theta(z)$$

if $R$ is a sufficiently large positive number. For all $t > 1$ the difference of two determinations of $\phi_n^2$ is given by the formula

$$\phi_n^2(t) - \phi_n^2(1 + (t-1)e^{2\pi i}) = \frac{P_n}{2\pi i} \int_C y_0(z) e^{-2n\pi i z + t\psi_\theta(z)} d\psi_\theta(z)$$

where $C$ is a contour consisting of the half line from $R - iR - \infty$ to $R - iR$, the segment $[R - iR, R + iR]$ and the half line from $R + iR$ to $R + iR + \infty$ (cf. Fig.2). This formula shows that the Stokes multipliers $P_n$ are determined by the nature of the singularity at $t = 1$ of $\phi_n^2$.

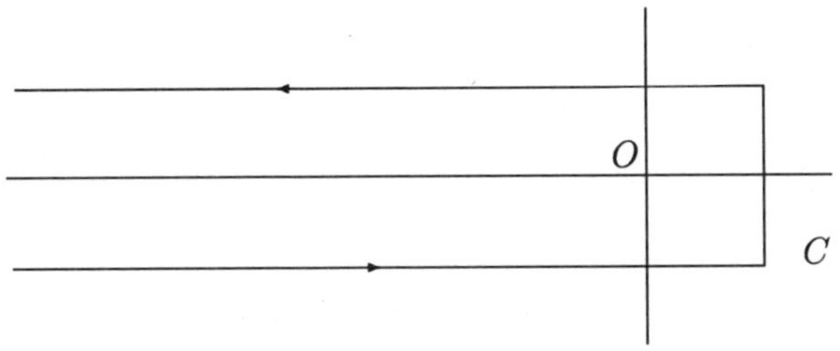


Figure 2

# References

[1] B.L.J. Braaksma and W.A. Harris jr. On an open problem in the theory of linear difference equations. *Nieuw Arch. Wisk. (3)*, 23:228 – 240, 1975.

[2] J. Ecalle. Cohesive functions and weak accelerations. to appear in : Journal d'An. Math.

[3] J. Ecalle. *Les Fonctions Résurgentes, tome III.* Publ. Math. d'Orsay, Université de Paris-Sud, Paris, 1985.

[4] G.K. Immink. Asymptotic expansions with error bounds for solutions of difference equations of 'level $1^+$'. In *Équations différentielles dans le champ complexe, Vol.1: Équations différentielles ordinaires*, pages 35–60, Strasbourg, 1988. Publ. I.R.M.A.

[5] G.K. Immink. On the summability of the formal solutions of a class of inhomogeneous linear difference equations. preprint, 1992.

# STABILITY OF SYMMETRIC PERIODIC SOLUTIONS OF A DIFFERENTIAL DELAY EQUATION

ANATOLI F. IVANOV *

*Department of Mathematics, University of Rhode Island*
*Kingston, RI 02881-0816, USA*

ABSTRACT

We prove a long standing conjucture that the unique slowly oscillating periodic solution of the equation $\dot{x}(t) = -\alpha x(t-1)[1 - x^2(t)], \alpha > \pi/2$, is asymptotically stable with the asymptotic phase. Similarly, the unique slowly oscillating periodic solution of the equation $\dot{x}(t) = -\alpha x(t-1)[1 + x^2(t)], 0 < \alpha < \pi/2$, is unstable.

## 1. Introduction

Consider differential delay equation

$$\dot{x}(t) = -\alpha x(t-1)[1 + \beta x^2(t)], \tag{1}$$

where $\alpha > 0, \beta \in \mathbf{R}$.

It is known that under certain assumptions on the parameters $\alpha, \beta$ Eq. (1) has periodic solutions of period 4 satisfying additional symmetry condition $x(t+2) = -x(t), t \in \mathbf{R}$ (see, for example, [4], [7] and further references therein). These solutions are called *special symmetric periodic solutions* (SSPS). In the case $\alpha > \pi/2, \beta < 0$ the SSPS has been proved to be unique [7]. Similar arguments apply to show the uniqueness in the case $0 < \alpha < \pi/2, \beta > 0$. The main objective of this paper is to study the stability of the SSPS of Eq. (1).

**Theorem 1.1.** (a) For every $\alpha > \pi/2$ and $\beta < 0$ the unique SSPS of Eq. (1) is asymptotically stable with the asymptotic phase.
(b) For every $0 < \alpha < \pi/2$ and $\beta > 0$ the unique SSPS of Eq. (1) is unstable.

Proof of the Theorem is based mainly on two ingredients:

(i) The negative feedback in Eq. (1) imposes strict restrictions on the number of Floquet multipliers outside the unit disc: there may be at most one such a multiplier. This is a straightforward corollary of a result in [1] derived for the equation $\dot{x}(t) = g(x(t-1)), g(-x) = g(x), g'(x) < 0$.

(ii) The Floquet multipliers can be described in terms of zeros of an analytic function called the *characteristic function*. The construction of the characteristic

* On leave from the Institute of Mathematics, Ukrainian Academy of Sciences, Kiev, Ukraine

function itself uses the fundamental matrix solution of a system of ordinary differential equations. This is an approach introduced in [8].

## 2. Preliminaries

Consider scalar differential delay equation

$$\dot{x}(t) = f(x(t), x(t-1)) \tag{2}$$

with the nonlinearity $f$ safisfying the following (symmetry) conditions

$$f(-u, v) = f(u, v) = -f(u, -v). \tag{3}$$

Assume in addition that $f(u, v)$ satisfies (locally) the negative feedback condition, i.e. there exists a neighbourhood $U$ of $(0,0)$ in $\mathbf{R}^2$ such that

$$v\,f(u, v) < 0 \quad \text{for all} \quad (u, v) \in U \quad \text{and} \quad v \neq 0. \tag{4}$$

### *2.1. Existence of Special Symmetric Periodic Solutions*

Definition 2.1. A periodic solution $x = p(t)$ of Eq. (1) is called special symmetric (SSPS) if: a) $p(t+2) = -p(t)$, and b) $p(-1) = 0, p(t) > 0$ in $(-1, 0]$.

We note that b) is purely a normalization assumption. It follows that any SSPS has period 4.

Consider a system of ordinary differential equations, defined by $f(u, v)$, of the following form:

$$\dot{u} = f(u, v), \quad \dot{v} = -f(v, u). \tag{5}$$

Conditions (3) and (4) imply that some open neighbourhood of the equilibrium $(0,0)$ of system (5) consists of a one-parameter family of closed (symmetric) trajectories. The relationship between SSPS of Eq. (2) and symmetric trajectories of system (5) is given by the following

**Theorem 2.1.** Eq. (2) has a SSPS if and only if system (5) has a symmetric trajectory of minimal period 4.

For the proof of Theorem 2.1 see [3], [4].

### *2.2. Stability of Special Symmetric Periodic Solutions*

The stability of a SSPS $x = p(t)$ is derived from the location of its Floquet multipliers in the complex plane C.

Consider corresponding variational equation along $x = p(t)$:

$$\dot{y}(t) = a(t)y(t) + b(t)y(t-1), \tag{6}$$

where $a(t) = f_u(p(t), p(t-1))$, $b(t) = f_v(p(t), p(t-1))$, $t \in \mathbf{R}$.

Let $V : C_{\mathsf{C}} \to C_{\mathsf{C}}$ be period 4 map along solutions of Eq. (6). Since $V$ is linear and compact the spectrum $\sigma(V)$ consists of eigenvalues of $V$ of finite multiplicity. Eigenvalues of $V$ are called *Floquet multipliers* of periodic solution $x = p(t)$.

Floquet multipliers are either real or occur in complex conjugate pairs. Differentiation of Eq. (2) shows that $1 \in \mathsf{C}$ is an eigenvalue with an eigenfunction $\dot{p}(t)$ restricted to a unit time interval.

**Lemma 2.2.** $a(t+2) = a(t), \quad b(t+2) = b(t), \quad t \in \mathbf{R}$.

*Proof.* The symmetry condition (3) implies that $f_u(u,v)$ is odd and $f_v(u,v)$ is even in both arguments. Also $p(t) = -p(t-2)$, $t \in \mathbf{R}$. Therefore,

$$a(t+2) = f_u(p(t+2), p(t+1)) = f_u(-p(t), -p(t-1)) = f_u(p(t), p(t-1)) = a(t)$$
$$b(t+2) = f_v(p(t+2), p(t+1)) = f_v(-p(t), -p(t-1)) = f_v(p(t), p(t-1)) = b(t).$$

*Corollary 2.3.* Let $U$ be period 2 map along solutions of equation $(ve)$. Then $V = U \circ U$.

*Corollary 2.4.* Let $\lambda = \mu^2$. Then $\lambda \in \sigma(V)$ if and only if $\mu$ or $-\mu \in \sigma(U)$.

Proof follows from the decomposition $V - \lambda I = (U + \mu I) \circ (U - \mu I)$ where $I$ is the identity operator.

Changing variables by

$$y(t) = v(t) \exp\left(\int_0^t a(s)\, ds\right), \tag{7}$$

reduces Eq. (6) to

$$\dot{v}(t) = B(t) v(t-1), \quad B(t) = b(t) \exp\left(-\int_{t-1}^t a(s)\, ds\right). \tag{8}$$

Since both $a(t)$ and $b(t)$ are 2-periodic (Lemma 2.2) it is immediate that $B(t+2) = B(t)$, $t \in \mathbf{R}$.

**Lemma 2.5**

$$\int_0^2 a(t)\, dt = 0.$$

*Proof.*

$$\int_0^2 a(t)\, dt = \int_0^1 f_u(p(t), p(t-1))\, dt + \int_1^2 f_u(p(t), p(t-1))\, dt.$$

Note that $p(-t) = p(t), t \in \mathbf{R}$, and $p(1-t) = p(t-1), t \in [0,1]$. Therefore,

$$\begin{aligned} \textstyle\int_1^2 f_u(p(t), p(t-1))\, dt &= -\textstyle\int_0^1 f_u(p(t), p(t-1))\, dt = -\textstyle\int_0^1 f_u(p(-t), p(1-t))\, dt \\ &= -\textstyle\int_0^1 f_u(p(t), p(t-1))\, dt. \end{aligned}$$

Define $W$ to be period 2 map along solutions of equation (8).

**Lemma 2.6** $\sigma(W) = \sigma(U)$. Furthermore, $\varphi \in C_{\mathbf{C}}$ is an eigenfunction of $W$ if and only if $\psi(s) = \varphi(s)\exp\{\int_0^s a(\xi)\,d\xi\}$ is an eigenfunction of $U$.

*Proof.* Follows from (7) and Lemma 2.5.

From now on we assume that $b(t) < 0$ (which is the case for Eq. (1)).

Definition 2.2. A solution $y(t)$ of Eq. (6) is called *slowly oscillating* at $t$ if $y(s) \neq 0$ for $s \in [t-1, t]$, or $y(s)$ has exactly one zero in $[t-1, t]$ which is simple.

Change of variables (7) shows that a solution $y(t)$ of Eq. (6) is slowly oscillating at $t$ if and only if the corresponding solution $v(t) = y(t)\exp\{-\int_0^t a(s)\,ds\}$ of Eq. (8) is slowly oscillating at $t$.

Since $1 \in \sigma(U)$ with the eigenfunction $\dot{p}(t)|_{[-1,0]}$, which is slowly oscillating, then $1 \in \sigma(W)$ with a slowly oscillating eigenfunction $\dot{p}(t)\exp\{-\int_0^t a(s)\,ds\}|_{[-1,0]}$.

For a Floquet multiplier $\mu \in \sigma(W)$ let $G(\mu) \subset C_{\mathbf{R}}$ be a linear space of solutions of Eq. (8) which are real parts of complex-valued solutions generated by the eigenspace defined by eigenvalues $\mu$ and $\bar{\mu}$. Therefore, the slowly oscillating periodic solution generated by $v = \dot{p}(t)\exp\{-\int_0^t a(s)\,ds\}|_{[-1,0]}$ belongs to the space $G(1)$. It follows from Lemmas 5,6,7 of [1] that the set

$$\bigcup{}_{\mu\in\sigma(W),|\mu|\geq 1} G(\mu)$$

consists of slowly oscillationg solutions. Also, Corollary 1 of [1] gives that

$$\dim \bigcup{}_{\mu\in\sigma(W),|\mu|\geq 1} G(\mu) \leq 2.$$

Therefore, there may exists at most one eigenvalue of $W$ outside the unit circle which must be real. In view of Lemma 2.6 we arrive at

**Lemma 2.7.** If $\mu \in \sigma(W)$ satisfies $|\mu| > 1$ then $\mu$ is real with the multiplicity $m(\mu) = 1$.

The spectrum $\sigma(W)$ can be characterized in terms of the fundamental matrix solution of a system of ordinary differential equations. This characterization was suggested in [8]. We describe the results we need here refering to [1], [5], [8] for details.

For a nonzero $\mu \in \mathbf{C}$ consider system of ordinary differential equations given by

$$\dot{u} = B(t)z, \quad \dot{z} = \frac{1}{\mu}A(t)u. \tag{9}$$

Let $S(t,\mu) = \begin{pmatrix} u_1(t,\mu) & u_2(t,\mu) \\ z_1(t,\mu) & z_2(t,\mu) \end{pmatrix}$ be the fundamental matrix solution of (9) satisfying $S(0,\mu) = I = \begin{pmatrix} 1 & 0 \\ 0 & 1 \end{pmatrix}$.

Define $r(\mu) = \det\left[S(-1,\mu) - \begin{pmatrix} 0 & 1 \\ 1/\mu & 0 \end{pmatrix}\right]$.

$r(\mu)$ is called the characteristic function.

**Theorem 2.8.** $0 \neq \mu \in \sigma(W) \Longleftrightarrow r(\mu) = 0$.

For the proof see [8], [2], [3].

The characteristic function has the properties:

a) $\lim_{|\mu|\to\infty} r(\mu)/\mu = -1$, ([8], Proposition 3.5; [1], Lemma 11);

b) $r(-1) = 0, r'(-1) = u_1(-1,-1)\int_{-1}^{0} A(s)[u_2(s,-1)]^2\, ds$, ([1], subsection 7.II).

It follows, in view of Lemma 2.7, that in the case $u_2(-1,-1) > 0$ all nontrivial eigenvalues of $W$ are strictly inside the unit disc, and in the case $u_2(-1,-1) < 0$ $W$ has exactly one real (negative) eigenvalue greater than 1 in absolute value (in either case $-1 \in \sigma(W)$ is a simple eigenvalue).

Introducing polar coordinates by

$$(u(t), z(t)) = \rho(t)(\cos\theta(t), \sin\theta(t))$$

decouples system (9) to

$$\dot{\theta} = (1/\mu)A(t)\cos^2\theta - B(t)\sin^2\theta, \quad \dot{\rho} = [B(t) + (1/\mu)A(t)]\, r\cos\theta\sin\theta.$$

The condition $u_1(-1,-1) > 0$ is equivalent to $\theta(-1) \in (-\pi/2, 0)$ for the solution $\theta(t)$ of the initial value problem:

$$\dot{\theta} = -A(t)\cos^2\theta - B(t)\sin^2\theta, \quad \theta(0) = 0.$$

Similarly, $u_1(-1,-1) > 0$ is equivalent to $\theta(-1) \in (-\pi, -\pi/2)$. For details see [1] and [5] respectively. It has been shown that if $B(t)$ is nonincreasing in $[-1, 0]$ then $\theta(-1) \in (-\pi/2, 0)$ ([1], subsection 7.V), and if $B(t)$ is nondecreasing in $[-1, 0]$ then $\theta(-1) \in (-\pi, -\pi/2)$ ([5], proof of Theorem 3.1).

## 3. Proof of Main Result

Consider equation (1) again

$$\dot{x}(t) = -\alpha x(t-1)[1 + \beta x^2(t)].$$

It is known that in the case $\beta < 0$ Eq. (1) has a SSPS for every $\alpha > \pi/2$. Its uniqueness has been shown in [7]. In the case $\beta > 0$ Eq. (1) has a SSPS for every $0 < \alpha < \pi/2$ [4]. Its uniqueness can be shown in a way similar to arguments in [7].

Let $x = p(t)$ be a SSPS of (1). Consider corresponding variational equation:

$$\dot{y}(t) = a(t)y(t) + b(t)y(t-1),$$

where $a(t) = -2\alpha\beta p(t)p(t-1)$, $b(t) = -\alpha[1 + \beta p^2(t)]$.

Corresponding change of variables (7) gives the equation

$$\dot{v}(t) = B(t)v(t-1),$$

with $B(t) = -\alpha[1 + \beta p^2(t)] \exp\left(\alpha\beta \int_{t-1}^{t} p(s)p(s-1)\, ds\right)$.

Direct differentiation shows:

$$\frac{d}{dt}[B(t)] = -2\alpha\beta p(t)p(t-1)[1 + \beta p^2(t)] \exp\left(\alpha\beta \int_{t-1}^{t} p(s)p(s-1)\, ds\right).$$

Since $p(t-1) < 0$ and $p(t) > 0$ for $t \in (-1, 0]$ it follows that $B(t)$ is nonincreasing for $\beta < 0$ and nondecreasing for $\beta > 0$.

**Acknowledgement.** This research has been supported in part by the Alexander von Humboldt Stiftung.

# References


[1] S.-N. Chow and H.-O. Walther, *Characteristic multipliers and stability of symmetric periodic solutions of* $\dot{x}(t) = g(x(t-1))$, Trans. Amer. Math. Soc. **307**(1988), 127–142.

[2] P. Dormayer, *The stability of special symmetric periodic solutions of* $\dot{x}(t) = \alpha f(x(t-1))$ *with small amplitude*, Nonlinear Analysis, Theory, Methods, & Applications, **14**(1990), 701–715.

[3] P. Dormayer and A.F. Ivanov, *Stability of symmetric periodic solutions with small amplitude of* $\dot{x}(t) = f(x(t), x(t-1))$, Preprint, 1992.

[4] A.F. Ivanov and A.N. Sharkovsky, *On periodic solutions of certain classes of differential-difference equations*, In: Oscillation and Stability of Solutions of Differential Functional Equations, Institute of Mathematics of the Ukrainian Academy of Sciences, 1982, 25–34 (in Russian).

[5] A.F. Ivanov, B. Lani–Wayda, and H.-O. Walther, *Unstable hyperbolic periodic solutions of differential delay equations*, Recent Trends in Differential Equations, WSSIAA **1**(1992), 301-316.

[6] J.L. Kaplan and J.A. Yorke, *Ordinary differential equations which yield periodic solutions of differential delay equations*, J.Math.Anal.Appl., **48**(1974), 317–324.

[7] R.D. Nussbaum, *Uniqueness and nonuniqueness for periodic solutions of* $\dot{x}(t) = -g(x(t-1))$, J.Diferential Equations, **34**(1979), 25–54.

[8] H.-O. Walther, *Bifurcation from periodic solutions in functional differential equations*, Math.Z., **182**(1983), 269–289.

# A new method for a system of two nonlinear equations without Poincaré condition

Masahiro Iwano
Department of Mathematics, Chuo University
1-13-27 Kasuga, Bunkyo-ku, JAPAN
and
Department of Mathematics and Statistics
Western Michigan University
Kalamazoo, MI 49008-5152 USA

## ABSTRACT

Stable domains and stability theorems are important subjects in the local theory of nonlinear equations at singular points in the complex domain. The pioneer work of Prof. M. Hukuhara in 1931 on the first order equation $x\dfrac{dy}{dx} = f(x,y)$ with a negative rational $f_y(0,0)$, which is a special case without Poincaré condition, is the starting point of many results of the author in the past. Recently, the same result was obtained in quite a different method by the author. This new method plays a fundamental role in recent studies of Painlevé equations at the irregular type singularities by K Takano, S. Yoshida and S. Simomura. It also can be applied to the following system of two nonlinear equations at an irregular type singularity without Poincaré condition:

$$x^2\frac{dy}{dx} = (\mu + \alpha x)y + f(x,y,z), \quad x^2\frac{dz}{dx} = (-\nu + \beta x)z + g(x,y,z), \quad (\mu > 0, \quad \nu > 0).$$

**1. Introduction.** We consider a differential equation of the form

$$x^{\sigma+1}\frac{dy}{dx} = F(x,y), \qquad F(0,0) = 0, \tag{A}$$

where $\sigma$ is a nonnegative integer; $x$ is a complex indedpendent variable; $y$ is an $n$-vector and $F(x,y)$ is an $n$-vector whose entries are holomorphic functions of $(x,y)$ at the origin of the space $\mathbb{C}^{n+1}$. The origin of the complex $x$-plane $\mathbb{C}$ is a singular point which is called *a Briot-Bouquet type singular point*. The equation

(A) has been studied by many authors since Briot-Bouquet. A main purpose of those authors is to construct analytical expressions of bounded solutions depending on several arbitrary constants in a full or a sectorial neighbourhood of the origin. A solution, which is defined in a domain $\mathcal{D} \in \mathcal{C}$ with $O \in \bar{\mathcal{D}}$, is said to *be bounded in* $\mathcal{D}$ if, for any $x_0 \in \mathcal{D}$, there exists a peicewise smooth curve in $\mathcal{D}$, starting from $x_0$ and extending to the origin, such that $x$ tends to the origin along this curve, the solution converges to zero.

Assume $\sigma > 0$. Let $F_y(x,y)$ be the Jacobi matrix of the $n$-vector $F(x,y)$ with respect to $y$ and $\{\lambda_1, \cdots, \lambda_n\}$ be the eigenvalues of $F_y(0,0)$. If the origin is outside the convex hull $\Pi$, containing the points $\lambda_j \in \mathcal{C}$, the equation (A) is said to *satisfy Poincaré condition at the origin* . Otherwise, the origin must be either in the interior or on the boundary of $\Pi$. Then, the system (A) does *not satisfy Poincaré condition at the origin* . In the case of $\sigma = 0$, the convex hull $\Pi$ must contain, besides the $\lambda_j$, the point 1 also.

In 1931, Professor Masuo Hukuhara(Professor, Emeritus, The University of Tokyo) created a theory which enables us to study analytical meanings of formal solutions, in the same manner, whether they are convergent or not. This method is called Hukuhara's fixed point technique. This is a very useful and powerful tool when we want to construct analytical expressions of bounded solutioons at a singular point. By using this method, M. Hukuhara [1, 2] obtained almost complete results about the first order nonlinear equation of the form

$$x\frac{dy}{dx} = f(x,y), \qquad f(0,0) = 0. \tag{B}$$

**2. Review of Hukuhara's work.** Among Hukuhara's work about equation (B), the most remarkable results are those for the cases of the vanishing $\lambda$ and a negative rational $\lambda$ , wehere $\lambda = f_y(0,0)$. Equation (B) for these two cases does not satisfy Poincaré condition at $x = 0$. The work for a negative rational $\lambda$ is the starting point of the present author's work. It is easy to verify that, in the case of $\lambda \notin \mathcal{N}$ (the set of all positive integers), there exists a particular solution $\phi(x)$ holomorphic at $x = 0$. Put $y = z + \phi(x)$ and rewrite $y$ for $z$. The resulting equation can be written as

$$x\frac{dy}{dx} = y(\lambda + g(x,y)), \qquad g(0,0) = 0, \tag{1}$$

where $\lambda$ is a nagative rational and $g(x,y)$ is a holomorphic function of $(x,y)$ at $(0,0)$.

Hukuhara's method consists of two parts. The first part is to find a formal transformation which formally changes (1) to an equation in a form as simple as possible. The formal transformation is obtained by the limit of sequence of polynomial type transformations. The convergence of the sequence as formal power series is due to the topology which is called a topological ring. The transformed equation is called *a simplified* or *a reduced equation* which is, in general, able to be solved by quadratures or by parametrical representations. The second part is to give analytical meaning(convergence or asymptoticity) to the formal transformation. If we substitute a solution of the simplified equation into the transformation, the resulting expression represents a solution of (1) as long as it is valid.

Let $\lambda = -\frac{\mu}{\nu}$, where $\mu$ and $\nu$ are mutually prime natural numbers. There is, as was done by M. Hukuhara, a formal transformation

$$y = z\left(1 + \Sigma' p_{jk} x^j z^k\right), \quad j + k \geq 1, \quad p_{jk} \in \mathcal{C}, \tag{2}$$

which formally changes (1) to an equation of the form

$$x\frac{dz}{dx} = z\left( -\frac{\mu}{\nu} + \alpha x^{m\mu} z^{m\nu} + \beta x^{2m\mu} z^{2m\nu} \right), \qquad \alpha, \beta \in \mathcal{C}. \tag{3}$$

Here we assume that $\alpha \neq 0$ and, consequently, $m \in \mathcal{N}$. In the case of $\alpha = 0$, the quantity $\beta$ is also zero. This case had been already done earlier than M. Hukuhara.

It is important to notice that the single power series $\sum p_{jk} x^j$ (for each fixed $k$) and $\sum p_{jk} z^k$ (for each fixed $j$) are convergent when $|x|$ and $|z|$ are small. Prof. M. Hukuhara obtained the following result.

THEOREM 1. *(M. Hukuhara [1,2]) Let $\epsilon$ be a preassinged sufficiently small positive constant and $r_0, r_1$ be sufficiently small positive numbers. Then, there exists a transformation $y = z\Phi(x,z)$ which changes (1) to (3) with the properties that:*
*(i) $\Phi(x,z)$ is a holomorphic function of $(x,z)$ in a domain of the $(x,z)$-space of the form*

$$\begin{cases} 0 < |x| < r_0, \quad 0 < |z| < r_1, \\ \left|m\mu \arg x + m\nu \arg z + \arg\alpha \mp \frac{\pi}{2}\right| < \pi - \epsilon. \end{cases} \tag{4}$$

*(ii) $\Phi(x,z)$ admits asymptotic expansion to the power series (2) (excluded the factor $z$) in the sense that, for any $N \in \mathcal{N}$, an inequality of the form*

$$\left|\Phi(x,z) - \left(1 + \sum_{(N)} p_{jk} x^j z^k\right)\right| \leq K_N \left|x^\mu z^\nu\right|^N$$

*holds, where the summation* $\sum_{(N)}$ *is to be take over all arrangements* $(j,k)$ *of nonnegative integers* $j$ *and* $k$ *such that* $\min\{\frac{j}{\mu},\frac{k}{\nu}\} < N$, $K_N$ *being a positive constant depending on* $N$. *(iii) Choose any point* $(x_0, z_0)$ *in domain (4) and denote by* $z = Z(x)$ *a holomorphic solution of (3) satisfying initial condition* $z = z_0$ *at* $x = x_0$. *Then, as long as the values of the functions* $(x, Z(x))$, *considered as points in the* $(x,z)$*-space, stay in domain (4), the expression* $Z(x)\Phi(x, Z(x))$ *represents a solution of equation (1).*

The domain (4) is a domain of the product space of the Riemann surfaces of $\log x$ and $\log z$. The construction of the function $\Phi(x,z)$ by M.Hukuhara is based on a theorem concerning continuity and differentiability of solutions with respect to initial values. It was strongly expected by the author that $\Phi(x,z)$ is expressible by a function of the form $\Psi(x^{m\mu}z^{m\nu}, x, z)$, where $\Psi(w,x,z)$ is a holomorphic function $(w,x,z)$ in a domain of the form

$$\begin{cases} 0 < |w| < r_2, & \left|\arg w + \arg\alpha \mp \frac{\pi}{2}\right| < \pi - \epsilon, \\ |x| < r_0, & |z| < r_1 \end{cases} \tag{6}$$

with $r_2$ sufficiently small.

**3. Stable domains.** Euqation(3) cann't be integrated by quadratures. But, the solution has a parametrical representation. To do see it, let $w = x^{m\mu}z^{m\nu}$ be an auxilliary parameter. Then, we can derive from (3) the linear system

$$(m\nu)(\alpha+\beta w)w^2\frac{dx}{dw} = x, \quad (m\nu)(\alpha+\beta w)w^2\frac{dz}{dw} = (\lambda + aw + \beta w^2)z. \tag{7}$$

This system is integrated by quadratures.

Let $z = Z(x)$ be a solution of (3) satisfying an initial condition $z = z_0$ at $x = x_0$ and $(x,z) = (X_1(w), Z_1(w))$ be a solution of (7) with initial condition $(x,z) = (x_0, z_0)$ at $w = w_0$. It is easy to verify that if $w_0 = x_0^{m\mu}z_0^{m\nu}$ the identity $w \equiv X_1(w)^{m\mu}Z_1(w)^{m\nu}$ holds. Hence, it turns out that the relation $Z(x) = Z_1(X_1^{-1}(x))$ holds if the branches of the both sides are properly chosen, where $w = X_1^{-1}(x)$ is an inverse function of $x = X_1(w)$.

After the study of many years, the author found a nice stable domain and the corresponding curve, which will be used for the path of integration. For the sake of simplicity, we assume that $\beta = 0$ and $\alpha \in \mathcal{R}$(the set of all real numbers). *A stable domain* (associated with solutions of (7)) is to be defined as a domain in the $(w,x,z)$-space of the form

$$\begin{cases} 0 < |w| < r_0\omega(\arg w), \quad |x| < r_1, \quad |z| < r_1\chi(\arg w), \\ \left|\arg w \mp \frac{\pi}{2}\right| < \pi - \epsilon. \end{cases} \tag{8}$$

$\epsilon$ is a preassigned sufficiently small positive constant, $r_0$ and $r_1$ are sufficiently small constants. The functions $\omega(\tau)$ and $\chi(\tau)$ are strictly positive valued continuous functions of $\tau$:

$$\omega(\tau) = \begin{cases} \dfrac{|\cos\tau|}{\sin\epsilon}, & \text{for } \dfrac{\pi}{2} \le \left|\tau \mp \dfrac{\pi}{2}\right| \le \pi - \epsilon, \\ \dfrac{1}{\sin\epsilon}, & \text{for } \left|\tau \mp \dfrac{\pi}{2}\right| < \dfrac{\pi}{2}, \end{cases} \tag{9,}$$

$$\chi(\tau) = \begin{cases} |\cos\tau|^{1/m\nu}, & \text{for } \dfrac{\pi}{2} \le \left|\tau \mp \dfrac{\pi}{2}\right| \le \pi - \epsilon, \\ 1, & \text{for } \left|\tau \mp \dfrac{\pi}{2}\right| < \dfrac{\pi}{2}. \end{cases} \tag{10}$$

Denote by $\mathcal{D}_w$ the domain of $w$ which is expressed as the $w$-component of domain (8). For any $w_0 \in \mathcal{D}_w$, a curve $\Gamma(w_0)$, starting from $w_0$ and extending to the origin, consists generally of two parts $\Gamma'$ and $\Gamma''$ .
When $\left|\arg w_0 \mp \frac{\pi}{2}\right| \le \frac{\pi}{4}$, the $\Gamma(w_0)$ consists of the part $\Gamma'$ only. The variable point $w = w_1(\sigma)$ on $\Gamma'$ is expressed by the formula

$$\frac{1}{w} = \frac{A_0}{A_0^2 + B_0^2} + \sigma - i\frac{B_0}{A_0^2 + B_0^2}e^{\kappa\sigma}, \quad 0 \le \sigma < \infty. \tag{11}$$

Here, $A_0 + iB_0 = w_0 (i = \sqrt{-1})$ and $\kappa$ is a certain positive quantity independent of $w_0$. This curve is tangent to the imaginary axis at the origin of the $w$-complex plane in the part of $\Re w > 0 \text{and} \Im w > 0$.

When $\frac{\pi}{4} < \left|\arg w_0 \mp \frac{\pi}{2}\right| < \pi - \epsilon$, the $\Gamma(w_0)$ consists of the two parts. Let $\theta_0 = \arg w_0$. The variable point $w = w_2(\tau)$ on $\Gamma'$ is given by

$$w = \left(\frac{|w_0|\cos\tau}{\cos\theta_0}\right)e^{i\tau}, \ quad \ \ \theta_0 \le \tau < \frac{\pi}{4} \ \text{ or } \ \frac{3\pi}{4} < \tau \le \theta_0. \tag{12}$$

And, at $\tau = \frac{\pi}{4}$ or $\tau = \frac{3\pi}{4}$ , this curve must be switched to a curve $\Gamma'$ of the form (11) with $A_0 = B_0 = \dfrac{|w_0|}{2\cos\theta_0} > 0$ or $-A_0 = B_0 = \dfrac{|w_0|}{-2\cos\theta_0} > 0$.

The meaning of stable domains will be clarifed in the theorem below:

THEOREM 2. *(M. Iwano[3]). For an arbitrary point $(w_0, x_0, z_0)$ in domain (8), define the solution $(X_1(w), Z_1(w))$ of system (7). Then, as $w$ travels on the curve $\Gamma(w_0)$, the values of the functions $(w, X_1(w), Z_1(w))$, considered as points* *in the $(w,x,z)$-space, stay in the domain (8). Moreover, the functions $|X_1(w)|$ and $|Z_1(w)|$ steadily decrease to zero as $w$ goes to the origin along $\Gamma'$.*

To construct the function $\Psi(w,x,z)$, we have to introduce a special equation. To do this, put, for example,

$$P(x,z) = 1 + \Sigma_{(3m)} p_{jk} x^j z^k \quad (\text{for} \quad \min\{\frac{j}{\mu}, \frac{k}{\nu}\} < 3m). \tag{13}$$

As mentioned above, this partial sum of the power series appearing in (2) is convergent. Apply successively two transformations of the form

$$y = YP(x,Y), \tag{14.1}$$

$$Y = \frac{Z(x)}{1 - \left(x^{m\mu} Z(x)^{m\nu}\right) U} \tag{14.2}$$

where $Z(x)$ is a general solution of (3) with $Z(x_0) = z_0$. The resulting equation will be written in a form

$$x\frac{dU}{dx} = (m\nu\alpha)\left(x^{m\mu} Z(x)^{m\nu}\right)^2 J(x^{m\mu} Z(x)^{m\nu}; x, Z(x), U), \tag{15}$$

where $J(w;x,z,u)$ is a holomorphic and bounded function at the origin of the $(w;x,z,u)$-space.

Corresponding to this equation, we consider the equation

$$\frac{dU_1}{dw} = J(w; X_1(w), Z_1(w), U_1), \tag{16}$$

which is formally obtained from (15) by the substitutions

$$\begin{cases} x^{m\mu} Z(x)^{m\nu} \to w, \quad x\dfrac{d}{dx} \to (m\nu\alpha) w^2 \dfrac{d}{dw}, \\ x \to X_1(w), \quad Z(x) \to Z_1(w), \quad U \to U_1. \end{cases} \tag{17}$$

It is now easy to prove the following

THEOREM 3. *(M. Iwano [3]) There exists a function $U_1(w,x,z)$ holomorphic in $(w,x,z)$ for a domain of the form (8) such that $U_1(x, X_1(w), Z_1(w))$ represents a solution of equation (16) as long as the values of $(x, X_1(w), Z_1(w))$, considered as points in the $(w,x,z)$-space, stay in domain (8). And, the function $U(x,Z(x)) = U_1(x^{m\mu}Z(x)^{m\nu}, x, Z(x))$ is a solution of equation (15) whenever the values of the* *functions $(x, Z(x))$, considered as points in the $(x, z)$-space, belong to a domain of the form (4).*

Thanks to this theorem, the function $\Psi(w, x, z)$ in question is easily obtained from (14.1) and (14.2). In the case of $\beta \neq 0$, to construct a stab le domain, we need to apply the change of variables, as was done in M. Iwano [3], which reduces the equation to an equation of $\beta = 0$. In the case of a complex $\alpha$, we have only to rotate the curve $\Gamma(w_0)$ by the angle equal to $\arg \alpha$ clockwise.

**4. A system of two nonlinear differential equations without Poincaré condition.** Now that a stability theorem like Theorem 3 has been established, it is not so difficult to study a system of the form

$$x^2 \frac{dy}{dx} = (\mu + \alpha x)y + xf(x, y, z), \qquad x^2 \frac{dz}{dx} = (-\nu + \beta x)z + xg(x, y, z). \quad (18)$$

Here, $\mu > 0$, $\nu > 0$ and $\alpha > 0$, $\beta > 0$. The functions $f(x, y, z)$ and $g(x, y, z)$ are holomorphic in $(x, y, z)$ at (0, 0, 0) and their Taylor series expansions in powers of $(y, z)$ begin with terms of degree at least 2.

We can construct an analytical expression of a general solution of (18) by the use of the following theorem.

THEREOM 4. *(M. Iwano [4]) Let $\epsilon > 0$ be a preassigned sufficiently small constant, and $r_0$ and $r_1$ be small positive constants. Then, there exists a transformation $(y, z) = (\Phi(x, u, v), \Psi(x, u, v))$ which changes equations (18) to the linear equations*

$$x^2 \frac{du}{dx} = (\mu + \alpha x)u, \qquad x^2 \frac{dv}{dx} = (-\nu + \beta x)v, \quad (19)$$

*for a domain of the form*

$$\left| \arg x \mp \frac{\pi}{2} \right| < \pi - \epsilon, \quad 0 < |x| < r_0, \quad |u| < r_1, \quad |v| < r_1. \quad (20)$$

*The functions $\Phi$ and $\Psi$ are holomorphic and bounded in $(x, u, v)$ for (20) and they admit uniformly convergent the Taylor series expansions in powers of $(u, v)$ whose coefficients are asymptotically devedlopable to power series in $x$ as $x$ tends to the origin through $|\arg x \mp \pi/2| < \pi - \epsilon$.*
*In particular, the double power series $\Phi(0, u, v)$ and $\Psi(0, u, v)$ in $(u, v)$, which can be obtained from the Taylor expansions of $\Phi(x, u, v)$ and $\Psi(x, u, v)$ by letting $x$ tend to 0, are convergent for $|u| < r_1$, $|v| < r_1$.*

To prove this theorem, we need to establish a stability theorem. So, in the course of the proof, domain (20) must be replaced by a stable domain of the form

$$\begin{cases} 0 < |x| < r_0' \omega(\arg x), \quad |u| < r_1' \chi_\alpha(\arg x), \quad |v| < r_1' \chi_\beta(\arg x), \\ \left| \arg x \mp \dfrac{\pi}{2} \right| < \pi - \epsilon. \end{cases} \tag{21}$$

Here $\omega(\tau)$ is given in (9) and the $\chi_\delta(\tau)(\delta = \alpha, \beta)$ are of the form (10) where the exponent $\dfrac{1}{m\nu}$ is replaced by the quantity $\delta$.

Theorem 4 played an important role in the study of Painlevé equations. K. Takano[8] was successful in constructing an analytical expression of a 2-parameter family of bounded solutions at the point of infinity for Painlevé equation V. The equation involves three parameters. However, in Theorem 4, the quantities $\alpha$ and $\beta$ are assumed to be positive. Therefore, Takano's expansion is valid only if these parameters are confined to some range in the three parameter-space. Fortunately, it was shown later by S. Shimomura [7] that Theorem 4 is still valid if

$$\mu > 0, \quad \nu > 0 \ \text{ and } \ \alpha\nu + \beta\mu > 0 \ \text{ or } \ \Re(\alpha\nu + \beta\mu) > 0. \tag{22}$$

It turned out that the expansion due to Takano is valid without any artificial restriction among the three parameters.

The result in Theorem 4 is extended to a nonlinear $n$-system of the form

$$x^2 \frac{dy}{dx} = (\mathbf{1}_n(\mu) + x\mathbf{1}_n(\alpha))y + xf(x, y), \qquad f(0,0) = 0.$$

Here, $x$ is a complex variable; $y$ is an $n$-vector; $\mathbf{1}_n(\mu)$ and $\mathbf{1}_n(\alpha)$ stand for diagonal matrices respectively with diagonal entries $\{\mu_j\}(= \mu)$ and $\{\alpha_j\}(= \alpha)$ ; $f(x, y)$ is an $n$-vector with entries holomorphic in $(x, y)$ at the origin of the space $\mathcal{C}^{n+1}$ and no terms of degree less than 2 appear in the Taylor series expansion in powers of $y$.

We assume that all $\mu_j \in \mathcal{R} \setminus \{O\}$ and there is a positive number $\kappa$ satisfying

$$\mu_j + \kappa \cdot \Re\alpha_j > 0 \qquad \text{for all} \quad j. \tag{24}$$

It is easy to verify that this condition is equivalent to Shimomura's when $n = 2$. We have the folloiwng theorem.

THEOREM 5. *(M. Iwano [5]) There is a holomorphic and bounded function* $\Phi(x, u)$ *of* $(x, u)$ *for a domain of the form*

$$\left| \arg x \mp \frac{\pi}{2} \right| < \pi - \epsilon, \quad 0 < |x| < r_0, \quad \| u \| = \max_{j=1}^{n} |u_j| < r_1 \tag{25}$$

*with the properties that*
*(i) The transformation* $y = \Phi(x, u)$ *changes (23) to a system of the form*

$$x^2 \frac{du}{dx} = (\mathbf{1}_n(\mu) + x\mathbf{1}_n(\alpha))u + x\Sigma^{(\aleph)} u^q g_q(x). \tag{26}$$

*The symbol* $u^q$ *is a usual scalar expression.*
*(ii) Every entry of an n-vector* $g_q(x)$ *is a monomial of the form* $x^{\ell} g^0_{q,j}$. *If a j-the entry* $g^0_{q,j}$ *is nonvanishing, the n-tupe q and the corresponding exponent* $\ell$ *must satisfy the relations* $\mu_j = q\mu$ *and* $\alpha_j = \ell + q\alpha$.

Owing to the assumption (24), the set of all $n$-tuples $q$ appearing in $\Sigma^{(\aleph)}$ is a finite set and, consequently, the simplified system (26) is integrated by quadratures. In the course of the proof, the domain (25) should be replaced by a stable domain similar to (21).

On the other hand, S. Yoshida [9, 10] has devised a system of two nonlinear differential equations which is useful for constructing a 2-parameter family of bounded solutions near all the irregular type singularities of Painlevé equations except for V.

**5. A new method.** The equations (18) and (23) contain a factor $x$ in the nonlinear terms. The factor $x$ has seemed to be indespensable for the author in estimating the ratio of growth of certain functions along the path of integration. Such an estimation is always necessary in the proof of the existence of solutions. So, a system (18) without the factor $x$ is of some interest to the author from a mathematical point of view. On the other hand, S. Yoshida [9, 11] already studied such a system b y the use of a singular transformation. T. Koitabashi[6] studied the same equations by assuming Siegel's condition. Recently the author has devised a new method which overcomes some difficulty.

We consider the differential equations of the form

$$x^2 \frac{dy}{dx} = (\mu + \alpha x)y + f(x, y, z), \quad x^2 \frac{dz}{dx} = (-\nu + \beta x)z + g(x, y, z) \tag{27}$$

with the assumptions that:
(i) $\mu > 0, \quad \nu > 0$ and the ratio $\mu : \nu$ is incommensurable,
(ii) $\alpha\nu + \beta\mu > 0$ or $\Re(\alpha\nu + \beta\mu) > 0$,
(iii) $f(x, y, z)$ and $g(x, y, z)$ have the same meaning as in (18).

The assumption (ii) is equivalent to the condition that there exists a positive number $\kappa$ satisfying the inequalities

$$\mu + \kappa \cdot \Re\alpha > 0, \quad -\nu + \kappa \cdot \Re\beta > 0. \tag{28.1}$$

As can be easily verified, there is a formal transformation of the form

$$y = u + \sum p_{jk}(x) u^j v^k, \quad z = v + \sum q_{jk}(x) u^j v^k, \quad (\text{for} \quad j + k \geq 2), \tag{29}$$

which formally changes equations (27) to the linear equations

$$x^2 \frac{du}{dx} = (\mu + \alpha x)u, \quad x^2 \frac{dv}{dx} = (-\nu + \beta x)v. \tag{30.1}$$

Here the coefficients $p_{jk}(x)$ and $q_{jk}(x)$ are holomorphic and bounded functions of $x$ for a domain of the form

$$\left| \arg x \mp \frac{\pi}{2} \right| < \pi - \epsilon, \quad 0 < |x| < r' \tag{31}$$

and admit asymptotic expansions in powers of $x$ as $x$ tends to 0 through (31).

The main task is to show the convergence of the formal transformation (29). It is to be noted that the single power series of two kinds

$$\begin{cases} \left( \sum_j p_{jk}(x) u^j, \sum_j q_{jk}(x) u^j \right) & (\text{for each fixed} \quad k), \\ \left( \sum_j p_{jk}(x) v^k, \sum_j q_{jk}(x) v^k \right) & (\text{for each fixed} \quad j) \end{cases}$$

are uniformly convergent for $x$ in domain (31) when $|u|$ or $|v|$ is small.

To prove the convergence of (29), we have to make particular truncatedc equations. To do this, for any $N \in \mathcal{N}$, we put

$$P_{(N)}(x, u, v) = u + \sum_{(N)}^{1} p_{jk}(x) u^j v^k, \quad Q_{(N)}(x, u, v) = v + \sum_{(N)}^{2} q_{jk}(x) u^j v^k. \tag{32}$$

Here $\sum\limits_{(N)}^{1}$ is the summation over all arrangements $(j, k)$ of nonnegative integers $j$ and $k$ satisfying $\min\{j - 1, k\} < 2N$, while $\sum\limits_{(N)}^{2}$ is that satisfying $\min\{j, k - 1\} < 2N$. The both power series are convergent for $x$ in domain (31) when $|u|$ and $|v|$ are small. Make the change of variables

$$y = P_{(N)}(x, \eta, \zeta), \quad z = Q_{(N)}(x, \eta, \zeta). \tag{33}$$

The transformed equations will be written as

$$\begin{cases} x^2\dfrac{d\eta}{dx} = (\mu+\alpha x)\eta + \eta^{2N+1}\zeta^{2N}F(x,\eta,\zeta), \\ x^2\dfrac{d\zeta}{dx} = (-\nu+\beta x)\zeta + \eta^{2N}\zeta^{2N+1}G(x,\eta,\zeta). \end{cases} \tag{34}$$

$F(x,\eta,\zeta)$ and $G(x,\eta,\zeta)$ are holomorphic and bounded functions of $(x,\eta,\zeta)$ for a domain of the form

$$\left|\arg x \mp \frac{\pi}{2}\right| < \pi - \epsilon, \quad 0 < |x| < r_0', \quad |\eta| < r_1', \quad |\zeta| < r_1'.$$

$r_0'$ and $r_1'$ may depend on $N$.

Denote by $(u,v) = (U(x),V(x))$ a holomorphic solution of (30.1) satisfying an arbitrary initial condition $(u,v) = (u_0,v_0)$ at $x = x_0$. But, $x_0$ must be in domain (31). Set

$$\eta = \frac{U(x)}{1-Y}, \qquad \zeta = \frac{V(x)}{1-Z}. \tag{35}$$

Then, the system (34) is changed to a system of the form

$$\begin{cases} x^2\dfrac{dY}{dx} = (U(x)V(x))^{2N}F_N(x,U(x),V(x),Y,Z), \\ x^2\dfrac{dZ}{dx} = (U(x)V(x))^{2N}G_N(x,U(x),V(x),Y,Z), \end{cases} \tag{36}$$

where $F_N(x,u,v,Y,Z)$ and $G_N(x,u,v,Y,Z)$ are holomorphic and bounded functions of $(x,u,v,Y,Z)$ in a domain of the form

$$\begin{cases} \left|\arg x \mp \dfrac{\pi}{2}\right| < \pi - \epsilon, \quad 0 < |x| < r_0'', \quad |u| < r_1'', \quad |v| < r_1'', \\ |Y| < r_2'', \quad |Z| < r_2'', \end{cases} \tag{37}$$

$r_j''$ being sufficiently small positive constants depending on $N$.

We consider the differential equation

$$x^2\frac{dw}{dx} = (\lambda+\gamma x)w, \quad \text{with} \quad \lambda = \mu - \nu \qquad \text{and} \gamma = \alpha + \beta, \tag{30.2}$$

which is satisfied by the function $w = U(x)V(x)$. Notice that

$$\lambda + \kappa \cdot \Re\gamma = (\mu + \kappa \cdot \Re\alpha) + (-\nu + \kappa \cdot \Re\beta) > 0. \tag{28.2}$$

Change the role of independent variable from $x$ to $w$. From (30.1) and (30.2) it follows that

$$w\frac{dx}{dw}=\frac{x^2}{\lambda+\gamma x},\quad w\frac{du}{dw}=\frac{\mu+\alpha x}{\lambda+\gamma x}u,,\quad w\frac{dv}{dw}=\frac{-\nu+\beta x}{\lambda+\gamma x}v \tag{38}$$

Let $(u,v,w)=(U(x),V(x),W(x))$ be a holomorphic general solution of (30.1) – (30.2) satisfying an arbitrary initial condition $(u,v,w)=(u_0,v_0,w_0)$ at $x=x_0$, where $x_0$ must be in domain (31). Denote by $(X_1(w),U_1(w),V_1(w))$ a holomorphic general solution of (38) with an arbitrary initial condition $(x,u,v)=(x_0,u_0,v_0)$ at $w=w_0$, where $x_0$ is in domain (31). Then, we have $U(x)\equiv U_1(W(x)),V(x)\equiv V_1(W(x))$. If $u_0v_0=w_0$, the identity $U(x)V(x)\equiv W(x)$ holds.

In view of (30.2), we consider, instead of (36), the equations

$$\begin{cases}\dfrac{dY_1}{dw}=\dfrac{w^{2N-1}}{\lambda+\gamma X_1(w)}F_N(X_1(w),U_1(w),V_1(w),Y_1,Z_1),\\[2mm] \dfrac{dZ_1}{dw}=\dfrac{w^{2N-1}}{\lambda+\gamma X_1(w)}G_N(X_1(w),U_1(w),V_1(w),Y_1,Z_1).\end{cases} \tag{39}$$

We can assert that:

PROPOSITION 1. *(Fundamental existence theorem) The equations (39) possess a solution* $\{\Phi_{(N)}(X_1(w),U_(w),V_1(w),w),\Psi_{(N)}(X_1(w),U_1(w),V_1(w),w)\}$ *as long as the values of the functions* $(X_1(w),U_1(w),V_1(w),w)$*, considered as points of the* $(x,u,v,w)$*-space, belong to a domain of the form*

$$\begin{cases}0<|x|<r_0^{(N)},\quad |u|<r_1^{(N)},\quad |v|<r_1^{(N)},\quad \dfrac{|w|^N}{|x|}<1,\\[2mm] \left|\arg x\mp\dfrac{\pi}{2}\right|<\pi-\epsilon,\end{cases} \tag{40}$$

$r_j^{(N)}$ *being sufficiently small positive constants depending on* $N$.
*(i)* $\Phi_{(N)}(x,u,v,w)$ *and* $\Psi_{(N)}(x,u,v,w)$ *are holomorphic functions of* $(x,u,v,w)$ *in the domain (40) and satisfy there*

$$\left|\Phi_{(N)}(x,u,v,w)\right|\le K_N|w|^N,\quad \left|\Psi_{(N)}(x,u,v,w)\right|\le K_N|w|^N,$$

*where* $K_N$ *is a certain constant depending on* $N$.
*(ii) A solution of (39) satisfying the order condition* $Y_1=O(|w|^N)$, $Z_1=O(|w|^N)$ *is unique.*

In the course of the proof, the domain (40) should be replaced by a stable domain similar to (21). It is easy to verify that:

PROPOSITION 2. *The pair* $\{\Phi_{(N)}(x,U(x),V(x),U(x)V(x)), \Psi_{(N)}(x,U(x),V(x),U(x)V(x))\}$ *is a solution of (36) whenever the values of functions* $(x,U(x),V(x),U(x)V(x))$*, considered as points in the* $(x,u,v,w)$*-space, stay in domain (40).*

In view of (32), (33) and (35), we define the function $\hat{\Phi}^{(N)}(x,u,v,w)$ by

$$\hat{\Phi}^{(N)}(x,u,v,w) = P_{(N)}\left(x, \frac{u}{1-\Phi_{(N)}(x,u,v,w)}, \frac{u}{1-\Psi_{(N)}(x,u,v,w)}\right)$$

and the function $\hat{\Psi}^{(N)}(x,u,v,w)$ in a similar way.

We can assert that:

PROPOSION 3. *(i) The pair* $\{\hat{\Phi}^{(N)}(x,U(x),V(x),U(x)V(x)), \hat{\Psi}^{(N)}(x,U(x),V(x),U(x)V(x))\}$ *is a solution of equations (27) which is independent of* $N$ *whenever it is valid.*
*(ii) We denote this solution by* $\{\Phi(x,U(x),V(x)), \Psi(x,U(x),V(x))\}$*. Then, the functions* $\Phi(x,u,v)$ *and* $\Psi(x,u,v)$ *are holomorphic and bounded in* $(x,u,v)$ *for a domain of the form*

$$\left|\arg x \mp \frac{\pi}{2}\right| < \pi - \epsilon, \quad 0 < |x| < r_0, \quad |u| < r_1, \quad |v| < r_1,$$

*where* $r_0$ *and* $r_1$ *are sufficiently small.*

By Cauchy's theorem, the functions $\Phi(x,u,v)$ and $\Psi(x,u,v)$ admit the Taylor series expansions in $(u,v)$ for (43), which coincide with the double power series given by (29). Thus we have the theorem.

THEOREM 6. *There is a transformation* $(y,z) = (\Phi(x,u,v), \Psi(x,u,v))$ *which changes the equations (27) to the linear equations (32.1) for* $(x,u,v)$ *in a domain of the form (43). As long as the values of the functions* $(x,U(x),V(x))$*, considered as points in the* $(x,u,v)$*-space belong to domain (43), the pair* $\{\Phi,\Psi\}$ *of the functions* $\Phi(x,U(x),V(x))$ *and* $\Psi(x,U(x),V(x))$ *represents a 2-parameter family of bounded solutions of the lsystem (27).*

**Remark.** In this theorem, it cann't be expected that $\Phi(0,u,v)$ and $\Psi(0,u,v)$ admit convergent double power series expansions.

## References

[1] M. Hukuhara, Sur les points singuliers d'une équation différentièlle ordinaire du premier order, IV. Proc. Phys.-Math. Soc. Japan, **20**(1938),pp.865-907.

[2] M. Hukuhara, T. Kimura, $M^{me}$ T. Matuda, Équations différentièlles ordinaires du premier ordre dans le champ complex, 7(1961), Publ. Math. Soc. Japan.

[3] M. Iwano, On general solution of a first-order nonlinear differential equation of the form $xdy/dx = y(\lambda + f(x,y))$ with negative rational $\lambda$. Ann. Mat. Pura Appl. Serie **126**(1980), pp.19-80.

[4] M. Iwano, On a general solution of a nonlinear 2-system of the form $x^2dw/dx = \Lambda w + xh(x,w)$ with a constant diagonal matrix $\Lambda$ of signature (1, 1), Tohoku Math. Jour. **32**(1980), pp.453-486.

[5] M. Iwano, On an $n$-parameter family of solutions of a nonlinear $n$-system with an irregular type singularity, Ann. Mat. Pura Appl. Serie 4, **140**(1985), pp.57-132.

[6] T. Koitabashi, On a 2-parameter family of solutions to a nonlinear 2-system with an irregular singularity of signature (1,1), Kumamoto Jour. Math. 5(1992), pp.21-48.

[7] S. Shimomura, Analytic integration of some nonlinear ordinary differential equations and the fifth Painlevé equation in the neighbourhood of an irregular singular points, Funkcialaj Ekva. **26**(1983), pp.301-338.

[8] K. Takano, A 2-parameter family of solutions of Painlevé equation (V) near the point at infinity, Ibid.26(1983), pp.79-113.

[9] S. Yoshida, A general solution of a nonlinear 2-system without Poincaré condition at an irregular singular point, Ibid. **27**(1983) pp.367-391.

[10] S. Yoshida, 2-parameter family of solutions for Painlevé equations (I) ~(V), Ibid.**28**(1985), pp.233-248.

[11] S. Yoshida, A general solution of a nonliner 2-system without Poincaré condition at an irregular singular point (II), Ibid.35(1992), pp.405-427.

# UNIQUENESS OF LIMIT CYCLES OF CERTAIN PLANAR VECTOR FIELDS WITH CODIMENSION TWO SINGULARITIES*

MAOAN HAN
Department of Mathematics
Shandong Mine Institute
Taian, Shandong 271019, China

and

KATIE JIANG
Department of Mathematics
GMI Engineering & Management Insitute
Flint, MI 48504-4898

## ABSTRACT

Hopf and heteroclinic bifurcations in certain planar vector fields with singularities of codimension two are studied. A necessary and sufficient condition is given for the uniqueness of limit cycles for these vector fields. This problem is solved strictly for the first time in the present paper, and our method is also of great advantage.

## 1. The Introduction

A family of three dimensional systems with two parameters can be reduced into a two-dimensional normal form by restoring the cubic terms. (Guckenheimer & Holmes [5]):

$$\dot{r} = \mu_1 r + arz + cr^3 + dr^2 z$$

$$\dot{z} = \mu_2 + br^2 - z^2 + er^2 z + fz^3 \tag{1.1}$$

where c,d,e, and f are real constants, $r\geq 0$, $a\neq 0$, $b = \pm 1$ and $\mu_1$ and $\mu_2$ are small parameters. Furthermore, certain coordinate changes can be introduced so that all cubic terms other

*This work was supported by the Chinese Natural Science Funds.

than $z^3$ in Eq(1.1) can be eliminated (Wiggins [7]). Therefore, without loss of generality, we will analyze the following normal form

$$\dot{r} = \mu_1 r + arz$$
$$\dot{z} = \mu_2 + br^2 - z^2 + fz^3 \tag{1.2}$$

We let

$$r = \varepsilon u,\ z = \varepsilon v,\ \mu_1 = \varepsilon^2 \nu_1,\ \mu_2 = \varepsilon^2 \nu_2$$

and rescale the time $t \longrightarrow \varepsilon t$, so that Eqs.(1.1) and (1.2) now become

$$\dot{u} = auv + \varepsilon\left(\nu_1 u + cu^3 + duv^2\right)$$
$$\dot{v} = \nu_2 + bu^2 - v^2 + \varepsilon\left(eu^2 v + fv^3\right) \tag{1.3}$$

and

$$\dot{u} = auv + \varepsilon \nu_1 u$$
$$\dot{v} = \nu_2 + bu^2 - v^2 + \varepsilon f v^3 \tag{1.4}$$

respectively. We can suppose $\nu_2 = \pm 1$. The Eq.(1.4) may have a limit cycle for $\varepsilon > 0$ and small if $b = -\nu_2 = \pm 1$ and $ab < 0$. In this paper we are interested in the case of $b = -\nu_2 = -1$ and $a > 0$, so that Eq.(1.4) becomes

$$\dot{u} = auv + \varepsilon \nu_1 u$$
$$\dot{v} = 1 - u^2 - v^2 + \varepsilon f v^3. \tag{1.5}$$

For $\varepsilon = 0$, we have

$$\dot{u} = auv$$
$$\dot{v} = 1 - u^2 - v^2 \tag{1.6}$$

which has a family of periodic orbits encircling the center at $(u,v) = (1,0)$ and limiting on the heteroclinic loop $F(u,v) = 0$, where $F(u,v) = \frac{a}{2} u^{2/a}\left(1 - \frac{u^2}{1+a} - v^2\right)$ is the first integral of Eq.(1.6). As mentioned in [5] and [7], F is not a Hamiltonian (unless a = 2), but it can be made into Hamiltonian by multiplying the right side of Eg.(1.5) by the integrating factor $u^{2/a-1}$ to obtain

$$\dot{u} = au^{2/a} v + \varepsilon v\, u^{2/a}$$
$$\dot{v} = u^{2/a-1} - u^{2/a+1} - u^{2/a-1} v^2 + \varepsilon f u^{2/a-1} v^3. \tag{1.7}$$

For $\varepsilon=0$, Eq. (1.7) is a Hamiltonian with the Hamiltonian function

$$H(u,v) = \frac{a}{2} u^{2/a}v^2 - \frac{a}{2} u^{2/a} + \frac{a}{2(1 + a)} u^{2/a +2}$$

and the family of periodic orbits $\Gamma$ of Eq.(1.6) is given by the compact component part of the curves:

$$H(u,v) = h, \quad h_0 = - \frac{a^2}{2(1 + a)} < h < 0.$$

Obviously, $\Gamma_{h_0}$ and $\Gamma_0$ correspond to the center at $(u,v) = (1,0)$ and the heteroclinic loop of Eq.(1.6) respectively, and the Melnikov functions are given by (cf. Wiggins [7]):

$$M(h,\nu_1) = af\oint_{\Gamma_h} u^{4/a-1}v^4dt - \nu_1 \oint_{\Gamma_h}\left(-u^{4/a+1}+ v^{4/a-1}+ u^{4/a-1}v^2\right)dt$$

Therefore, $M(h,\nu_1) = 0$, if and only if

$$\nu_1 = \frac{af\oint_{\Gamma_h} u^{4/a-1}v^4 dt}{\oint_{\Gamma_h} \left(-u^{4/a +1} + v^{4/a -1} + u^{4/a -1}v^2\right)dt} \equiv g(h).$$

Chow, Li and Wang [4] have proven that the function $g(h)$ is monotonic in $h$ on $(h_0,0)$. The monotonicity of $g$ implies that for any small $\varepsilon_0>0$ and $\varepsilon_1>0$, there exist $\varepsilon_2 = \varepsilon_2(\varepsilon_0,\varepsilon_1)$ and

$\tilde{g}(h,\varepsilon) = g(h) + O(\varepsilon)$, $h_0 - \varepsilon_0 \le h \le \varepsilon$, and $0<\varepsilon<\varepsilon_2$, such that for $0 < \varepsilon < \varepsilon_2$, Eq.(1.5) has a unique limit cycle in the region of

$\cup\left\{\Gamma_h | h_0 - \varepsilon_0 \le h \le \varepsilon_1\right\}$ if and only if $\nu_1 = \tilde{g}(h,\varepsilon)$ for for some $h\in[h_0 - \varepsilon_0,\varepsilon_1]$ and $\varepsilon\in(0,\varepsilon_2]$. However, in order to prove that the uniqueness of limit cycles in a neighborhood of the region $\cup\left\{\Gamma_h | h_0 \le h \le 0\right\}$, we need to prove the uniqueness problem of limit cycles bifurcating from $\Gamma_{h_0}$ and $\Gamma_0$ separately in order for us to have

$$\inf_{\varepsilon_0} \varepsilon_2(\varepsilon_0,\varepsilon_1) > 0 \quad \text{and} \quad \inf_{\varepsilon_1} \varepsilon_2(\varepsilon_0,\varepsilon_1) > 0 \tag{1.8}$$

usually, it is easy to show the first inequality in the Eq. (1.8) under certain nondegenerate conditions (by using Poincaré normal forms), but it is rather difficult to show the second one. So far, no one has shown the uniqueness of limit cycles bifurcation from the heteroclinic cycle $\Gamma_0$, although one conjects that it should follow from the monotonicity of g(h), and this problem has been neglected by all of [2,3,4,5,7]. In this paper, we will show the uniqueness of limit cycles for system Eq.(1.5) in the whole half plane u>0 easily by applying a result from Zhifen Zhang [8].

As shown in Broer and Vegter [1], the heteroclinic bifurcation value g(0) is closely related to the Silnikov bifurcation and an explicit expression of g(0) has not been given by [3,5,7] unless a=2. Perhaps, the reason is that the Eq.(1.7) is not suitable for the calculation of g(0). In this paper, we will obtain an explicit expression of g(0) for any a>0 by introducing a coordinate change to Eq.(1.5), which is different from [2]. In the end, we will correct a mistake made in the calculation of the Hopf bifurcation value in [5].

## 2. Hopf Bifurcation

In this section, we consider the Hopf bifurcation of Eq.(1.5) near the point (1,0). It is easy to see that Eq.(1.5) has a unique focus at the point $A_\varepsilon(u(\varepsilon),v(\varepsilon))$, where

$$u(\varepsilon) = \left(1 - \frac{\varepsilon^2\nu_1^2}{a^2} + \frac{f\nu_1^3}{a^3}\varepsilon^4\right)^{1/2}, \qquad v(\varepsilon) = -\frac{\varepsilon\nu_1}{a}.$$

Removing $A_\varepsilon$ to the origin in Eq.(1.5), we have

$$\begin{aligned}\dot{u} &= Bv + F(u,v,\varepsilon)\\ \dot{v} &= Cu + Dv + G(u,v,\varepsilon)\end{aligned} \tag{2.1}$$

where $B = au(\varepsilon)$, $C = -2u(\varepsilon)$, $D = -2v(\varepsilon) + 3fv^2(\varepsilon)\varepsilon$,

$F(u,v,\varepsilon) = auv$, $G(u,v,\varepsilon) = -u^2 - (1-3f\,v(\varepsilon)\varepsilon)v^2 + \varepsilon f v^3$.

The eigenvalues of the linear part of Eq.(2.1) have the real part of

$$\alpha(\varepsilon,\nu_1) = -v(\varepsilon) + \frac{3}{2}fv^2(\varepsilon)\varepsilon = \frac{\varepsilon\nu_1}{a}\left(1 + \frac{3f\nu_1}{2a}\varepsilon^2\right). \tag{2.2}$$

Thus, for $\varepsilon > 0$ and small, $\alpha(\varepsilon,\nu_1) = 0$ if and only if $\nu_1 = 0$. Let $\nu_1 = 0$ and set $\omega = \sqrt{-BC} = \sqrt{2a}\, u(\varepsilon)$, then, by making the the coordinate changes of

$$u = \frac{\omega}{c} x = -\frac{1}{2}\sqrt{2a}\, x, \quad v = y,$$

we obtain from (2.1)

$$\dot{x} = -\omega y + axy,$$
$$\dot{y} = \omega x - \frac{a}{2} x^2 - y^2 + \varepsilon f y^3. \tag{2.3}$$

The Eq.(2.3) has the following normal form in polar coordinates

$$\dot{r} = a_1 r^3 + O(r^5), \quad \dot{\theta} = \omega + O(r^2)$$

where, from the formula (3.4.11) of [5], $a_1 = \frac{3}{8} f\varepsilon$. Therefore, from Eq.(2.2) the normal form of Eq.(2.1) has the following Poincaré map

$$P(x,\nu_1,\varepsilon) = x + \alpha(\varepsilon,\nu_1)x + [a_1 + O(\alpha(\varepsilon,\nu_1))]x^3 + O(x^5)$$
$$= x + \varepsilon\left[\frac{\nu_1}{a}\left(1 + \frac{3}{2}\frac{f\nu_1}{a}\varepsilon^2\right)x + \left(\frac{3f}{8} + O(\nu_1)\right)x^3 + O(x^5)\right]$$

since $P(x,\nu_1,\varepsilon) \equiv x$ for $\varepsilon = 0$. Thus, we conclude that if $f \neq 0$ then for $\varepsilon > 0$ and $|\nu_1| > 0$ and both $\varepsilon$ and $\nu_1$ small, Eq.(1.5) has a unique limit cycle in a neighborhood of the focus $A_\varepsilon$ if and only if $\nu_1 f < 0$.

## 3. Heteroclinic Bifurcation

In this section, we will discuss the heteroclinic bifurcation near the heteroclinic cycle

$$u^{2/a}\left(1 - \frac{u^2}{1+a} - v^2\right) = 0.$$

By making the coordinate change of $\tilde{u} = u^{2/a}$, and reusing u instead of $\tilde{u}$, we have from Eq.(1.5)

$$\dot{u} = 2uv + \frac{2\nu_1}{a}\varepsilon u$$
$$\dot{v} = 1 - u^a - v^2 + \varepsilon f v^3. \tag{3.1}$$

Then for $\varepsilon = 0$ the Eq.(3.1) is Hamiltonian with Hamiltonian function

$$H(u,v) = u\left(1 - v^2 - \frac{1}{1+a}\, u^a\right). \tag{3.2}$$

Let $\Gamma_h$ denote the compact connected closed curve defined by

$$H(u,v) = h, \quad 0 < h < \frac{a}{1+a}.$$

Then the Melnikov function of Eq.(3.1) is given by

$$M(h,\nu_1) = \oint_{\Gamma_h}\left[2uvfv^3 - \frac{2\nu_1}{a}(1-u^a-v^2)u\right]dt = \oint_{\Gamma_h} fv^3du - \frac{2}{a}\nu_1 udv \tag{3.3}$$

note that along $\Gamma_h$

$$v = \pm\sqrt{1 - \frac{1}{1+a}\, u^a - \frac{h}{u}} \equiv \pm\, v(u,h),\ u_1(h) \le u \le u_2(h). \tag{3.4}$$

where $v\big(u_i(h),h\big) = 0$, $0 < u_1(h) < 1 < u_2(h) < (1 + a)^{1/a}$. That is, $\big(u_i(h),0\big)$ $i = 1,2$ are the intersection points of $\Gamma_h$ with the u-axis. By intergration by parts, we have

$$\oint_{\Gamma_h} v^3du = 2\int_{u_1(h)}^{u_2(h)} v^3(u,h)\,du = -3\oint_{\Gamma_h} v^2udv. \tag{3.5}$$

Similarly, $\displaystyle\oint_{\Gamma_h} udv = -\oint_{\Gamma_h} vdu,\ \oint_{\Gamma_h} u^{1+a}dv = -(1+a)\oint_{\Gamma_h} u^a v\, du.$ (3.6)

Then by using Eqs.(3.4) and (3.6) and the fact that $\displaystyle\oint_{\Gamma_h} dv = 0$,

we deduce that

$$\oint_{\Gamma_h} v^2udv = \oint_{\Gamma_h}\left(1 - \frac{u^a}{1 + a} - \frac{h}{u}\right)udv = \oint_{\Gamma_h} udv + \oint_{\Gamma_h} u^a v\, du. \tag{3.7}$$

Therefore, it follows from Eqs.(3.3), (3.5) and (3.7) that

$$M(h,\nu_1) = M_1(h)\left[\frac{2\nu_1}{a} + 3f - 3fQ(h)\right] \tag{3.8}$$

where

$Q(h) = M_2(h)\big/M_1(h)$, and $\displaystyle M_1(h) = \oint_{\Gamma_h} vdu,\ M_2(h) = \oint_{\Gamma_h} u^a v\, du.$

Thus, for $0 < h < \frac{a}{1+a}$, we have $M(h,\nu_1) = 0$ if and only if $\nu_1 = \frac{3af}{2}\left(Q(h)-1\right) \equiv G(h)$. Our goal is to compute $G(0)$, for this we derive the following integral formula

$$\int x^a (Ax^a+B)^P dx = \frac{1}{A(a+aP+1)}\left[x(Ax^a+B)^{P+1} - B\int (Ax^a+B)^P dx\right]. \quad (3.9)$$

by the integration by parts. Observe that $\lim_{h\to 0} u_1(h) = 0$, $\lim_{h\to 0} u_2(h) = (1+a)^{1/a}$. By Eqs.(3.9) and (3.4) we then have

$$M_2(0) = 2\int_0^{(1+a)^{1/a}} u^a\left(1 - \frac{u^a}{1+a}\right)^{1/2} du$$

$$= \frac{4(1+a)}{3a+2}\int_0^{(1+a)^{1/a}} \left(1 - \frac{u^a}{1+a}\right)^{1/2} du = \frac{2(1+a)}{3a+2} M_1(0).$$

This yields

$$Q(0) = M_2(0)/M_1(0) = \frac{2(1+a)}{3a+2}, \quad (3.10)$$

thus, $G(0) = \frac{-3a^2 f}{2(3a+2)}$. Noting that $\frac{\partial M}{\partial \nu_1} = \frac{2}{a} M_1(h) > 0$ from the Melnikov's method [5], there exists a smooth function $\nu_1 = K(\varepsilon)$ defined for $\varepsilon>0$ small and with $K(0) = \frac{-3a^2 f}{2(3a+2)} = G(0)$ such that the Eq.(3.1) has a heteroclinic cycle if and only if $\nu_1 = K(\varepsilon)$.

## 4. The Main Results

Theorem 4.1: Let $f\neq 0$ and $\varepsilon>0$, then Eq.(1.5) possesses at most one limit cycle in the region of $u>0$. If the limit cycle does exist, it is stable (unstable) if $f<0$ ($f>0$) respectively.

<u>Proof</u>: It is easy to see that the coordinate changes of $y = \ln u$, $x = -v - \frac{\varepsilon\nu_1}{a}$ carry the Eq.(1.5) to the following

$$\dot{x} = e^{2y} - F(x), \quad \dot{y} = -ax \quad (4.1)$$

where $F(x) = -\varepsilon f x^3 - \left(1 + \frac{3f\nu_1}{a}\varepsilon^2\right)x^2 - \left(\frac{2}{a} + \frac{3f\nu_1\varepsilon^2}{a^2}\right)\nu_1\varepsilon x +$

$$+\left(1 - \frac{\nu_1^2\varepsilon^2}{a^2} - \frac{f\nu_1^3\varepsilon^4}{a^3}\right).$$

Obviously, a necessary condition for the existence of a limit cycle of Eq.(4.1) is $F(0)>0$. In this case, removing the unique singular point $\left(0, \frac{1}{2}\ell n F(0)\right)$ of Eq.(4.1) to the origin, we have the following equations

$$\begin{aligned} \dot{x} &= \left(e^{2y} - 1\right)F(0) - \left(F(x) - F(0)\right) \\ \dot{y} &= -ax \end{aligned} \tag{4.2}$$

where $F(x) - F(0) = -\varepsilon f x^3 - \left(1 + \frac{3f\nu_1}{a}\varepsilon^2\right)x^2 - \left(\frac{2}{a} + \frac{3f\nu_1\varepsilon^2}{a^2}\right)\nu_1\varepsilon x.$

Let us prove that Eq.(4.2) has no limit cycles if

$$f\left(\frac{2}{a} + \frac{3f\nu_1}{a^2}\varepsilon^2\right)\nu_1 \geq 0. \tag{4.3}$$

For this purpose, we consider the system

$$\begin{aligned} \dot{x} &= \left(e^{2y} - 1\right)F(0) + \left(1 + \frac{3f\nu_1}{a}\varepsilon^2\right)x^2 \\ \dot{y} &= -ax \end{aligned} \tag{4.4}$$

The vector fields of Eq.(4.4) are symmetric with respect to the y-axis. Thus, the Eq.(4.4) has a family of periodic orbits encircling the origin. On the other hand, a direct computation yields that

$$\left.\frac{dx}{dy}\right|_{Eq.(4.2)} - \left.\frac{dx}{dy}\right|_{Eq.(4.4)} = -\frac{\varepsilon}{a}\left[fx^2 + \left(\frac{2}{a} + \frac{3f\nu_1\varepsilon^2}{a^2}\right)\nu_1\right].$$

which will be either positive or negative under Eq.(4.3), therefore, Eq.(4.2) cannot have any limit cycle if Eq.(4.3) holds. Further, we suppose that

$$f\left(\frac{2}{a} + \frac{3f\nu_1\varepsilon^2}{a^2}\right)\nu_1 < 0 \tag{4.5}$$

Since, $\left(\frac{F'(x)}{ax}\right)' = \frac{\varepsilon}{ax^2}\left[-3fx^2 + \left(\frac{2}{a} + \frac{3f\nu_1\varepsilon^2}{a^2}\right)\nu_1\right]$, it follows from

Eq.(4.5) that $f\left(\frac{F'(x)}{ax}\right)' < 0$ for all $x \neq 0$.

Therefore, from Zhifen Zhang [8], Eq.(4.2) has at most one limit cycle under Eq.(4.5) and its stability is determined by the sign of $f$ if it exists. This completes the proof of theorem 4.1.

Remark 4.1: This theorem is true for all real $\varepsilon>0$, not just for $\varepsilon>0$ and small.

Remark 4.2: We can also obtain the uniqueness of a limit cycle for Eq.(1.4) in the case of $b = -\nu_2 = 1$ and $a<0$.

By applying the results from Section 2,3 and Theorem 4.1, we have the following theorem:

Theorem 4.2: Let $f\neq 0$, then there exists a smooth function $K(\varepsilon)$ for $0\le\varepsilon<\varepsilon_0$ with $K(0) = -\frac{3a^2f}{2(3a+2)}$, such that Eq.(1.5) has a unique limit cycle if and only if $fK(\varepsilon)<\nu_1 f<0$, and has a heteroclinic cycle if and only if $\nu_1 = K(\varepsilon)$.

Remark 4.3: For the special case of $a = 2$, our result coincides with Guckenheimer & Holmes' [5].

Remark 4.4: As we have seen, our methods in discussing Eq.(1.1) are all different from those in [2,3,4,5,7]. We also would like to point out that our methods have a great advantage in the following aspects: As we have mentioned at the beginning of this paper, the Eq.(1.1) comes from a three-dimensional system. However, if we restore all terms in the three-dimensional system, then we obtain a time periodic perturbation of Eq.(1.1). We expect that the periodic system should possess invariant tori. Note that our change of variables $\tilde{u} = u^{2/a}$, which makes Eq.(3.1) Hamiltonian, does not change the time parameter. This makes it possible for us to investigate the existence of invariant tori. On the contrast, the change of variables from Eq.(1.5) to Eq.(1.7) does change the time parameter. Because of that, the perturbation of Eq.(1.7) is no longer time periodic. Based on this paper, we will study the invariant tori for the complete three-dimensional system in the forthcoming paper.

## References

1. Broer, H.W and G. Vegter, "Subordinate Silnikov bifurcations near some singularities of vector fields having two codimension" Ergod. Th. & Dynam. Sys. 1984, 4, pp 509-525.

2. Carr, J., S.N. Chow and J.K. Hale, "Abelian Integrals and Bifurcation Theory", Journal of Diff. Eqs., 59 1985, pp 413-436.

3. Chow, S.N., J.K. Hale, Method of Bifurcation Theory, Springer-Verlag, New York, 1982.

4. Chow, S.N., C.Z. Li and D. Wang, "Uniqueness of periodic orbits of some vector fields with codimension two singularities" J. o Diff. Eqs. 77, 1989, pp 231-253.

5. Guckenheimer, J. and P. Holmes, Nonlinear Oscillation, Dynamical System and Bifurcation of Vector Fields, Spring-Verlag, New York, 1983.

6. Roussarie, R., "On the number of limit cycles which appear by perturbat of separatrix loop of planar vector fields", Bul. Soc. Bras. Mat. 17, 1986, pp 67-101.

7. Wiggins, S., Introduction to Applied Nonlinear Dynamical Systems and Chaos, Springer-Verlag, New York, 1990.

8. Zhang, Z.F., "Proof of the Uniqueness Theorem of Limit Cycles of Generalized Lienard Equations" Applicable Analysis, 23, 1986, pp 63-76.

# Parallelization of Initial Value Problem Codes for MIMD-DM Architectures

J. Kapenga
Western MIchigan Umiversity

and

I. Rattan
Central Michigan University

**Abstract**

Parallelization of numerical solutions for a class of IVPs, which are large and computationally intensive, is considered. It is indicated that a straightforward parallelization of existing IVP codes can sometimes be effective.

The problem we will address is the first order nonlinear system,

$$\frac{d}{dx}\bar{y} = \bar{f}(\bar{y},t)\ ,\ \ \bar{y}(0) = \bar{b}, \tag{1}$$

where $\bar{y}$ and $\bar{f}$ are of dimension N and $\bar{y}(T_f)$ is the desired solution. When a computed approximation for a vector is used, we will use a tilde in place of a bar over the vector, such as $\tilde{y}$ as an approximation of $\bar{y}$. We reserve $P$ to represent the number of processors to be used and $i$ will be an index into the processors.

Given $P \leq N$, we conformally partition $\bar{y}$, $\bar{f}$ and $\bar{b}$ as

$$\bar{y} = \begin{pmatrix} \bar{y}_0 \\ \bar{y}_1 \\ \vdots \\ \bar{y}_{P-1} \end{pmatrix} \quad \bar{f} = \begin{pmatrix} \bar{f}_0(\bar{y},t) \\ \bar{f}_1(\bar{y},t) \\ \vdots \\ \bar{f}_{P-1}(\bar{y},t) \end{pmatrix} \quad \bar{b} = \begin{pmatrix} \bar{b}_0 \\ \bar{b}_1 \\ \vdots \\ \bar{b}_{P-1} \end{pmatrix} \tag{2}$$

where the first $N\ mod\ P$ components are of dimension $\lceil N/P \rceil$ and the last $P - (N\ mod\ P)$ components are of dimension $\lfloor N/P \rfloor$. We denote the dimension of component $i$ as $N_i$.

$$\frac{d}{dx}\bar{y}_i = \bar{g}_i(\bar{y}_i,t)\ ,\ \ \bar{y}_i(0) = \bar{b}_i \tag{3}$$

$$\bar{g}_i(\bar{y}_i,t) \ = \ \bar{g}_i(\bar{y}_0, \bar{y}_1, \cdots, \bar{y}_{i-1}, \bar{y}_i, \bar{y}_{i+1}, \cdots, \bar{y}_{P-1}, t) \ = \ \bar{f}_i(\bar{y}) \tag{4}$$

We are interested in problems where $N$ is large, compared to $P$ and $\bar{f}$ is non-trivial to evaluate. We assume the partitioning above is carried out so that all $\bar{f}_i$ are approximately all equally costly to evaluate.

Applied problems in this class include: galactic dynamics, quantum molecular and some classical chemical dynamics (5,12), some electronic circuit simulations (4), as well as systems resulting in the methods of lines being applied to singular perturbation problems or PDEs.

## 1 Supercomputer Background

Classical vector supercomputers, led by the CRAY family, have been at the cutting edge of technology since their introduction in the mid 1970s. They have pushing clock rates to the limits available and required massive engineering to keep their large hand built circuts cool. Because of the cutting edge technology employed, they have always been very expensive. A single CRAY processor might have a peak performance of 600 Million FLOating Point instructions a Second (MFLOPS) and cost $20,000,000., that is 30 FLOPS per dollar.

Now compare this with a current state of the art Very Large Scale Integrated circuit (VLSI) CPU chip, such as a DEC Alpha, Intel i860, HP PA series or MIPS R4400. These chips keep their circuits on an single chip and require no exotic cooling. They produce a peak floating point rating of about 40-100 MFLOPS. The cost of the CPU chip itself is under $1000. That is 40,000-100,000 FLOPS per dollar. Putting 1000 VLSI microprocessors together in a single machine can potentially provide 100 times more performance than a classic vector supercomputer at one tenth the cost.

By the end of the 1980s this new approach for building future supercomputers was accepted by all major high performance computer manufactures as the most likely replacement for the classic supercomputer. This computer architecture is called Multiple Instruction Multiple Data (MIMD). There are two main classes on MIMD machines, shared memory (MIMD-SM) and distributed memory (MIMD-DM). The true MIMD-SM machines are effectively limited to about 30 processors, while commercial MIMD-DM machines have been built with many thousands of processors. We will address the MIMD-DM architecture.

In MIMD-DM machines each processor has its own memory and programs must coordinate the computation explicitly by sending and receiving data with other processors. A processor may only compute with data that is in its local memory.

## 2 The Performance of MIMD Supercomputers

To consider the parallelization of ODE solvers on MIMD-DM machines we will need to have models for both the floating point performance of the processors and a model for the communication time required for sending messages between processors. We will give three sets of parameters for a communication model in Figure 1. These are 1989, 1995 and 2001 technology. The model we take will be a communication cost of $c_l + c_b b$ seconds for transferring a message of size $b$ bytes between memory of any two processors, assuming no contention.

Starting with 1988 technology, the nCUBE-2 processor is a non-pipelined processor with a floating point unit able to do approximately 4 MFLOPS for add, subtract, multiply or divide. It uses a hypercube connection network topology to connect the processors. Hardware supported cut-through wornhole routing is used to transfer messages between processors. This implies that messages can be transferred between any two processors in the machine, as long as there is no contention in the communication network, with a time that is nearly independent of the distance between the processor in the network. This last property is shared by most current MIMD-DM machines and is the justification for our simple communication cost model.

The nCUBE-3 is scheduled for release in 1994. it has a scaler floating point unit capable of 100 MFLOPS and communication latency will be reduced by a factor of 10, while transfer bandwidth should go up by a factor of 4. This will be in the same range as the processors on several other commercial MIMD-DM supercomputers.

Estimating what will be available in VLSI microprocessors in 2001 is somewhat more risky. However it is fairly safe to expect another 10 fold increase in both floating point performance and communications.

As long as there is no contention, these communication models, work for most current machines, wormhole routed systems like the nCUBE, mesh connected machines such as Intel

| Year | Node Performance in $1/FLOPS$ | Communication Speed in seconds ($b$ is bytes) |
|---|---|---|
| | $c_f$ | $c_s + c_b b$ |
| 1989 | .00000025 (scaler) | $0.0001 + 0.0000004 * b$ |
| 1995 | .000000001 (scaler) | $0.00001 + 0.0000001 * b$ |
| 2001 | .0000000001 (vector) | $0.000001 + 0.00000001 * b$ |

Figure 1: Projected Performance for MIMD Supercomputers

Paragons and transputers and more exotic communication networks, like the KSR1 and CM5.

From these estimates one can infer that sometime between 1995 and 2001 there will be single chip microprocessors that will out perform any supercomputer now made. That is the driving force behind the interest in MIMD-DM architectures.

The challenge in using MIMD-DM supercomputers is to produce programs which can take advantage of their potential. Unlike the vector supercomputers, there currently are not effective parallel compilers which map serial code onto MIMD-DM machines. Writing a MIMD-DM program requires explicit care in the placement of data and processing on the processors of the machine. However, is not really too different from the care a CRAY programmer must take when he reorganizes his code to get maximum vectorization.

One aid in writing programs for MIMD-DM machines is the use of a Single Program Multiple Data (SPMD) programming style, where all the processors execute the same program. They may do somewhat different things by using their processor id as part of branching and indexing. A common approach used is to partition the data across the processors in an manner that both balances the computational loads on the processors and allows all processors to execute the same functions in an iterated Communicate/Compute loop. Note that the processors do not have to be executing the same portions of the program at the same time (as in a Single Instruction Multiple Data (SIMD) computer). They will all loosely synchronize at the communicate section of the iteration.

For this approach to be a reusable style of programming a set of well understood and efficient global communication primitives is required. These primitives should also scale, that is they should work with any number of available processors.

We will discuss a few global communication primitives, that are commonly used. We refer to these primitives as topologically independent because they do not explicitly refer to the network topology of the machine they are running on. Hence programming with these primitives helps abstract one away from many details of the parallel machine.

Assume there are $P$ nodes (processors) numbered $1, 2, \cdots P-1$, and let one node $i$ be designated as the root.

- Single Node Broadcast.
  START: node $i$ (the root) has data $d_i$.
  STOP: node $j$ has data $d_i$ for all nodes $j$.

- Single Node Accumulate.
  For a commutative and associative operator $< op >$
  START: node $j$ has data $d_j$ for all nodes $j$.
  STOP: node $i$ (the root) has the result, $d_0 < op > d_1 < op > \cdots < op > d_{P-1}$.

| Communication Primitive | Bound | Simpler Bound |
|---|---|---|
| Single Node Broadcast | $\lfloor \log P \rfloor (c_s + c_b B)$ | $(\log P)(c_s + c_b B)$ |
| Single Node Accumulate | $\lfloor \log P \rfloor (c_f F + c_s + c_b B)$ | $(\log P)(c_f F + c_s + c_b B)$ |
| Single Node Scatter | $\lfloor \log P \rfloor c_s + 2^{\lfloor \log P \rfloor} c_b B$ | $(\log P) c_s + P c_b B$ |
| Single Node Gather | same as Single Node Scatter | same as Single Node Scatter |
| Multiple Node Broadcast | $\lfloor \log P \rfloor 2c_s + (P + 2^{\lfloor \log P \rfloor}) c_b B$ | $2(\log P) c_s + 2P c_b B$ |

Figure 2: Global Communication Bounds for the Linear Array

- Single Node Scatter.
  START: node $i$ (the root) has data $d_0, d_1, \cdots, d_{P-1}$.
  STOP: node $j$ has data $d_j$ for all nodes $j$.
- Single Node Gather.
  START: node $j$ has data $d_j$ for all nodes $j$.
  STOP: node $i$ (the root) has data $d_0, d_1, \cdots, d_{P-1}$.
- Multiple Node Broadcast.
  START: node $j$ has data $d_j$ for all nodes $j$.
  STOP: node $j$ has data $d_0, d_1, \cdots, d_{P-1}$ for all nodes $j$.

Now to give a general form for the communication costs of these primitives we note that all current machines can be assumed to have a linear array embedded within their topology. While the linear array is not sufficient for many dense linear algebra operations, it is sufficient for bounding the above primitives. Optimal global communication routines on a specific machine may be considerably faster (1). The point is these upper bounds are expected to hold on any reasonable machine.

Given a linear array we can form a binary tree with the middle node as the root, recursively bisecting every half to get the left and right children. This tree is $\lfloor \log P \rfloor$ deep (all log s will be base 2). It also has the property that at every level all the paths between parents and children are distinct (they share no edges) and hence, communications moving up and down the tree in parallel have no contention.

We now make the assumptions that all data $d_i$ is of size $B$ bytes and the operator $< op >$ requires $F$ FLOPS. We also assume a node can transmit or receive two messages at the same time (otherwise the communication costs double, however in this case using a U-tree on a linear array reduces the costs back down to nearly these bounds). Then by using the above binary tree and the linear non-contention communication model, we have bounds for the global communication primitives are given in Figure 2.

## 3 Asynchronous Relaxation

The majority of the research into the problem of solving large systems of differential equations has been in PDEs, rather than ODEs. This work has been driven by the available computer technology. Through the 1970s and 1980s most efforts were directed at vectorization of solution methods on classical CRAY type vector supercomputers, which was the mainstay of the high performance scientific computing community. Several methods for parallelizing IVP solvers have been explored (2,3,9,10,14) One approach to the parallel numerical solution of IVPs on MIMD-DM machines is asynchronous relaxation (AR) (1,6,11). To apply AR to (1) we assign

Communicate: single node scatter processor $i$ gets $\bar{b}$, $t_f$ and $K$
$\tilde{y}^{(0)}(0) = \bar{b}$ ;
k = 0 ;
while(All Solutions Have not Converged) {
    Compute: processor $i$ computes $\tilde{y}_i^{(k+1)}(t)$ such that $0 \leq t \leq T$ from (3) and (4),
        Holding all $\tilde{y}_j^{(k)}$ such that $j \neq i$ fixed.
    Communicate: multiple node broadcast of $\tilde{y}_i^{(k+1)}$ returns $\tilde{y}^{(k+1)}$.
    $k = k + 1$
}

Figure 3: SPMD Asynchronous Relaxation

each processor $i$ the $i$th problem in Equation (3) and apply the Single Program Multiple Data (SPMD) procedure given in Figure 3.

In SPMD programs each processor executes an identical program. The processors perform different actions based of their processor id, for which we reserve the variable $i$. The SPMD program style is very powerful, and simplifies parallelization when it can be applied.

In presenting such an SPMD program we require all computations to be on variables available in the processors local memory and that access to values computed in other processors must be indicated with an explicitly shown communication.

Note that on each iteration $k$ processor $i$ will find a solution over the entire interval for a relaxed problem of dimension $N_i$, without communication using a standard serial ODE IVP solver. Also note the multinode broadcast is not of a single vector of size $N$, but rather a vcector of functions over the interval. There must be some representation (such as a spline) of the functions chosen and used. Also note the actual solution method applied in the computational step is not specified, any serial method can be applied.

The question is, does $\bar{y}^{(k)}$ converge to a solution of Equation (1)? This problem and conditions which insure convergence are addressed in (1,6,9). We note this approach is used even as a serial method, for solving ODEs that arise from some electronic circuit simulations is common practice (4). In particular for those applications, Parallel AR looks very promising.

In practice the Parallel AR methods would be applied over a sequence of subintervals, rather than the whole interval at one time. The Parallel AR method has advantages over the simple conformal decomposition we will see in the next section in that:

- It can balance the load between even a few processors when $N$, the number of equations, is about the same size as $P$.
- There is the potential for balancing the load between processors when some solutions are stable (and converge quickly) and others are not.
- It can be applied to problems with very easy to evaluate (sparse) $\bar{f}$ functions and still get effective parallelization.

Some problems with AR are that:

- Though AR may work, potentially a serial AR program may be very slow on a given problem, compared to other serial methods. So getting a good parallel AR of a parallelization of the serial AR program means it can run in near $1/P$th of the time of the serial AR program, which can still result in a slow program.
- The number of iterations that must be applied for convergence may be very large.
- Convergence may not take place at all. even for systems which could be solved by the serial solver used in the computation step if it was applied to the entire $N$ dimensional system.
- the communication requires a large amount of data be exchanged and stored by each processor, not a single vector of $N$ scalers.

## 4 Euler's Method

Now we start to discuss a simple parallelization of common ODE solver methods, which parallelizes the work required at each step of the serial method. We call this the conformal decomposition method. This results in the parallel algorithm computing exactly the same values as the serial algorithm would. Thus the questions of stability and convergence are answered by the theory from the serial method being parallelized. This is more appealing than the AR method, if it can be effectively applied to the problem.

To keep things simple we start by examining Euler's method (7,8). The basic method uses a fixed step size $h = T_f/K$ and computes approximate values of $\bar{y}$ at $t = jh$ for $j = 1,\ 2\ \cdots\ K$ in a forward manner, starting from $\tilde{y}(0) = \bar{b})$ and ending with the computed approximate solution $\tilde{y}(T_f)$. The iteration used is,

$$\tilde{y}(t+h) = \tilde{y}(t) + h * \bar{f}(\tilde{y}(t), t) \tag{5}$$

Now suppose we want processor $i$ to compute the update for $\tilde{y}_i$ at each step of the iteration. That is, it computes,

$$\tilde{y}_i(t+h) = \tilde{y}_i(t) + h * \bar{f}_i(\tilde{y}(t), t) \tag{6}$$

Note that that each processor requires the entire $N$ vector $\tilde{y}$ for the iteration and evaluates $N_i$ rows of $\bar{f}$ as well as $N_i$ additions and multiplications.

This leads to the parallel SPMD Euler algorithm in Figure 4. Assume that there are 4 bytes in floating point and integer values, then with our previous assumptions on global communication costs, the costs for this SPMD Parallel Euler Method are as follows (ignoring the initialization and assuming $\lfloor N/P \rfloor$ is approximately $N/P$:

- Space on each processor
  $N$ locations for $\tilde{y}(t)$ and $N_i$ locations for $\tilde{y}_i(t+h)$.
- Communication time: The $K$ multinode broadcasts require
  $K2(c_s \log P + c_b N)$
- Computation time: For each of $K$ steps
  Assume the parallel evaluations of $f_i$ requires $comp(N/P)$.
  Assume the serial evaluation of $f$ requires $comp(N)$.
  Let the constant $c$ be such that $comp(N/P) = c\frac{1}{p}comp(N)$
  (ideally $c = 1$ for perfect parallelization of this step.)

Communicate: single node scatter processor $i$ gets $\bar{b}$, $t_f$ and $K$
$\tilde{y}(0) = \bar{b}$ ;
$h = T_f/K$ ;
$t = 0$ ;
while($t < T_f$) {
    Compute: processor $i$ computes $\tilde{y}_i(t+h)$ from $\tilde{y}(t)$
    Communicate: multiple node broadcast of $\tilde{y}_i(t+h)$ returns $\tilde{y}(t+h)$ ;
    $t = t + h$ ;
}

Figure 4: SPMD Parallel Euler Method

- Speed Up
  The parallel time (computation + communication) of the parallel algorithm divided by the time for the serial algorithm is $\frac{c}{P} + \frac{2(c_s \log P + c_b N)}{comp(N)}$

This is quite reasonable, if $c$ is near 1 and we want 50% utilization on a 128 processor machine, which has the compute and communicate speed of today's machines, we require the serial evaluation of $\bar{f}$ to take about .1 second.

## 5 Runge Kutta Methods.

In moving toward more practical method than Euler's method, consider the classical Runge Kutta formula (7,8). As in Euler's method, classical Runge Kutta uses a fixed step size $h = T_f/K$ and computes approximate values of $\bar{y}$ at $t = jh$ for $j = 1,\ 2\ \cdots\ K$ in a forward manner, starting from $\tilde{y}(0) = \bar{b}$) and ending with the computed approximate solution $\tilde{y}(T_f)$. The iteration used is,

$$\bar{k}_1 = h * \bar{f}(\tilde{y}(t), t) \quad (7)$$

$$\bar{k}_2 = h * \bar{f}(\tilde{y}(t) + \frac{1}{2}\bar{k}_1, t + \frac{1}{2}h) \quad (8)$$

$$\bar{k}_3 = h * \bar{f}(\tilde{y}(t) + \frac{1}{2}\bar{k}_2, t + \frac{1}{2}h) \quad (9)$$

$$\bar{k}_4 = h * \bar{f}(\tilde{y}(t) + \bar{k}_3, t + h) \quad (10)$$

$$\tilde{y}(t+h) = \tilde{y}(t) + \frac{1}{6}(\bar{k}_1 + 2\bar{k}_2 + 2\bar{k}_3 + \bar{k}_4) \quad (11)$$

Again suppose we want processor $i$ to compute the update for $\tilde{y}_i$ at each step of the iteration as well as conformal parts of each $\bar{k}_i$. Note that that each processor requires the entire $N$ vector $\tilde{y}$ as well as each $\bar{k}_i$ (although these can each be overwritten as used).

This leads to the parallel SPMD Program program for the classical Runge Kutta algorithm in Figure 5. For the classical Runge Kutta method the Adjust step size phase is empty.

It should be clear that the same analysis of the performance of the SPMD Euler Method holds for the SPMD classical Runge Kutta method. Furthermore, extending the classical Runge Kutta method to a practical method, such as RK45, only adds a few more Communicate / Compute steps in each loop iteration (two for RK45) and an actual operation in the Adjust

Communicate: single node scatter processor $i$ gets $\bar{b}$, $t_f$ and $K$
$\tilde{y}(0) = \bar{b}$ ;
$h = T_f/K$ ;
$t = 0$ ;
while($t < T_f$) {
    Compute: processor $i$ computes $\bar{k}_{1,i}$ from $\tilde{y}(t)$ ;
    Communicate: multiple node broadcast of $\bar{k}_{1,i}$ returns $\bar{k}_1$ ;
    Compute: processor $i$ computes $\bar{k}_{2,i}$ from $\tilde{y}(t)$ and $\bar{k}_1$;
    Communicate: multiple node broadcast of $\bar{k}_{2,i}$ returns $\bar{k}_2$ ;
    Compute: processor $i$ computes $\bar{k}_{3,i}$ from $\tilde{y}(t)$ and $\bar{k}_2$;
    Communicate: multiple node broadcast of $\bar{k}_{3,i}$ returns $\bar{k}_3$ ;
    Compute: processor $i$ computes $\bar{k}_{4,i}$ from $\tilde{y}(t)$ and $\bar{k}_3$;
    Communicate: multiple node broadcast of $\bar{k}_{4,i}$ returns $\bar{k}_4$ ;
    Compute: processor $i$ computes $\tilde{y}_i(t+h)$ ;
    Communicate: multiple node broadcast of $\tilde{y}_i(t+h)$ returns $\tilde{y}(t+h)$ ;
    Adjust step size.
    $t = t + h$ ;
}

Figure 5: SPMD classical Runge Kutta

step size phase (8). This adjust step size step can be done completely in parallel with virtually no extra communication an extra by adding an additional piece of computed data to the multinode broadcast for $\tilde{y}(t+h)$.

## 6 Predictor Corrector Methods

Implicit methods have several advantages over the explicit methods such as Runge Kutta, and must be considered for stiff problems (7,8,15). Perhaps the most common implicit methods are of the Predictor Corrector type.

In codes of Predictor Corrector Methods, an implicit equation needs to be solved for the next $\bar{y}(t+h)$. This is usually structured as a predictor step (PE) and one or more corrector steps (CE).

The parallelization of the PE phase is in general identical to the Runge Kutta parallelization. The parallelization of the CE phase is dependent on what is being used as a corrector. There are three common types, functional iteration, Newton's method with explicit derivatives and Newton's method without explicit derivatives. The CE steps are employed either a fixed number of times or until some convergence criteria is met. This difference does not effect the parallelization in a significant manner.

Functional Iteration CE steps use an evaluation of $\bar{f}$ at each iterate. They parallelize exactly like an Euler step.

Newton's method with explicit derivatives must be able to compute the Jacobian

$$F' = F'(\tilde{y}^{(k)}(t+h)) \tag{12}$$

matrix explicitly and use it in solving a system of the form

$$F'\bar{d}^{(k)} = \bar{e}^{(k)} \tag{13}$$

If the independent evaluation of the conformal pieces of of $\bar{f}$ is effective across the $P$ processors, then it is most likely that letting processor $i$ evaluate a conformal set of rows of $F'$ will be an effective parallelization. As to the parallelization of the solution of the linear system, several effective methods are known, both direct and iterative. These tend to provide excellent parallelization in the cases we are interested in ($N$ several times larger than $P$). (1,13,14).

In the Newton Method with explicit derivatives case, the total parallelization will tend to be even more effective than that of the Runge Kutta methods, because of the much larger amount of work required in forming $F'$ and solving the linear system. We note that most practical codes do not reevaluate $F'$ at each iteration, but only when it seems needed. This causes no additional problems in parallelization.

Newton's Method without explicit derivatives requires the program to supply an approximation for $F'$ as needed, based on values of $\bar{f}$, A very effective parallelization follows if the full $F'$ approximation is evaluated using the second order difference approximations from of the values

$$\bar{f}(\bar{y}^{(k)}(t+h) + e\bar{e}_i, t+h), e = -\gamma, 0, \gamma \tag{14}$$

where $e_i$ is 1 in the $i$th position and 0 elsewhere. Some codes that use approximations to $F'$ do not explicitly compute the above points and only update the $F'$ when needed. The choice of update time appears to pose no problem, the actual parallelization of $F'$ updates that don't use the full set of points above must be considered on an algorithm by algorithm basis, and looks like a very interesting problem to investigate.

Again note that the adaptation of step size poses no additional problems for parallelization. The parallelization of implicit methods is even more effective than the parallelization of the explicit methods because of the extra high grained work that must be done. This work will dominate the communication even more than in the case of explicit methods.

## 7 Summary

We have indicated some conditions on problems when parallelized IVP codes using the conformal decomposition can be effective and indicated the codes for implicit methods should parallelize even better than codes for explicit methods. This class of problems includes some of applied interest, which are computationally demanding enough to warrant attention with respect to parallelization.

Some basic conditions that insure effective conformal parallelization are;

- The evaluation of any $\bar{f}_i$ takes about $N/P$ of the time for the full evaluation of $\bar{f}$, as a serial computation.

- For machines with the same ratio of computation to communication speed as current machines, it requires about $1000PlogP + PN$ floating point operation in the evaluation of $\bar{f}$ by a serial program.

The effective parallelization of advanced implicit methods for large IVP systems is even more promising than that of explicit methods. In that case the condition of the cost in evaluating $\bar{f}$ can be relaxed. In the case of a Newton Corrector using explicit derivatives, the condition may be replaced by something of the form $N > cP$ where $c$ is a small constant (perhaps 10 on current machines). This follows because the CE step now dominates the computational cost.

The practical parallelization of the solution of small systems of ODE IVP problems is much more difficult to address in general. Although relaxation methods have been proposed, little practical experience is available.

The analysis and implementation of conformally decomposed parallel implicit methods appears to be worthy of further investigation.

## References

(1) Parallel and Distributed Computation, Numerical Methods, D. Bertsekas and J. Tsitsiklis, Prentice Hall, 1989

(2) The Potential for Parallelization in Ordinary Differintial Equations, C. Gear, Report No. UIUCDCS-R-86-1246, University of Illinois at Urbana-Champaign, 1986

(3) Parallel Methods for Ordinary Differential Equations, C. Gear, Report No. UIUCDCS-R-87-1369, University of Illinois at Urbana-Champaign, 1987

(4) Relaxation-Based Electrical Simulation, A. Newton and A. Sangiovanni-Vincentelli, IEEE Trans. on Electronic Devices, ED-30, pp 1184-1207, 1983

(5) Distributed Molecular Dynamics Using the PVM System, H. Simonsen and J. Amundsen, Proc. of the Sixth SIAM Conf. on Parallel Processing for Sci. Comp., SIAM, pp 183-186, 1993

(6) Asynchronous Relaxiation for the Numerical Solution of Ordinary Differential Equations by Parallel Computers. D. Mitra, SIAM J. Sci. Stat. Comput., No 8, pp 43-58, 1987

(7) Numerical Methods and Software, D. Kahaner, C. Moler adn S. Nash, Prentice Hall, 1977

(8) The Numerical Analysis of Ordinary Differential Equations, J. Butcher, Wiley, 1987

(9) Parallel Methods for ODEs With Improved Absolute Stability Boundaries, P. Worland, J. Parallel and Dist. Comp., no. 18, pp 25-32, 1993

(10) Parallel Algorithms for Initial Value Problems: Parallel Shooting, B. Khalaf and D. Hutchenson, Parallel Computing, no. 18, pp 661-673, 1992

(11) Parallel Methods for Integrating Ordinary DIfferential Equations, J. Nievergelt, Comm ACM, 7(12), pp 731-733, 1964

(12) Stress and Strain Analysis in Molecular Dynamics Modeling, P. Scagnetti, T. Bifano, R. Nagem and G. Sandri, Proc. of the Sixth SIAM Conf. on Parallel Processing for Sci. Comp., SIAM, pp 203-210, 1993

(13) LAPACK for Distriibuted Memory Architectures, The Next Generation, J. Demmel, J,. Dongarra, R. van de Geijn, and D. Walker. Proc. of the Sixth SIAM Conf. on Parallel Processing for Sci. Comp., SIAM, pp 183-186, 1993

(14) Scientific Computing, An Introduction with Parallel Processing, G. Gulob and J. Ortega, Academic Press, 1993

(15) L-stable Parallel One-Block Methods for Stiff ODEs, P. Chartier and B. Philippe, Proc. of the Sixth SIAM Conf. on Parallel Processing for Sci. Comp., SIAM, pp 743-746, 1993

# ON HYPERGEOMETRIC FUNCTIONS OF CONFLUNT TYPE OF SEVERAL VARIABLES

HIONOBU KIMURA
*Department of Mathematical Sciences, University of Tokyo, Tokyo 153, Japan*

YOSHISHIGE HARAOKA
*Department of Mathematics, Kumamoto University, Kumamoto 860, Japan*

and

KYOICHI TAKANO
*Department of Mathematics, Kobe University, Kobe 657, Japan*

## 1. Introduction

In this talk, we give a generalization of hypergeometric functions of conflunet type to the functions of several variables. Our generalization relates to Aomoto-Gelfand hypergeometric function ([A],[G]) as the confluent hypergeometric function of Kummer does to the Gauss hypergeometric function. Before proceeding to the definition of our functions, we recall the explicit form of integral representations of the Gauss hypergeometric function and well known hypergeometric functions of confluent type, and the differential equations satisfied by them.

1) Gauss hypergeometric function: The power series

$$F(a,b,c;x) = \sum_{k=0}^{\infty} \frac{(a)_k(b)_k}{(c)_k(1)_k} x^k, \quad (a)_k = a(a+1)\cdots(a+k-1)$$

converges in $|x| < 1$ and the defined function is called *Gauss hypergeometric function.* The function $F(a,b,c;x)$ satisfies the famous Gauss hypergeometric equation

$$x(1-x)\frac{d^2\Phi}{dx^2} + \{c-(a+b+1)x\}\frac{d\Phi}{dx} - ab\Phi = 0 \tag{1.1}$$

and has an integral representation

$$F(a,b,c;x) = \frac{\Gamma(c)}{\Gamma(a)\Gamma(c-a)} \int_0^1 u^{a-1}(1-u)^{c-a-1}(1-xu)^{-b} du, \tag{1.2}$$

where $\Gamma(z)$ is the gamma function of $z$.

2) Kummer's confluent hypergeometric function: We make a change of variables and parameters

$$x = \epsilon\xi, \quad b = 1/\epsilon, \quad \Psi(\xi) = \Phi(\epsilon\xi).$$

Then the equation for $(\xi, \Psi)$ is

$$\xi(1-\epsilon\xi)\frac{d^2\Psi}{d\xi^2} + \{c - \epsilon(a+\epsilon^{-1}+1)\xi\}\frac{d\Psi}{d\xi} - a\Phi = 0 \tag{1.3}$$

and the coefficients of $d^2\Psi/d\xi^2$, $d\Psi/d\xi$ and $\Psi$ depend holomorphically on $\epsilon$ at $\epsilon = 0$. Putting $\epsilon = 0$ in Eq. (1.3) and writing again $x, \Phi$ in place of $\xi, \Psi$ in the equation thus obtained, we have

$$x\frac{d^2\Phi}{dx^2} + (c-x)\frac{d\Phi}{dx} - a\Phi = 0. \tag{1.4}$$

The above limit process gives also an integral representation of solutions of Eq. (1.4)

$$\Phi(x) = \text{const.} \times \int_\Delta u^{a-1}(1-u)^{c-a-1}\exp(xu)du \tag{1.5}$$

by choosing an appropriate path in $\mathbb{C}$.

3) Bessel function: In (1.4), put $a = \nu - 1/2$, $c = 2\nu + 1$ and $\Psi(x) = e^{-ix}x^\nu\Phi(2ix)$. Writing again $\Phi$ instead of $\Psi$ we obtain the Bessel equation:

$$x^2\frac{d^2\Phi}{dx^2} + x\frac{d\Phi}{dx} + (x^2 - \nu^2)\Phi = 0. \tag{1.6}$$

Its solutions are given by

$$\Phi(x) = \text{const.} \times \int_\Delta u^{-\nu-1}\exp(\frac{1}{2}(u - 1/u)x)du. \tag{1.7}$$

4) Hermite function:

$$\frac{d^2\Phi}{dx^2} - 2x\frac{d\Phi}{dx} - 2n\Phi = 0. \tag{1.8}$$

The solutions are given by

$$\Phi(x) = \text{const.} \times \int_{\Delta} u^{-n-1} \exp(2ux - u^2) du. \tag{1.9}$$

5) Airy function:

$$\frac{d^2\Phi}{dx^2} - x\Phi = 0. \tag{1.10}$$

The solutions are given by

$$\Phi(x) = \text{const.} \times \int_{\Delta} \exp(\frac{1}{3}u^3 - xu) du. \tag{1.11}$$

In the following sections, we introduce the generalized confluent hypergeometric function and show that the classical hypergeometric functions of confluent type, reviewed above, are obtained as special cases of our function.

## 2. Confluent hypergeometric systems

Let $r$ and $n$ $(> r)$ be positive integers and let $Z_{r,n}$ be the set of $r \times n$ complex matrices of maximal rank. The element is denoted by $z = (z_{ip})_{0 \le i \le r-1, 0 \le p \le n-1} \in Z_{r,n}$.

There is given a composition $\lambda = (\lambda_0, \dots, \lambda_{\ell-1})$ of $n$, i.e. the sequence of positive integers satisfying $\lambda_0 + \dots + \lambda_{\ell-1} = n$ and $\lambda_k$ $(k = 0, \dots, \ell - 1)$ are not necessarily arrayed in descending or ascending order. We associate a composition $\lambda$ of $n$ with a diagram defined as the set of points $(k, m) \in \mathbb{Z}^2$ such that $0 \le m \le \lambda_k - 1$. In drawing such diagram we adopt the convention that the first coordinate $k$ (the row index) increases as one goes downwards, and the second coordinate $m$ (the column index) increases as one goes from left to right, and we place a square at each point $(k, m) \in \mathbb{Z}^2$. The number of squares of the diagram $\lambda$ will be called the weight of $\lambda$ and will be denoted by $|\lambda|$.

**Definition 2.1.** There are given a diagram $\lambda = (\lambda_0, \cdots, \lambda_{\ell-1})$ of weight $n$ and a constant column vector $\alpha = {}^t(\alpha^{(0)}, \dots, \alpha^{(\ell-1)})$, $\alpha^{(k)} = (\alpha_0^{(k)}, \alpha_1^{(k)}, \dots, \alpha_{\lambda_k-1}^{(k)})$ $(k = 0, \dots, \ell - 1)$, of dimension $n$ satisfying $\sum_{k=0}^{\ell-1} \alpha_0^{(k)} = -r$. *The confluent hypergeometric system of type* $\lambda$ (CHG system for short) is the system of linear partial differential equations:

$$\begin{cases} L_m^{(k)} u = \alpha_m^{(k)} u, & k = 0, \dots, \ell - 1;\ m = 0, \dots, \lambda_k - 1 \\ M_{ij} u = -\delta_{ij} u, & i, j = 0, \dots, r - 1 \\ \square_{ij,pq} u = 0 & i, j = 0, \dots, r - 1;\ p, q = 0, \dots, n - 1 \end{cases} \tag{$M_\lambda$}$$

where

$$L_m^{(k)} = \sum_{q=0}^{r-1} \sum_{\substack{\lambda_1+\cdots+\lambda_{k-1}+m \\ \leq p \leq \lambda_1+\cdots+\lambda_k-1}} z_{q,p-m} \frac{\partial}{\partial z_{qp}},$$

$$M_{ij} = \sum_{p=0}^{n-1} z_{ip} \frac{\partial}{\partial z_{jp}},$$

$$\Box_{ij,pq} = \frac{\partial^2}{\partial z_{ip} \partial z_{jq}} - \frac{\partial^2}{\partial z_{iq} \partial z_{jp}},$$

$\delta_{ij}$ being the Kronecker's symbol. We denote the system by $M_\lambda(\alpha)$ or $M_\lambda(z;\alpha)$, if it is necessary to indicate the constants $\alpha$ or, moreover, the independent variables $z = (z_{ip})$.

**Theorem 2.2.** *The generalized confluent hypergeometric system $M_\lambda$ is holonomic.*

This theorem implies that the space of solutions for $M_\lambda$ at a generic point of $Z_{r,n}$ is isomorphic to a finite dimensional complex vector space. Any solution of the system $M_\lambda$ is called *the generalized confluent hypergeometric function of type* $\lambda$ (CHG function for short).

Before stating the properties of the CHG functions of type $\lambda$, let us explain several terminologies.

Let $J(m)$ be the Jordan group of size $m$, namely, a maximal abelian Lie subgroup of $GL(m, \mathbb{C})$ defined by

$$J(m) := \left\{ c = \sum_{i=0}^{m-1} c_i \Lambda^i; \quad c_i \in \mathbb{C}, \quad c_0 \neq 0 \right\},$$

where $\Lambda = \Lambda_m = (\delta_{i+1,j})_{0 \leq i,j \leq m-1} \in M(m, \mathbb{C})$. The group $J(m)$ is isomorphic to the group $\mathbb{C}^\times \times \mathbb{C}^{m-1}$, where $\mathbb{C}^{m-1}$ is equipped with the natural additive structure. In fact, this isomorphism is establised by associating $c = \sum_{i=0}^{m-1} c_i \Lambda^i$ with $\left(c_0, \theta_1(c/c_0), \ldots, \theta_{m-1}(c/c_0)\right)$. Here

$$\theta_i(x) := \theta_i(x_1, \ldots, x_i), \quad i \geq 1,$$

are polynomials of $x = (x_1, x_2, x_3, \ldots)$ defined by

$$\sum_{i=1}^{\infty} \theta_i(x) t^i = \log(1 + x_1 t + x_2 t^2 + \cdots).$$

We note that if the weight of $x_k$ is defined to be $k$, $\theta_i(x)$ ($i \geq 1$) is a weighted homogeneous polynomial of $x_1, \ldots, x_i$ of weight $i$. It follows from the explicit form of the isomorphism $J(m) \to \mathbb{C}^\times \times \mathbb{C}^{m-1}$ that a character $\chi : \tilde{J}(m) \to \mathbb{C}^\times$ of the universal covering group $\tilde{J}(m)$ of $J(m)$ is written as

$$\chi(c) = c_0^{\alpha_0} \exp\left(\sum_{i=1}^{m-1} \alpha_i \theta_i(c/c_0)\right), \quad c = \sum_{i=0}^{m-1} c_i \Lambda^i$$

for some $\alpha = {}^t(\alpha_0, \cdots, \alpha_{m-1}) \in \mathbb{C}^m$. This character will be denoted by $\chi(\,\cdot\,;\alpha)$ to indicate the dependence on $\alpha$.

For a diagram $\lambda = (\lambda_0, \ldots, \lambda_{\ell-1})$ of weight $n$, we consider an abelian Lie subgroup $H_\lambda = J(\lambda_0) \times \cdots \times J(\lambda_{\ell-1}) \subset GL(n, \mathbb{C})$ of dimension $n$ and its universal covering group $\tilde{H}_\lambda$. Let $\alpha = {}^t(\alpha^{(0)}, \ldots, \alpha^{(\ell-1)}) \in \mathbb{C}^n$ be the constant column vector given in Definition 2.1. Then $\chi_\lambda(\,\cdot\,;\alpha) : \tilde{H}_\lambda \to \mathbb{C}^\times$ defined by $\chi_\lambda(c;\alpha) = \prod_k \chi(c^{(k)};\alpha^{(k)})$ for

$$c = (c^{(0)}, \cdots, c^{(\ell-1)}) \in \tilde{H}_\lambda, \quad c^{(k)} = \sum_{i=0}^{\lambda_k - 1} c_i^{(k)} \Lambda_{\lambda_k}^i \in \tilde{J}(\lambda_k)$$

is a character of $\tilde{H}_\lambda$.

As one of the important properties for the CHG functions, we have

**Proposition 2.3.** *A generalized confluent hypergeometric function $\Phi_\lambda(z;\alpha)$ of type $\lambda$ satisfies*

$$\begin{cases} \Phi_\lambda(zc;\alpha) = \Phi_\lambda(z;\alpha)\chi_\lambda(c;\alpha) & \text{for } c \in H_\lambda. \\ \Phi_\lambda(gz;\alpha) = (\det g)^{-1}\Phi_\lambda(z;\alpha) & \text{for } g \in GL(r,\mathbb{C}) \ . \end{cases} \tag{2.1}$$

The first and the second properties in (2.1) are called the *$H_\lambda$-homogeneity* and the *$GL(r, \mathbb{C})$-homogeneity* for $\Phi_\lambda(z;\alpha)$, respectively.

*Remark 2.4.* The Eqs. $L_m^{(k)} u = \alpha_m^{(k)} u$ and $M_{ij} u = -\delta_{ij} u$ are infinitesimal expressions for the $H_\lambda$-homogeneity and $GL(r, \mathbb{C})$-homogeneity for CHG function, respectively.

*Remark 2.5.* In the case where the composition of $n$ is $\lambda = (1, \ldots, 1)$ the system $M_\lambda$ coincides with the Aomoto-Gel'fand's hypergeometric system ([G]) and in the case where $\lambda = (n)$, the system $M_\lambda$ has been already introduced in [GRS].

## 3. Integral representation

The CHG system $M_\lambda$ admits solutions in an integral representation. Define a biholomorphic mapping

$$\iota_\lambda : \ H_\lambda \to \prod_{k=0}^{\ell-1} (\mathbb{C}^\times \times \mathbb{C}^{\lambda_k - 1}) \subset \mathbb{C}^n$$

by $\iota_\lambda(c) = (c_0^{(0)}, \dots, c_{\lambda_0-1}^{(0)}, \dots, c_0^{(\ell-1)}, \dots, c_{\lambda_{\ell-1}-1}^{(\ell-1)})$ for $c = \bigoplus_{k=0}^{\ell-1} \sum_{i=1}^{\lambda_k-1} c_i^{(k)} \Lambda_{\lambda_k}^i \in H_\lambda \subset GL(n, \mathbb{C})$. For $t = (t_0, \dots, t_{r-1}) \in \mathbb{C}^r$ and for $z = (z_0, \dots, z_{n-1}) \in Z_{r,n}$, we set

$$\langle t, z \rangle = (\langle t, z_0 \rangle, \dots, \langle t, z_{n-1} \rangle), \quad \langle t, z_j \rangle = \sum_{i=0}^{r-1} t_i z_{ij} \quad (0 \le j \le n-1).$$

Let $\omega$ be the $(r-1)$-form defined by

$$\omega := \sum_{i=0}^{r-1} (-1)^i t_i dt_0 \wedge \cdots \wedge dt_{i-1} \wedge dt_{i+1} \wedge \cdots \wedge dt_{r-1}.$$

**Proposition 3.1.** *For an appropriate $(r-1)$-dimensional cycle $\Delta$ in $\mathbb{C}^r$, the integral*

$$\Phi_\lambda(z; \alpha) = \int_\Delta \chi_\lambda(\iota_\lambda^{-1}(\langle t, z \rangle); \alpha)\, \omega$$

*gives a solution of the CHG system $M_\lambda$.*

## 4. Finite group symmetries

There is a trivial symmetry for the CHG systems. To describe it we change the manner of indexing of $z \in Z_{r,n}$ as

$$z = (z^{(0)}, \dots, z^{(\ell-1)}), \quad z^{(k)} = (z_0^{(k)}, z_1^{(k)}, \dots, z_{\lambda_k-1}^{(k)}) \quad (k = 0, \dots, \ell-1).$$

Let $\mathfrak{S}_\ell$ be the symmetric group acting on the set $\{0, \dots, \ell-1\}$ and define the action of $\sigma \in \mathfrak{S}_\ell$ on $(\lambda, \alpha, z)$ by

$$\begin{aligned} \sigma \cdot \lambda &= (\lambda_{\sigma(0)}, \dots, \lambda_{\sigma(\ell-1)}), \\ \sigma \cdot \alpha &= (\alpha^{(\sigma(0))}, \dots, \alpha^{(\sigma(\ell-1))}), \\ \sigma \cdot z &= (z^{(\sigma(0))}, \dots, z^{(\sigma(\ell-1))}). \end{aligned}$$

**Proposition 4.1.** *Let $\lambda = (\lambda_0, \dots, \lambda_{\ell-1})$ be a diagram of weight $n$ and let $\sigma \in \mathfrak{S}_\ell$. Then the change of variables $(\alpha, z) \mapsto (\alpha', z') = (\sigma \cdot \alpha, \sigma \cdot z)$ takes the CHG system $M_\lambda(\alpha)$ to the system $M_{\sigma \cdot \lambda}(\sigma \cdot \alpha)$.*

This trivial result says that if the diagram $\mu$ is obtained from $\lambda$ by permuting its rows, the systems $M_\lambda$ and $M_\mu$ are essentially the same. Note that, even though this result looks trivial, it gives, for example, a well known non trivial transformation formula for the Kummer's confluent hypergeometric function.

## 5. CHG functions on $Z_{2,4}$ and the classical confluent hypergeometric functions

We assert that the CHG functions on $Z_{2,4}$ with various composition of weight 4 give the classical hypergeometric functions of confluent type given in Introduction. The compositions (or rather, partitions) of weight 4 are listed as

$$(1,1,1,1), \quad (2,1,1), \quad (2,2), \quad (3,1), \quad (4).$$

We restrict $z \in Z_{2,4}$ on an appropriate one dimensional subspace $z = z(x)$, where $x$ denotes the coordinate of this one dimensional subspace.

1) $\lambda = (1,1,1,1)$. Set

$$z(x) = \begin{pmatrix} 1 & 1 & 0 & 1 \\ 0 & -x & 1 & -1 \end{pmatrix}, \quad \alpha = (b-c, -b, a-1, c-a-1).$$

Then $\langle t, z(x) \rangle = (t_0, t_0 - xt_1, t_1, t_0 - t_1)$ and

$$\begin{aligned} \Phi_\lambda(z(x);\alpha) &= \int t_0^{b-c}(t_0 - xt_1)^{-b} t_1^{a-1}(t_0-t_1)^{c-a-1}(t_0 dt_1 - t_1 dt_0) \\ &= \int u^{a-1}(1-u)^{c-a-1}(1-xu)^{-b} du. \end{aligned}$$

Here and hereafter we set $u = t_1/t_0$. This is the Gauss hypergeometric function.

1) $\lambda = (2,1,1)$. Set

$$z(x) = \begin{pmatrix} 1 & 0 & 0 & 1 \\ 0 & -x & 1 & -1 \end{pmatrix}, \quad \alpha = (-c, -1, a-1.c-a-1).$$

Then $\langle t, z(x) \rangle = (t_0, -t_1 x, t_1, t_0 - t_1)$ and

$$\begin{aligned} \Phi_\lambda(z(x);\alpha) &= \int t_0^{-c} \exp\left(\frac{t_1}{t_0} x\right) t_1^{a-1}(t_0-t_1)^{c-a-1}(t_0 dt_1 - t_1 dt_0) \\ &= \int e^{xu} u^{a-1}(1-u)^{c-a-1} du. \end{aligned}$$

This is Kummer's confluent hypergeometric function.

3) $\lambda = (2,2)$. Set

$$z(x) = \begin{pmatrix} 1 & 0 & 0 & x \\ 0 & x & 1 & 0 \end{pmatrix}, \quad \alpha = (\nu - 1, 1/2, -\nu - 1, -1/2).$$

Then

$$\Phi_\lambda(z(x);\alpha) = \int t_0^{\nu-1}\exp\left(\frac{1}{2}\frac{t_1}{t_0}x\right)t_1^{-\nu-1}\exp\left(-\frac{1}{2}\frac{t_0}{t_1}x\right)(t_0dt_1 - t_1dt_0)$$
$$= \int \exp\left(-\frac{x}{2}\left(u - \frac{1}{u}\right)\right)du.$$

This is the Bessel function.

4) $\lambda = (3,1)$. Set

$$z(x) = \begin{pmatrix} 1 & 0 & 0 & 0 \\ 0 & 1 & x & 1 \end{pmatrix}, \quad \alpha = (n-1, 0, 2, -n-1).$$

Then

$$\Phi_\lambda(z(x);\alpha) = \int t_0^{n-1}\exp\left(2\theta_2(t_1/t_0, xt_1/t_0)\right)t_1^{-n-1}(t_0dt_1 - t_1dt_0)$$
$$= \int \exp(2ux - u^2)u^{-n-1}du.$$

This is the Herminte function.

5) $\lambda = (4)$. Set

$$z(x) = \begin{pmatrix} 1 & 0 & 0 & 1 \\ 0 & 1 & 0 & -x \end{pmatrix}, \quad \alpha = (-2, 0, 0, 1).$$

Then

$$\Phi_\lambda(z(x);\alpha) = \int t_0^{-2}\exp\left(\theta_3(t_1/t_0, 0, 1 - xt_1/t_0)\right)(t_0dt_1 - t_1dt_0)$$
$$= \int \exp\left(\frac{1}{3}u^3 - ux\right)du.$$

This is the Airy function.

## 6. Confluence for CHG systems

**Definition 6.1.** Let $\lambda$ and $\mu$ be diagrams of weight $n$. The diagram $\mu$ is said to be *adjacent* to $\lambda$ if $\mu$ is obtained from $\lambda$ by making some two rows of $\lambda$ into a single row whose length is the sum of those of the two rows of $\lambda$. We denote this relation by the symbol $\lambda \to \mu$.

**Example 6.2.** For the diagrams of weight 4, the following figure describes all the relations of adjacency among them modulo permutations of rows in the diagrams.

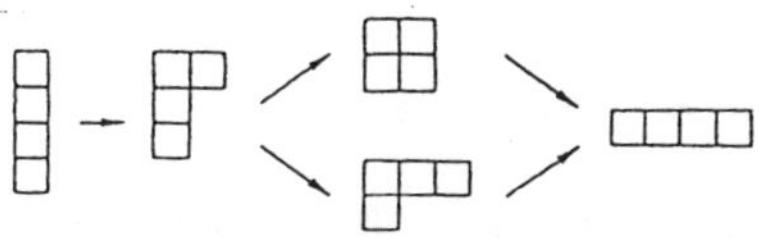

Let $\mu = (\mu_0, \dots, \mu_{\ell-2})$ be adjacent to $\lambda = (\lambda_0, \dots, \lambda_{\ell-1})$ of weight $n$. Taking account of Proposition 4.1 and of the remark which follows it, we may assume that the rows in the diagram $\lambda$, which are put togather to make $\mu$, are arrayed consecutively in $\lambda$, namely, $\mu_1 = \lambda_1, \dots,\ \mu_{h-1} = \lambda_{h-1},\ \mu_h = \lambda_h + \lambda_{h+1},\ \mu_{h+1} = \lambda_{h+2}, \dots,\ \mu_{\ell-2} = \lambda_{\ell-1}$ for some $0 \le h \le \ell - 2$. We define a holomorphic mapping $g = g_{\lambda\to\mu} : \mathbb{C}\setminus\{0\} \to GL(n, \mathbb{C})$ by $g(\epsilon) = (g_{ij}(\epsilon))_{0\le i,j\le \ell-1}$, where $g_{ij}(\epsilon)$ is a $\lambda_i \times \lambda_j$ matrix given by

$$
\begin{aligned}
&g_{ii}(\epsilon) = I_{\lambda_i}, \qquad i \ne h+1, \quad g_{ij}(\epsilon) = 0, \quad \text{for } i \ne j \text{ and } (i,j) \ne (h, h+1),\\
&\begin{pmatrix} g_{h,h+1}(\epsilon) \\ g_{h+1,h+1}(\epsilon) \end{pmatrix}\\
&= D_{\lambda_h+\lambda_{h+1}}(\epsilon) \begin{pmatrix} \binom{0}{0} & \binom{0}{1} & \cdots & \binom{0}{\lambda_{h+1}-1} \\ \binom{1}{0} & \binom{1}{1} & \cdots & \binom{1}{\lambda_{h+1}-1} \\ \vdots & \vdots & \cdots & \vdots \\ \binom{\lambda_h+\lambda_{h+1}-1}{0} & \binom{\lambda_h+\lambda_{h+1}-1}{1} & \cdots & \binom{\lambda_h+\lambda_{h+1}-1}{\lambda_{h+1}-1} \end{pmatrix} D_{\lambda_{h+1}}(\epsilon^{-1}),
\end{aligned}
$$

$D_m(\epsilon)$ denoting $\mathrm{diag}(1, \epsilon, \epsilon^2, \dots, \epsilon^{m-1})$. Remark that $\det g(\epsilon) = \epsilon^{\lambda_h\lambda_{h+1}}$. Define $H_{\lambda\to\mu,\epsilon}$ by

$$H_{\lambda\to\mu,\epsilon} := \left\{\iota_\lambda^{-1}(\iota_\mu(c)g(\epsilon)) \in H_\lambda\ ;\ c \in H_\mu\right\},$$

where $g(\epsilon) = g_{\lambda\to\mu}(\epsilon)$. Then we have

**Theorem 6.3 (confluence for $H_\lambda$ ).** *It holds that*

$$\lim_{\epsilon\to 0} Ad_{g(\epsilon)} H_{\lambda\to\mu,\epsilon} = H_\mu,$$

*more precisely, for every $c \in H_\mu$,*

$$Ad_{g(\epsilon)}\iota_\lambda^{-1}(\iota_\mu(c)g(\epsilon)) = g(\epsilon)\iota_\lambda^{-1}(\iota_\mu(c)g(\epsilon))g(\epsilon)^{-1} \to c$$

*as $\epsilon \to 0$.*

**Theorem 6.4 (confluence for character).** *For every $c \in H_\mu$ and every column vector $\beta \in \mathbb{C}^n$, we have*

$$\lim_{\epsilon \to 0} \chi_\lambda\left(\iota_\lambda^{-1}(\iota_\mu(c)g(\epsilon)); g(\epsilon)^{-1}\beta\right) = \chi_\mu(c;\beta).$$

*Therefore, for $t \in \mathbb{C}^r$ and $z \in Z_{r,n}$, we have*

$$\lim_{\epsilon \to 0} \chi_\lambda\left(\iota_\lambda^{-1}(\langle t, zg(\epsilon)\rangle); g(\epsilon)^{-1}\beta\right) = \chi_\mu\left(\iota_\mu^{-1}(\langle t, z\rangle);\beta\right).$$

As an immediate consequence of Proposition 3.1 and Theorem 6.4, we obtain the

**Theorem 6.5 (confluence for CHG system).** *The system $M_\lambda(zg(\epsilon); g(\epsilon)^{-1}\beta)$ tends to the system $M_\mu(z;\beta)$ as $\epsilon \to 0$.*

**Remark 6.6.** If $z \in Z_{r,n}$ is a nonsingular point of the system $M_\mu$, then $zg(\epsilon) \in Z_{r,n}$ is also a nonsingular point of the system $M_\lambda$.

**Example 6.7.** Let us observe that the confluence of two singular points of the equation (1.1) stated in the introduction is a simple example of our general process of confluence. Let $\lambda = (\lambda_0, \lambda_1, \lambda_2, \lambda_3) = (1,1,1,1)$ and $\mu = (\mu_0, \mu_1, \mu_2) = (2,1,1)$ be two compositions of 4. We consider $\mu$ to be adjacent to $\lambda$ in the sense that $\mu_0 = \lambda_0 + \lambda_1, \mu_1 = \lambda_2, \mu_2 = \lambda_3$. Set

$$z(x) = \begin{pmatrix} 1 & 1 & 0 & 1 \\ 0 & -x & 1 & -1 \end{pmatrix} \in Z_{2,4}$$

$$w(\xi) = \begin{pmatrix} 1 & 0 & 0 & 1 \\ 0 & -\xi & 1 & -1 \end{pmatrix} \in Z_{2,4}.$$

Then, the equation (1.1) ( or (1.3)) is the restriction of the system $M_\lambda(z; {}^t(-c+b, -b, a-1, c-a-1))$ ( or $M_\mu(w; {}^t(-c,-1,a-1,c-a-1))$ ) on the 1-dimensional space $z = z(x)$ ( or $w = w(\xi)$ ). Note that

$$g(\epsilon) = g_{\lambda\to\mu}(\epsilon) = \begin{pmatrix} 1 & 1 & & \\ & \epsilon & & \\ & & 1 & \\ & & & 1 \end{pmatrix}$$

$$w(\xi)g(\epsilon) = \begin{pmatrix} 1 & 1 & 0 & 1 \\ 0 & -\epsilon\xi & 1 & -1 \end{pmatrix}$$

$$g(\epsilon)^{-1}\, {}^t(-c,-1,a-1,c-a-1) = {}^t(-c+\frac{1}{\epsilon}, -\frac{1}{\epsilon}, a-1, c-a-1).$$

From Theorem 6.5, it follows that

$$\lim_{\epsilon\to 0} M_\lambda(wg(\epsilon); {}^t(-c+\frac{1}{\epsilon}, -\frac{1}{\epsilon}, a-1, c-a-1)) = M_\mu(w; {}^t(-c,-1,a-1,c-a-1)).$$

Remark that the equation (1.3) is the restriction of $M_\lambda(wg(\epsilon);{}^t(-c+\frac{1}{\epsilon},-\frac{1}{\epsilon},a-1,c-a-1))$ on $w=w(\xi)$ and that

$$M_\lambda(z(x);{}^t(-c+b,-b,a-1,c-a-1)) = M_\lambda(w(\xi)g(\epsilon);{}^t(-c+\frac{1}{\epsilon},-\frac{1}{\epsilon},a-1,c-a-1))$$

if and only if

$$x=\epsilon\xi, \quad b=\frac{1}{\epsilon}.$$

Thus the confluence process from (1.1) to (1.3) is equivalent to that from $M_\lambda(z(x);{}^t(-c+b,-b,a-1,c-a-1))$ to $M_\mu(w(\xi);{}^t(-c,-1,a-1,c-a-1))$.

## References


[A] K. Aomoto, *Les équation aux différences linéaires et les intégrales des functions multiformes*, J. Fac. Sci. Univ. Tokyo, Sec. IA **22** (1975), 271–297.

[G] I. M. Gelfand, *General theory of hypergeometric functions*, Dokl. Akad. Nauk. SSSR **288** (1986), 14–48; English translation Soviet Math. Dokl. **33** (1986), 9–13.

[GG] I. M. Gelfand, and S. I. Gelfand, *Generalized hypergeometric equations*, Dokl. Akad. Nauk. SSSR **288** (1986), 279–283; English transl. in Soviet Math. Dokl. **33** (1986), 643–646.

[GRS] I. M. Gelfand, V. S. Retakh, and V. V. Serganova, *Generalized Airy funcitons, Schublrt cells, and Jordan groups*, Dokl. Akad. Nauk. SSSR **298** (1988), 17–21; English transl. in Soviet Math. Dokl. **37** (1988), 8–12.

[HK] Y. Haraoka and H. Kimura, *Contiguity relations of generalized confluent hypergeometric functions*, Proc. Japan Acad. **69** (1993).

[IKSY] K. Iwasaki, H. Kimura, S. Shimomura and M. Yoshida, *Vieweg, Wiesbaden* (1991).

[KHT1] H. Kimura, Y. Haraoka and K. Takano, *The generalized confluent hypergeometric functions*, Proc. Japan Acad. **68** (1992), 290–295.

[KHT2] H. Kimura, Y. Haraoka and K. Takano, *On confluences of the general hypergeometric systems*, Proc. Japan Acad. **69** (1993).

[KHT3] H. Kimura, Y. Haraoka and K. Takano, *On contiguity relations of the confluent hypergeometric systems*, Proc. Japan Acad. **69** (1993).

[O] K. Okamoto, *Isomonodromic defomation and Painlevé equations and the Garnier system*, J. Fac. Sci. Univ. Tokyo Sec. IA, Math. **33** (1986), 576–618.

[OK] K. Okamoto and H. Kimura, *On particular solutions of Garnier systems and the hypergeometric functions of several variables*, Quarterly J. Math. **37** (1986), 61–80.

# Connection Problem for Hypergeometric Systems

**Mitsuhiko Kohno**
*Department of Mathematics, Kumamoto University*
*Kumamoto, 860, Japan*

**Abstract**

It is well-known that hypergeometric systems (Fuchsian equations) of the form

$$(t-B)\frac{dX}{dt} = AX, \quad B = \mathrm{diag}(\lambda_1, \lambda_2, \ldots, \lambda_n), \quad A \in M_n(\mathbf{C})$$

are invariant under the *Euler* transformation and are reduced to Birkhoff canonical systems of rank 1 by the *Laplace* transformation. These facts have been used in the calculation of monodromy groups and in the derivation of relations between connection coefficients and Stokes multipliers. There is one more effective transformation, namely the *Mellin* transformation which connects difference equations with differential equations. Making full use of the transformation, we shall analyze connection problems for hypergeometric systems.

In order to analyze linear differential equations in the large, we often use the Euler transformation and the Laplace transformation. We here summarize the well-known formulas for such transformations.

Consider the Riemann-Liouville integral (an extension of differentiation):

$$I_\lambda^\mu \phi(t) = \frac{1}{\Gamma(\mu)} \int_\lambda^t (t-z)^{\mu-1}\phi(z)dz \qquad (\lambda = \text{finite or } \infty),$$

which is just the *Euler* transformation.

If $\phi(t)$ is holomorphic except for a regular singularity at $t = \lambda$ with the characteristic exponent $\rho$ ($\rho \notin \mathbf{Z}^-$), then the Riemann-Liouville integral $I_\lambda^\mu \phi(t)$ has again only a regular singularity at $t = \lambda$ with the characteristic exponent $\rho + \mu$ and moreover is holomorphic with respect to the parameter $\mu \in \mathbf{C}$.

We have the following properties:

$$\text{(E-i)} \qquad I_\lambda^\mu I_\lambda^{\mu'} \phi(t) = I_\lambda^{\mu+\mu'} \phi(t) \qquad (\rho + \mu' \notin \mathbf{Z}^-),$$

$$\text{(E-ii)} \qquad I_\lambda^{-n} \phi(t) = \frac{d^n}{dt^n} \phi(t) \qquad (n = 0, 1, \dots),$$

$$\text{(E-iii)} \qquad \phi(t) = \sum_{m=0}^{\infty} g(m)(t-\lambda)^{m+\rho} \qquad (|t-\lambda| < r)$$

$$\Rightarrow I_\lambda^\mu \phi(t) \equiv \Phi(t) = \sum_{m=0}^{\infty} \frac{\Gamma(m+\rho+1)}{\Gamma(m+\rho+\mu+1)} g(m)(t-\lambda)^{m+\rho+\mu}$$

$$\phi(t) = \sum_{s=0}^{\infty} h(s) t^{-s-\nu} \qquad (|t| > r)$$

$$\Rightarrow I_\infty^\mu \phi(t) = (-1)^\mu \sum_{s=0}^{\infty} \frac{\Gamma(s+\nu-\mu)}{\Gamma(s+\nu)} h(s) t^{-s-\nu+\mu}.$$

The integral

$$\mathcal{L}[\phi(t)] = \int_C e^{-tz} \phi(z) dz,$$

is called the *Laplace* transformation with the properties

$$\text{(L-i)} \qquad D^k \mathcal{L}[\phi(t)] = \mathcal{L}[(-t)^k \phi(t)] \qquad (D = d/dt),$$

$$\text{(L-ii)} \qquad t^k \mathcal{L}[\phi(t)] = \mathcal{L}[D^k \phi(t)],$$

where the path of integration $C$ is appropriately taken and here the remainder term is omitted.

Now we consider the hypergeometric system (Fuchsian equation)

$$(\mathrm{HS})_\mu \qquad \begin{cases} (t-B)X' = (A+\mu)X, \\ B = \mathrm{diag}(\lambda_1, \lambda_2, \dots, \lambda_n), \quad A \in M_n(\mathbf{C}), \quad \mu \in \mathbf{C}, \end{cases}$$

which has regular singularities at $t = \lambda_j$ $(j = 1, 2, ..., n)$ and $t = \infty$ in the whole complex $t$-plane.

Let us denote a solution of $(\mathrm{HS})_\mu$ near a regular singular point $t = \lambda$ by $X(t, \mu)$. Then, from the properties (E-i) and (E-ii), we can easily obtain the formula

$$(1) \qquad X(t, \mu) = I_\lambda^{\mu-\mu'} X(t, \mu').$$

In fact, putting

$$X(t,\mu) = I_\lambda^{\mu-\mu'}\phi(t)$$

and substituting it into $(\mathrm{HS})_\mu$, we have

$$\begin{aligned}
&(t-B)(I_\lambda^{\mu-\mu'}\phi(t))' \\
&= tI_\lambda^{\mu-\mu'}\phi'(t) - BI_\lambda^{\mu-\mu'}\phi'(t) \\
&= \tfrac{\mu-\mu'}{\Gamma(\mu-\mu'+1)}\int_\lambda^t (t-z)^{\mu-\mu'}\phi'(z)dz + \tfrac{1}{\Gamma(\mu-\mu')}\int_\lambda^t (t-z)^{\mu-\mu'-1}(z\phi'(z))dz - BI_\lambda^{\mu-\mu'}\phi'(t) \\
&= \underline{(\mu-\mu')I_\lambda^{\mu-\mu'+1}\phi'(t)} + \underline{I_\lambda^{\mu-\mu'}(t-B)\phi'(t)} \\
&= \underline{(A+\mu)I_\lambda^{\mu-\mu'}\phi(t)},
\end{aligned}$$

whence, if $\phi(t)$ satisfies the hypergeometric system $(\mathrm{HS})_{\mu'}$, then $X(t,\mu)$ becomes a solution of $(\mathrm{HS})_\mu$. The formula (1) just implies that hypergeometric systems (Fuchsian equations) are invariant under the Euler transformation.

| |
|---|
| Hypergeometric Systems ( Fuchsian Equations ) |
| $\Uparrow$ |
| Euler Transformation |
| $\Downarrow$ |
| Hypergeometric Systems ( Fuchsian Equations ) |

This fact, together with (E-iii), immediately yields several important results. Let $X(t,0)$ be a solution of $(\mathrm{HS})_0$, which is holomorphic in the Riemann surface $D(\lambda)$ of $\log(t-\lambda)$ over $\mathbf{C}\setminus\{\,\ell(\lambda_k-\lambda)+\lambda;\ \lambda_k\neq\lambda, \ell\geq 1\,\}$ and is expanded in terms of the power series like $\phi(t)$ in the formula (E-iii) near $t=\lambda$:

(2) $$X(t,0) = \sum_{m=0}^{\infty} g(m)(t-\lambda)^{m+\rho}.$$

Then, $X(t,\mu)$ defined by putting $\mu'=0$ in (1) becomes a solution of $(\mathrm{HS})_\mu$, which can be expanded as $\Phi(t)$ in (E-iii) near $t=\lambda$, and moreover admits the asymptotic behavior

(3) $$X(t,\mu) \sim \Phi(t) \qquad (\operatorname{Re}\mu\to\infty)$$

for any compact set of $D(\lambda)$. We also have directly from (1) and $(\mathrm{HS})_\mu$

$$(4)\qquad \begin{cases} X'(t,\mu) = X(t,\mu-1), \\ (A+\mu)X(t,\mu) = (t-B)X(t,\mu-1). \end{cases}$$

Combining this with the relation

$$X(t,-\rho-m) = I_\lambda^{-\rho-m}X(t,0) = I_\lambda^{-m}X(t,-\rho),$$

we can easily see that the coefficients of the power series (2)

$$g(m) = \frac{1}{\Gamma(m+\rho+1)}X(\lambda,-\rho-m)$$

satisfies the linear difference equation

$$(5)\qquad (B-\lambda)(m+\rho+1)g(m+1) = (m+\rho-A)g(m).$$

Furthermore, the combination of (3) and (4) has led to the derivation of the so-called extended Gauß formula, which played an important role in the calculation of the monodromy group of the hypergeometric system. (See **[1]**.)

On the other hand, from the properties (L-i) and (L-ii) we immediately see that

$$Y(t,\mu) = \mathcal{L}[\,X(t,\mu)\,]$$

satisfies the Birkhoff canonical system

$$(6)\qquad -t\frac{dY}{dt} = (\,A+\mu+1\,+\,t\,B\,)\,Y,$$

which has a regular singularity at $t=0$ and an irregular singularity of rank 1 at $t=\infty$.

| |
|---|
| Birkhoff Systems |
| ⇑ |
| Laplace Transformation |
| ⇓ |
| Hypergeometric Systems |

This equivalence verifies that connection coefficients of $(\mathrm{HS})_\mu$ are closely related with Stokes multipliers of the Birkhoff system (6). (See **[2]**.)

Now we shall consider the connection problem for the hypergeometric system

$$(\mathrm{HS})_0\qquad (t-B)X' = A\,X.$$

For simplicity, it is assumed that $\lambda_i \neq \lambda_j$ $(i \neq j\,;\, i,j = 1,2,...,n)$ and $A$ is diagonalizable, i.e.,

$$A \sim \operatorname{diag}(\nu_1, \nu_2, \ldots, \nu_n),$$

together with the condition:

$$\nu_i \not\equiv 0 \pmod{\mathbf{Z}}, \qquad \nu_i \not\equiv \nu_j \pmod{\mathbf{Z}} \qquad (i \neq j\,;\, i,j = 1,2,\ldots,n).$$

Denoting the diagonal elements of $A$ by $\rho_i$ $(i = 1,2,...,n)$, we moreover assume that

$$\rho_i \not\equiv 0 \pmod{\mathbf{Z}}, \qquad \rho_i \not\equiv \nu_j \pmod{\mathbf{Z}} \qquad (i,j = 1,2,\ldots,n).$$

Let $\lambda$ be one of the $\lambda_i$ and then $\rho$ denotes the corresponding $\rho_i$ or 0. In the below, changing the ordering of the $\lambda_i$, we may regard $\lambda = \lambda_n$. Then the connection problem is, for example, to derive exact connection coefficients $T_k$ in the expression of a linear combination

$$X(t) = \sum_{k=1}^{n} T_k X^k(t) \tag{7}$$

between any solution $X(t) \equiv X(t,0)$ near a finite regular singular point $t = \lambda$ and a fundamental set of solutions near $t = \infty$

$$X^k(t) = t^{\nu_k} \sum_{s=0}^{\infty} h_k(s)\, t^{-s} \qquad (k = 1,2,\ldots,n).$$

Concerning this problem, the Euler transformation

$$X(t,0) = I_\lambda^{-\mu'} X(t,\mu'),$$

in which $\mu'$ is taken as an eigenvalue of $A$, probably works well, since $X(t,\mu')$ becomes a solution of a reduced hypergeometric system of lower order or with fewer singular points. For example, in case $n = 2$, $X(t,\mu')$ is a solution of a linear differential equation of the first order. Also, in some cases, through the Laplace transformation, such a problem can be cleared up from relations of hypergeometric systems and Birkhoff systems. However, just from the knowledge explained above, the use of such transformations seems to yield only a tautology in most cases.

As is observed in (5), the coefficient $g(m)$ in the power series expansion (2) of $X(t)$ satisfies the linear difference equation, that is, $g(m)$ is a particular solution of (5) subject to the condition:

$$(B - \lambda)\rho g(0) = 0, \qquad g(-r) = 0 \qquad (r = 1,2,\ \ldots\ ). \tag{8}$$

We have already proved

**Proposition 1 (M. Kohno [3])** *A solution $g(m)$ of (5) can be expressed in the form*

$$g(m) = \sum_{k=1}^{n} \widehat{T}_k \, F_k(m) + E(m), \tag{9}$$

*where the $F_k(m)$ and $E(m)$ are particular meromorphic and entire solutions of (5), respectively. In particular, in the left half m-plane, the $F_k(m)$ are expanded in terms of the factorial series*

$$F_k(m) = \sum_{s=0}^{\infty} h_k(s)\gamma_k(m,s) \qquad (\, k = 1, 2, \ldots, n\,),$$

*where*

$$\gamma_k(m,s) = \frac{e^{\pi i(m+\rho)}\,\Gamma(m+s+\rho-\nu_k)}{\lambda^{m+s+\rho-\nu_k}\,\Gamma(m+\rho+1)\Gamma(s-\nu_k)}.$$

*The constants $\widehat{T}_k$ are calculated as*

$$\left\{ \begin{aligned} \mathrm{Res}\,[g(m)\,:\,-\rho+\nu_k\,] &= \widehat{T}_k\,\mathrm{Res}\,[F_k(m)\,:\,-\rho+\nu_k\,] \\ &= \widehat{T}_k\,h_k(0) \qquad (\, k = 1, 2, \ldots, n\,). \end{aligned} \right. \tag{10}$$

Combining this with the Barnes-integral representation

$$X(t) = -\,\frac{(t-\lambda)^{\rho}}{2\pi i} \int_C g(z)\left(\frac{\pi}{\sin \pi z}\right)(\lambda - t)^z dz,$$

we then have shown that the connection coefficients $T_k$ in (7) are given by

$$T_k = \widehat{T}_k \left\{ \frac{\pi\, e^{\pi i(\rho-\nu_k)}}{\sin \pi(\rho-\nu_k)} \right\} \qquad (\, k = 1, 2, \ldots, n\,).$$

In order to prove the above fact completely, we need to know the global behavior of $g(m)$ in the whole complex $m$-plane. To this end, we here have a very effective transformation, namely the *Mellin* transformation

$$g(m) = \int_C z^{m-1}\,\phi(z)dz, \tag{11}$$

which relates linear difference equations to linear differential equations.

| |
|---|
| Linear Difference Equations<br>⇑<br>Mellin Transformation<br>⇓<br>Linear Differential Equations |

It is easy to see that the Mellin transformation has the property:

$$(\mathrm{M-i}) \qquad (m)_p\, g(m) = (-1)^p \int_C z^{m+p-1} D^p \phi(z) dz,$$

where

$$(m)_p = m(m+1)\cdots(m+p-1), \quad (m)_0 = 1 \quad \text{and} \quad D = d/dz.$$

In the expression (M-i), the path of integration $C$ is assumed to be suitably chosen so that the remainder term is vanishing.

Now, applying (M-i) to (5), we immediately obtain the Fuchsian equation

$$(12) \qquad z\{1-(B-\lambda)z\}\phi'(z) = \{(\rho - A) - (B-\lambda)\rho\, z\}\phi(z).$$

Moreover, if we put

$$g(m) = \frac{\overline{g}(m)}{\Gamma(m+\rho+1)},$$

then we can rewrite (5) in the form

$$(B-\lambda)\overline{g}(m+1) = (m+\rho-A)\overline{g}(m).$$

In this case, by the Mellin transformation we obtain the Birkhoff system

$$(13) \qquad z\,\overline{\phi}'(z) = \{(\rho - A) - (B-\lambda)z\}\overline{\phi}(z).$$

Clearly, for an eigenvalue $\lambda$ of $B$ and an appropriately chosen $\rho$ of $A$, the differential systems (12) and (13) can be always reduced to lower ones.

We shall now investigate the Fuchsian system (12), which has regular singular points at

$$z = 0, \qquad z = \frac{1}{\lambda_j - \lambda} \quad (\lambda \neq \lambda_j), \qquad z = \infty.$$

Near $z = 0$, one can find a fundamental set of solutions

$$V_k(z) = z^{\rho - \nu_k} \sum_{s=0}^{\infty} v_k(s)\, z^s \qquad (k = 1, 2, \ldots, n),$$

where the coefficients $v_k(s)$ satisfy the difference equations

$$(14) \qquad (A - \nu_k + s)v_k(s) = (B-\lambda)(s-1-\nu_k)v_k(s-1).$$

On the other hand, denoting $\alpha_j = 1/(\lambda_j - \lambda)$, we can verify that near each regular singular point $z = \alpha_j$, there exists one nonholomorphic solution of the form

$$\Phi_j(z) = (z-\alpha_j)^{\rho_j} \sum_{s=0}^{\infty} \phi_{j,s}\, (z-\alpha_j)^s \qquad (j = 1, 2, \ldots, n-1).$$

According to the study of H. Galbrun, we can then construct a fundamental set of solutions of the system of linear difference equations (5) as follows:

$$G_j(m) = \int_{L_j} z^{m-1}\,\Phi_j(z)dz - \sum_{k=1}^{n} \frac{c_{j,k}}{e^{2\pi i(m+\rho-\nu_k)}-1} \int_{L_0} z^{m-1}\,V_k(z)dz$$

$$(j = 1, 2, \ldots, n-1),$$

where the constants $c_{j,k}$ denote the connection coefficients in the formula:

$$(15) \qquad (e^{2\pi i\rho_j} - 1)\,\Phi_j(z) = \sum_{k=1}^{n} c_{j,k}\,V_k(z)$$

and the path of integration $L_0$ and $L_j$ are taken as the figure in the below.

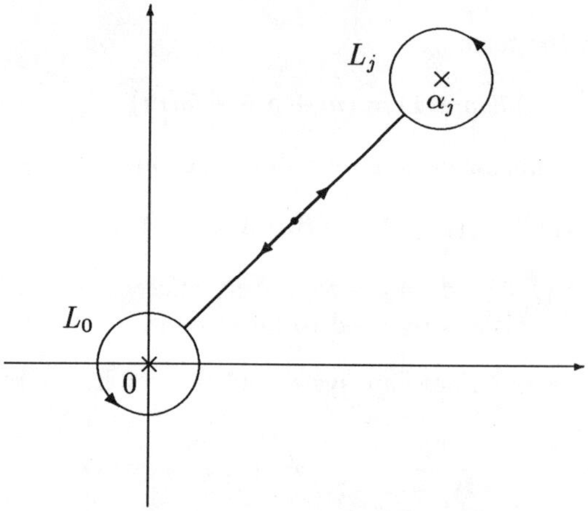


Figure 1: Path of integration $L_0$ and $L_j$

It is not difficult to see that each $G_j(m)$ is holomorphic in the right half $m$-plane and meromorphic in the whole $m$-plane with simple poles at

$$m = -\rho + \nu_k - r \qquad (r = 0, 1, 2, \ldots \;:\; k = 1, 2, \ldots, n)$$

and the residues at those poles are calculated as follows:

$$(16) \qquad \mathrm{Res}[G_j(m) : -\rho + \nu_k - r] = -c_{j,k}\,v_k(r).$$

Moreover, we have the following

**Proposition 2 (H. Galbrun [4])** *Assume that on the segment $\overline{O\alpha_j}$ connecting the origin with $\alpha_j$, there is no other $\alpha_\ell$ $(\ell \neq j)$. Then there holds the asymptotic relation*

$$G_j(m) \sim P_j(m)\, S_j(m) \quad as \quad m \to \infty \quad in \quad |\arg m| \leq \frac{\pi}{2} + \varepsilon, \tag{17}$$

*$\varepsilon$ being an arbitrarily small positive number, in which $P_j(m)$ and $S_j(m)$ are given as follows:*

$$\begin{aligned} P_j(m) &= \left\{ \alpha_j^{m+\rho_j} \frac{e^{\pi i(\rho_j+1)}}{m^{\rho_j+1}} \right\}, \\ S_j(m) &= E_0^{(j)} + \frac{E_1^{(j)}}{m} + \cdots + \frac{E_n^{(j)}}{m^n} + \cdots . \end{aligned}$$

Now, taking account of all results described so far, we shall return to the connection problem. The coefficient $g(m)$ of $X(t)$ can be now expressed in terms of the $G_j(m)$:

$$g(m) = \sum_{j=1}^{n-1} d_j\, G_j(m), \tag{18}$$

that is, the constants $d_j$ are determined so that the initial condition (8) is satisfied. This formula exactly corresponds to the expression (9).

Then, from the Barnes-integral and the asymptotic behavior (17), we can immediately prove that the convergent power series expression (2) of $X(t)$ holds in the open disk

$$|t - \lambda| < \min_j |\lambda - \lambda_j|.$$

and then $X(t)$ can be analytically continued to the domain

$$\bigcap_{j=1}^{n-1} \left| \arg\left(\frac{t-\lambda}{\lambda - \lambda_j}\right) \right| < \pi,$$

where the connection formula (7) holds.

As for the constants $\widehat{T}_k$, combiming (16) and (18) with (10), we have

$$\operatorname{Res}\left[g(m) \,:\, -\rho + \nu_k\right] = -\left(\sum_{j=1}^{n-1} d_j\, c_{j,k}\right) v_k(0) = \widehat{T}_k\, h_k(0).$$

Here it should be remarked that both $v_k(0)$ and $h_k(0)$ are eigenvectors corresponding to the eigenvalue $\nu_k$

$$(A - \nu_k)\, v_k(0) = (A - \nu_k)\, h_k(0) = 0.$$

So if we take $v_k(0) = h_k(0)$, we then consequently obtain

$$\widehat{T}_k = -\left(\sum_{j=1}^{n-1} d_j\, c_{j,k}\right) \qquad (\, k = 1, 2, \ldots, n\,).$$

This fact just implies that the connection coefficients $T_k$ for a Fuchsian equation can be represented in terms of the connection coefficients $c_{j,k}$ for a Fuchsian equation of lower order or with fewer singularities. We have thus established a *hierachy* of connection coefficients for Fuchsian equations.

**Examples**

(i) *Jordan-Pochhammer equation* In this case, through a transformation, the linear difference equation (5) can be reduced to the so-called Laplace difference equation ( an extension of hypergeometric difference equations )

$$\sum_{k=0}^{n-1}\{a_k(m+\rho-\theta)-b_k\}g(m+n-1-k)=0, \qquad a_k,\ b_k \in \mathbf{C}.$$

By the Mellin transformation, the associated differential equation (12) becomes a linear differential equation of the first order.

(ii) *Appell's hypergeometric functions $F_2,\ F_3,\ H_2, \ldots$* When we consider sectional differential equations of such hypergeometric functions, the associated differential equations (12) become linear differential equations of the first order or Gauß equation.

**Acknowledgements:** The author would like to take this opportunity to thank Professor Y. Sibuya, Professor M. Iwano and Professor W.A. Harris, Jr for guiding and encouraging him for a long time. During the preparation of this paper, the author was a guest researcher of the University of Minnesota supported by a grant from National Science Foundation.

## References

[1] K. Okubo. On the group of Fuchsian equations. Seminar Notes, Tokyo Metropolitan University, 1987.

[2] R. Schäfke. Über das globale Verhalten der Normallösungen von $x'(t) = (B + t^{-1}A)x(t)$ und zweier Arten von assoziierten Functionen. Math. Nachr. **121**, 123-145, 1985.

[3] M. Kohno. Global Analysis in Linear Differential Equations. Preprint.

[4] H. Galbrun. Sur la représentation des solutions d'une équation linéaire aux différences finies pour les grandes valeurs de la variable. Acta Math. **36**, 1-68, 1912.

# MONOTONE ITERATIVE METHOD FOR SECOND ORDER IMPULSIVE DIFFERENTIAL EQUATIONS

by

XINZHI LIU

Department of Applied Mathematics
University of Waterloo
Waterloo, Ontario N2L 3G1, Canada

ABSTRACT

This paper extends the monotone iterative method to impulsive differential equations. By modifying the notions of lower and upper solutions, extremal solutions of a boundary value problem of a second order impulsive differential equation are obtained as limits of monotone iterates.

## 1. Introduction

Many physical systems are subject to short-term perturpations which result in sudden changes of their state. For example, when a mass on a spring is given a blow by a hammer, it experiences a sharp change of velocity; and a pendulum in a mechanical clock undergoes a drastic increase of momentum every time it crosses its vertical position. The duration of these perturbations is negligible in comprison with that of the whole process. Thus they are often assumed to be in the form of impulses in the modelling process. Let the dynamics of the physical system between two successive impulses be described by a second order ordinary differential equation and the transition from the on-set of impulse to the begining of the subsequent continuous process be described by discrete maps. Then we get a second order impulsive differential equation. The monotone iterative method has been widely used in the treatment of certain nonlinear differential equations in recent years [3]. The basic idea of this method is that by using a suitable initial iteration one can construct a monotone sequence from an auxiliary quasilinear system, and this sequence converges monotonically to a solution of the nonlinear system. This method was extended to the study of the first order impulsive differential equations in [4]. We shall discuss, in this paper, the boundary value problems ($BVP$ for short) for a second order impulsive differential equation in which impulses occur at fixed times. With the help of lower and upper solutions, we shall employ the monotone iterative technique and obtain the existence of minimal and maximal solutions, which are limits of monotone sequences, of the $BVP$ with two point boundary conditions.

## 2. Preliminaries

Let us consider the following second order impulsive differential equation

$$\begin{aligned} -u'' &= f(t,u,u'), & t &\neq t_k, \\ u'(t^+) &= N(t,u'(t)), & t &= t_k, \\ u(t^+) &= I(t,u(t)), & t &= t_k, \end{aligned} \tag{2.1}$$

where $0 < t_1 < t_2 < \ldots < t_p < 1$, $f : J \times R \times R \to R$, $J = [0,1]$, and $N, I : J \times R \to R$ for each $k = 1, 2, \ldots, p$.

We denote by $\Omega$ the class of all functions $u : J \to R$, such that $u(t)$ is twice continuously differentiable for $t \neq t_k, 0, 1$, $u(t^-), u(t^+)$, $u'(t^-), u'(t^+)$ exist and $u(t^-) = u(t), u'(t^-) = u'(t)$ at $t = t_k, 0, 1$, where it is understood that at $t = 0$ we only mean right limit and at $t = 1$, left limit.

A function $v \in \Omega$ is said to be a lower solution of (2.1) if

$$\begin{aligned} -v'' &\leq f(t,v,v'), & t &\neq t_k, \\ v'(t^+) &\geq N(t,v'(t)), & t &= t_k, \\ v(t^+) &= I(t,v(t)), & t &= t_k\,. \end{aligned} \tag{2.2}$$

An upper solution of (2.1) is analogously defined by reversing the inequalities.

For any $v_0, w_0 \in \Omega$ such that $v_0(t) \leq w_0(t)$ on $J$, we define the conical segment $[v_0, w_0] = \{u \in \Omega : v_0 \leq u \leq w_0\}$.

Let us list the following assumptions for convenience:

$(A_0)$ $f : J \times R \times R \to R$ is continuous for $t \neq t_k$, $k = 1, 2, \ldots, p$,

$$\lim_{\substack{(t,x,y) \to (t_k,x_0,y_0) \\ t < t_k}} f(t,x,y) = f(t_k, x_0, y_0)$$

and

$$\lim_{\substack{(t,x,y) \to (t_k,x_0,y_0) \\ t < t_k}} f(t,x,y) = f(t_k^+, x_0, x_0) \text{ exists}\,.$$

$(A_1)$ $N(t,u)$ and $I(t,u)$ are continuous and nondecreasing in $u$ for $t = t_k$;

$(A_2)$ there exist $v_0, w_0 \in \Omega$ such that $v_0(t) \leq w_0(t)$ on $J$ with $v_0(t), w_0(t)$ being lower and upper solutions of (2.1), respectively;

$(A_3)$ $f(t,u,u')$ satisfies a Nagumo condition with respect to $v_0$ and $w_0$, see [1];

$(A_4)$ $f(t,u_1,u') - f(t,u_2,u') \geq -M(u_1 - u_2)$,
whenever $t \in J, v_0(t) \leq u_2 \leq u_1 \leq w_0(t)$ and $|u'| \leq C$, where
$C > \max[N, \max_J |v_0'|, \max_J |w_0'|]$, $N$ being the Nagumo constant and $M = M(C) > 0$.

In view of Theorem 3.1 in [2], we get easily the following existence result.

**Theorem 2.1.** Assume that $(A_0)$ to $(A_3)$ are satisfied. Then for any $\alpha, \beta \in R$ such that

$$v_0(0) \leq \alpha \leq w_0(0), \quad v_0(1) \leq \beta \leq w_0(1), \tag{2.3}$$

the $BVP$ (2.1) and (2.4) where

$$u(0) = \alpha, \quad u(1) = \beta, \tag{2.4}$$

has a solution $u(t)$ such that $v_0(t) \leq u(t) \leq w_0(t)$ on $J$ and $|u'(t)| \leq N$ for $t \neq t_k$, $t \in J$.

## 3. Monotone Iterative Method

We shall, in this section, consider the second order impulsive differential equation (2.1) subject to the boundary condition (2.4) and investigate the existence of extremal solutions. For this purpose, we shall employ the method of upper and lower solutions and the monotone iterative technique. This approach offers a constructive proof of existence which also provides numerical procedures for computation of solutions.

In order to develop the monotone iterative technique, we need to consider the quasi-linear impulsive differential equation

$$\begin{aligned} -u'' &= F(t,u,v'), & t &\neq t_k, \\ u'(t^+) &= N(t,u'(t)), & t &= t_k, \\ u(t^+) &= I(t,u(t)), & t &= t_k, \\ u(0) &= \alpha, u(1) = \beta, \end{aligned} \tag{3.1}$$

where for $\eta \in [v_0, w_0]$, $F(t,u,u') = f(t,\eta,q(u')) - M(u-\eta)$, and $q(u') = \max[-C, \min(u',C)]$.

In view of the definition of $q(u')$, it is easy to see that $(A_4)$ is equivalent to

$$(A_4^*)\ f(t,u_1,q(u')) - f(t,u_2,q(u')) \geq -M(u_1-u_2), \quad \text{for } M > 0,\ t \in J,$$

$$v_0(t) \leq u_2 \leq u_1 \leq w_0(t) \quad \text{and} \quad u' \in R.$$

Relative to the $BVP$ (3.1), we prove the following lemma.

**Lemma 3.1** Let the assumptions $(A_0)$ to $(A_4)$ hold. Then all the conditions $(A_0)$ to $(A_3)$ are also verified with respect to the $BVP$ (3.1) (i.e. $(A_0)$ to $(A_3)$ hold with $f$ replaced by $F$).

**Proof.** Clearly, $(A_0)$, $(A_1)$ hold with respect to $BVP$ (3.1). When $(A_2)$ to $(A_4)$ are satisfied, we have on $J$,

$$\begin{aligned} F(t,v_0,v_0') - f(t,v_0,v_0') &= f(t,\eta,q(v_0')) - M(v_0-\eta) - f(t,v_0,v_0') \\ &\geq -M(\eta-v_0) - M(v_0-\eta) = 0, \end{aligned}$$

since $|v_0| \le C$ and $q(v_0') = v_0'$. Hence we conclude that $v_0(t)$ is a lower solution of the $BVP$ (3.1). Similarly, we can show that $w_0(t)$ is an upper solution of the $BVP$ (3.1). Thus $(A_2)$ is satisfied with $f$ replaced by $F$.

When $(A_3)$ holds, we have

$$|f(t,u,u')| \le h(|u'|),\ t \in J, \quad v_0(t) \le u \le w_0(t) \quad \text{and} \quad u' \in R, \tag{3.2}$$

where $h \in C[R_+,(0,\infty)]$ and $\int^\infty \frac{s}{h(s)}ds = \infty$. Because of (3.2) and the definition of $f$, we get

$$|F(t,u,u')| \le h(|q(u')|) + M\gamma,$$

for $t \in [0,T], v_0(t) \le u \le w_0(t)$ and $u' \in R$, where $\gamma = \max_J |v_0(t) - w_0(t)|$. Defining $h_0(|u'|) = h(|q(u')|)$, we have

$$\int^\infty \frac{s}{h_0(s)}ds = \int^C \frac{s}{h(s)}ds + \int_C^\infty \frac{s}{h(c)}ds = \infty.$$

Also, since $\liminf_{s\to\infty} h_0(s) > 0$, there exists a $K > 0$ such that $M\gamma \le Kh_0(s)$ for all $s$ and hence

$$\int^\infty \frac{s}{h_0(s)+M\gamma}ds \ge \frac{1}{1+K}\int^\infty \frac{s}{h_0(s)}ds = \infty,$$

proving that $F$ also satisfies a Nagumo condition. The proof of the lemma is complete.

**Theorem 3.1.** Let $(A_0)$ to $(A_4)$ hold. Then there exists a solution $u$ of the $BVP$ (3.1) such that $v_0(t) \le u(t) \le w_0(t)$ on $J$ and $|u'(t)| \le N$ for $t \ne t_k$. Furthermore, the solution $u(t)$ is unique.

**Proof.** By Lemma 3.1, we have all the assumptions $(A_0)$ to $(A_3)$ satisfied relative to the $BVP$ (3.1). Now, employing Theorem 2.1, we obtain the existence of a solution $u = u(t)$ of the $BVP$ (3.1) with $v_0(t) \le u(t) \le w_0(t)$ on $J$ and $|u'(t)| \le N$ for $t \ne t_k$. It remains to show that $u(t)$ is unique.

Let us suppose that for any $\eta \in [v_0, w_0]$, the $BVP$ (3.1) admits two solutions $u_1(t), u_2(t)$ with $u_1, u_2 \in [v_0, w_0]$. We shall prove that $u_1(t) \equiv u_2(t)$ by showing (i) $u_1(t) \le u_2(t)$ and (ii) $u_2(t) \le u_1(t)$. If (i) is not true, then we have

$$\sup_{t\in J}[u_1(t) - u_2(t)] = \varepsilon > 0. \tag{3.3}$$

(1) We prove that $m(t) = u_1(t) - u_2(t)$ can not assume a positive local maximum at $t \ne t_k$. If this is not true, then $m(t)$ assumes a positive local maximum at some $t_0 \ne t_k, 0, 1$. Hence, $m'(t_0) = 0$ and $m''(t_0) \le 0$. Thus

$$\begin{aligned} -u_1''(t_0) &\ge -u_2''(t_0) = f(t_0, \eta(t_0), q(u_2'(t_0))) - M(u_2(t_0) - \eta(t_0)) \\ &> f(t_0\eta(t_0), q(u_1'(t_0)) - M(u_1(t_0) - \eta(t_0)) \\ &= F(t_0, u_1(t_0), u_1'(t_0)), \end{aligned}$$

which is a contradiction.

(2) If there is a $t_0 \in J$ such that $m(t_0) = u_1(t_0) - u_2(t_0) = \varepsilon$, then $t_0 = t_k$ for some $k = 1, 2, \ldots, p$ by (1). Hence $u_2'(t_k) \leq u_1'(t_k)$ and consequently, we have, since $I(t_k, u)$ and $N(t_k, u)$ are nondecreasing in $u$,

$$m(t_k^+) = u_1(t_k^+) - u_2(t_k^+) = \varepsilon, \tag{3.4}$$

and

$$m'(t_k^+) = u_1'(t_k^+) - u_2'(t_k^+) \geq 0. \tag{3.5}$$

In view of $(A_0)$, the assumptions on $f$ then give

$$-D_+u_1''(t_k^+) = F(t_k^+, u_1(t_k^+), u_1'(t_k^+)), \tag{3.6}$$

and

$$-D_+u_2''(t_k^+) = F(t_k^+, u_2(t_k^+), u_2'(t_k^+)), \tag{3.7}$$

where

$$-D_+u_1''(t_k^+) = lim\,inf_{h\to 0+}\ \frac{1}{h}[u_1'(t_k + h) - u_1'(t_k^+)],$$

and

$$-D_+u_2''(t_k^+) = lim\,inf_{h\to 0+}\ \frac{1}{h}[u_2'(t_k + h) - u_2'(t_k^+)].$$

By the definition of $\varepsilon, m(t)$ is nonincreasing to the right of $t_k$. As a result, $m'(t_k^+) \leq 0$. This, together with (3.5), implies

$$m'(t_k^+) = 0 \quad \text{and} \quad D_+m''(t_k^+) \leq 0. \tag{3.8}$$

Consequently, from (3.5),

$$\begin{aligned} -D_+u_1''(t_k^+) &\geq -D_+u_2''(t_k^+) = F(t_k^+, u_2(t_k^+), u_2'(t_k^+)) \\ &= f(t_k^+, \eta(t_k^+), q(u_1'(t_k^+))) - M(u_2(t_k^+) - \eta(t_k^+)) \\ &> f(t_k^+, \eta(t_k^+), q(u_1'(t_k^+))) - M(u_1(t_k^+) - \eta(t_k^+)) \\ &= F(t_k^+, u_1(t_k^+), u_1'(t_k^+)), \end{aligned}$$

which is a contradiction to (3.6).

(3) If $m(t) = u_1(t) - u_2(t) < \varepsilon$ for all $t \in J$, then

$$m(t_k^+) = u_1(t_k^+) - u_2(t_k^+) = \varepsilon, \tag{3.9}$$

for some $k = 1, 2, \ldots, p$, which in turn yields

$$u_1(t_k) > u_2(t_k). \tag{3.10}$$

Thus

$$\sup[u_1(t) - u_2(t), t_{k-1} < t \leq t_k] = \varepsilon_k > 0. \tag{3.11}$$

We then have the following two possibilities: 1° $u_1(t_0) - u_2(t_0) = \varepsilon_k$ for some $t_0 \in (t_{k-1}, t_k]$. By (i) $t_0 \notin (t_{k-1}, t_k)$. However, if $t_0 = t_k$, then $u_2'(t_k) \leq u_1'(t_k)$ and hence $u_2'(t_k^+) \leq u_1'(t_k^+)$. Now repeating the arguments used in case (2) shows that this is impossible.
2° $u_1(t_{k-1}^+) - u_2(t_{k-1}^+) = \varepsilon_k$. As in (3.10), we get $u_1(t_{k-1}) > u_2(t_{k-1})$. Repeating the same procedure we employed so far in (3) inductively, we either get a contradiction on the way or we have finally

$$m(t_1^+) = \sup[m(t) : t_1 < t \leq t_2] = \varepsilon_2 > 0, \tag{3.12}$$

and $m(t_1) > 0$. Hence

$$\sup[m(t) : 0 < t \leq t_1] = \varepsilon_1 > 0. \tag{3.13}$$

Since $m(0) \leq 0$ and $m(t) \neq \varepsilon_1$ for $t \in (0, t_1)$ by (1), we get

$$m(t_1) = \varepsilon_1 \quad \text{and} \quad m'(t_1) \geq 0, \tag{3.14}$$

and thus $m'(t_1^+) \geq 0$. However, this and (3.12) imply again a contradiction as in the case of 1°. Thus $u_1(t) \leq u_2(t)$ on $J$.

Repeating the above arguments for $u_2(t) - u_1(t)$, one can show that $u_2(t) \leq u_1(t)$ on $J$. Therefore, $u_1(t) \equiv u_2(t)$ and the proof of the lemma is complete.

Since, for every $\eta \in [v_1, w_0]$, the $BVP$ (3.1) admits a unique solution $u$, we can define the mapping $A$ by

$$A\,\eta = u. \tag{3.15}$$

The properties of $A$ that enable us to generate the monotone iterates that converge to the minimal and the maximal solutions of $BVP$ (2.1) and (2.4) are given in the next result.

**Theorem 3.2.** Under the assumptions of Theorem 3.1, the mapping $A$ defined by (3.15) satisfies

(i) $v_0 \leq Av_0$ and $w_0 \geq Aw_0$;

(ii) for $\eta_1, \eta_2 \in [v_0, w_0]$, $\eta_1 \leq \eta_2$ implies $A\eta_1 \leq A\eta_2$ (i.e. $A$ is a monotone operator on the segment $[v_0, w_0]$).

**Proof.** (i) Let $Av_0 = v_1$ and $Aw_0 = w_1$, where $v_1, w_1$ are the unique solutions of the $BVP$ (3.1) corresponding to $\eta = v_0$ and $\eta = w_0$, respectively. We shall only prove that $v_0 \leq v_1$ and similar arguments can be used to show $w_0 \geq w_1$.

Suppose $v_0(t) \leq v_1(t)$, $t \in J$ is not true. Then

$$\sup_{t \in J}[v_0(t) - v_1(t)] = \varepsilon > 0. \tag{3.16}$$

(1) We prove that $m(t) = v_0(t) - v_1(t)$ can not assume a positive local maximum at some $t_0 \neq t_k$. If this is not true, then $m(t)$ assumes a positive local maximum at some $t_0 \neq t_k, 0, 1$. Hence $m'(t_0) = 0$ and $m''(t_0) \leq 0$. Thus

$$\begin{aligned} -v_0''(t_0) &\geq -v_1''(t_0) = f(t_0, v_0(t_0), q(v_1'(t_0))) - M(v_1(t_0) - v_0(t_0)) \\ &= f(t_0, v(t_0), q(v_0'(t_0))) - M(v_1(t_0) - v_0(t_0)) \\ &> f(t_0, v_0(t_0), v_0'(t_0)), \end{aligned}$$

which is a contradiction.

(2) If there is a $t_0 \in J$ such that $m(t_0) = v_0(t_0) - v_1(t_0) = \varepsilon$, then $t_0 = t_k$ for some $k = 1, 2, \ldots, p$ by (1). Hence $v_1'(t_k) \leq v_0'(t_k)$ and consequently, we have

$$m(t_k^+) = v_0(t_k^+) - v_1(t_k^+) = \varepsilon, \tag{3.17}$$

and

$$m'(t_k^+) = v_0'(t_k^+) - v_1'(t_k^+) \geq 0. \tag{3.18}$$

In view of the assumptions on $f$, we have

$$-D_+ v_0''(t_k^+) \leq f(t_k^+, v_0(t_k^+), v_0'(t_k^+)), \tag{3.19}$$

and

$$-D_+ v_1''(t_k^+) = F(t_k^+, v_1(t_k^+), v_1'(t_k^+)). \tag{3.20}$$

By the definition of $\varepsilon$ and (3.18), we have $m'(t_k^+) = 0$ and $D_+ m''(t_k^+) \leq 0$. Thus

$$\begin{aligned} -D_+ v_0''(t_k^+) &\geq -D_+ v_1''(t_k^+) = F(t_k^+, v_1(t_k^+), v_1'(t_k^+)) \\ &= f(t_k^+, v_0(t_k^+), q(v_0'(t_k^+))) - M(v_1(t_k^+) - v_0(t_k^+)) \\ &> f(t_k^+, v_0(t_k^+), v_0'(t_k^+)), \end{aligned}$$

which is a contradiction to (3.19).

(3) If $m(t) = v_0(t) - v_1(t) < \varepsilon$ for all $t \in J$, then similar arguments to that used in the proof of Theorem 3.1 lead to a contradiction. This proves $v_0 \leq v_1$.

(ii) To prove that $A$ is a monotone operator on $[v_0, w_0]$, let us set $A\eta_1 = u_1$ and $A\eta_2 = u_2$, where $\eta_1, \eta_2 \in [v_0, w_0], \eta_1 \leq \eta_2$ and $u_1, u_2$ are the unique solutions of $BVP$ (3.1) with $\eta = \eta_1$ and $\eta = \eta_2$ respectively. We claim that $u_1(t) \leq u_2(t)$ whenever $\eta_1(t) \leq \eta_2(t)$, $t \in J$. If this is not true, then

$$\sup_{t \in J}[u_1(t) - u_2(t)] = \varepsilon > 0. \tag{3.21}$$

As before, we first prove that $m(t) = u_1(t) - u_2(t)$ can not assume a positive local maximum at $t \neq t_k$. If this is not true, then $m(t)$, assumes a positive local maximum at some $t_0 \neq t_k, 0, 1$. Hence $m'(t_0) = 0$ and $m''(t_0) \leq 0$. Thus

$$\begin{aligned}0 &\leq -u_1''(t_0) + u_2''(t_0) = f(t_0,\ \eta_1(t_0), q(u_1'(t_0))) - M(u_1(t_0) - \eta_1(t_0)) \\ &\quad -f(t_0, \eta_2(t_0), q(u_2'(t_0))) + M(u_1(t_0) - \eta_2(t_0)) \\ &\leq M(\eta_2(t_0) - \eta_1(t_0)) - M(u_1(t_0) - \eta_1(t_0)) + M(u_2(t_0) - \eta_2(t_0)) \\ &= -M(u_1(t_0) - u_2(t_0)) < 0,\end{aligned}$$

which is a contradiction.
Thus if there is a $t_0 \in J$ such that $m(t_0) = u_1(t_0) - u_2(t_0) = \varepsilon$, then $t_0 = t_k$ for some $k = 1, 2, \ldots, p$. Hence $u_2'(t_k) \leq u_1'(t_k)$ and consequently, we have,

$$m(t_k^+) = u_1(t_k^+) - u_2(t_k^+) = \varepsilon, \tag{3.22}$$

and

$$m'(t_k^+) = u_1'(t_k^+) - u_2'(t_k^+) \geq 0. \tag{3.23}$$

In view of the assumptions on $f$, we have,

$$-D_+u_1''(t_k^+) = F(t_k^+, u_1(t_k^+), u_1'(t_k^+)), \tag{3.24}$$

and

$$-D_+u_2''(t_k^+) = F(t_k^+,\ u_2(t_k^+),\ u_2'(t_k^+)). \tag{3.25}$$

By the definition of $\varepsilon$ and (3.23), we have $m'(t_k^+) = 0$ and $D_+m''(t_k^+) \leq 0$. Thus

$$\begin{aligned}0 &\leq -D_+m''(t_k^+) = f(t_k^+, \eta_1(t_k^+), q(u_1'(t_k^+))) - M(u_1(t_k^+) - \eta_1(t_k^+)) \\ &\quad -f(t_k^+, \eta_2(t_k^+), q(u_2'(t_k^+))) + M(u_2(t_k^+) - \eta_2(t_k^+)) \\ &\leq M(\eta_2(t_k^+) - \eta_1(t_k^+)) - M(u_1(t_k^+) - \eta_1(t_k^+)) + M(u_2(t_k^+) - \eta_2(t_k^+)) \\ &= -M(u_1(t_k^+) - u_2(t_k^+) < 0,\end{aligned}$$

which is a contradiction.
If $m(t) = u_1(t) - u_2(t) < \varepsilon$ for all $t \in J$, then a similar argument to that in the proof of Theorem 3.1 leads to a contradiction. Hence our claim $u_1(t) \leq u_1(t)$ on $J$ is proved. The proof is thus complete.

We can now state our main result in this section.

**Theorem 3.3** Let $(A_0)$ to $(A_4)$ hold. Then there exist monotone sequences $\{v_n(t)\}_{n=0}^{\infty}, \{w_n(t)\}_{n=0}^{\infty}$ such that $\lim_{n\to\infty} v_n(t) = \rho(t)$ and $\lim_{n\to\infty} w_n(t) = \gamma(t)$, monotonically and piecewise uniformly on $J$, where $\rho(t)$ and $\gamma(t)$ are the minimal and the maximal solutions of the $BVP$ (2.1) and (2.4) respectively.

**Proof.** By Theorem 3.1, we know that for any $\eta \in [v_0, w_0]$ of the $BVP$ (3.1) has a unique solution $u(t)$ such that $v_0(t) \leq u(t) \leq w_0(t)$ on $J$ and $|u'(t)| \leq \tilde{N}, t \neq t_k$, $\tilde{N}$ being the Nagumo constant that can be obtained for the function $F$. By Theorem 3.2, we see that the mapping $A$ defined by $A\eta = u$ generates monotone sequences $\{v_n(t)\}, \{w_n(t)\}$, where $Av_n = v_{n+1}$. In fact, we have on $J$,

$$v_0 \leq v_1 \leq \ldots \leq v_n \leq w_n \leq \ldots \leq w_1 \leq w_0.$$

Let $J_0 = [0, t_1], J_1 = (t_1, t_2], \ldots, J_{p-1} = (t_{p-1}, t_p], J_p = (t_p, 1]$. Then $J = \bigcup_{i=0}^{p} J_i$. Note that the $BVP$ (2.1) and (2.4) is equivalent to the system of integral equations

$$\begin{aligned} u(t) &= \int_{t_{i-1}}^{t_i} G_i(t,s) f(s,\ u(s),\ u'(s)) ds + \tfrac{u(t_i) - u(t_{i-1}^+)}{t_i - t_{i-1}} (t - t_{i-1}) + u(t_{i-1}^+), t \in J_i \\ i &= 1, 2, \ldots, p+1, \end{aligned} \tag{3.26}$$

where

$$G_i(t,s) = \begin{cases} (t_i - t)(s - t_{i-1})/(t_i - t_{i-1}), & \text{if} \quad t_{i-1} \leq s \leq t \leq t_i, \\ (t_i - s)(t - t_{i-1})/(t_i - t_{i-1}), & \text{if} \quad t_{i-1} \leq t \leq s \leq t_i, \end{cases}$$

$$u(t_{i-1}^+) = I(t_{i-1}, u(t_{i-1})),\ i = 2, 3, \ldots, p+1, t_0 = 0, t_{p+1} = 1, u(0^+) = \alpha, u(1) = \beta.$$

If we replace, in (3.26), $f$ by $F$, then (3.26) is equivalent to the $BVP$ (3.1). Thus it follows using standard arguments, see [3], that the sequence $\{v_n\}, \{w_n\}$ converge uniformly and monotonically to $\rho(t), \gamma(t)$ on $J_i, i = 0, 1, 2, \ldots, p$, respectively and $\rho, \gamma$ are solutions of the $BVP$ (2.1) and (2.4) satisfying $\rho,\ \gamma \in [v_0, w_0]$.

In order to prove that $\rho, \gamma$ are minimal and maximal solutions of $BVP$ (2.1) and (2.4), it is enough to show that if $u$ is any solution of $BVP$ (2.1) and (2.4) such that $u \in [v_0, w_0]$, then $v_n(t) \leq u(t) \leq w_n(t)$ on $J$ for all $n$. This is easy to prove by induction following an argument similar to that used to prove (1) in Theorem 3.2. Thus the proof of the theorem is complete.

## Acknowledgements

This research was supported by the Natural Sciences and Engineering Research Council of Canada which is gratefully acknowledged.

## References

1. Bernfeld, S. and Lakshmikantham, V., *An Introduction to Nonlinear Boundary Value Problems*, Academic Press, New York, 1974.

2. Hu, Shouchuan and Lakshmikantham, V., Periodic boundary value problems for second order impulsive differential systems, (to appear).

3. Ladde, G., Lakshmikantham, V., and Vatsala, A., *Monotone Iterative Technique for Nonlinear Differential Equations*, Pitman Publishing Company, 1985.

4. Liu, Xinzhi, Monotone iterative technique for impulsive different equations in a Banach space, Appl. Math. Phy. Sci., **24**, (1990) 183-191.

# ON REGULARIZING DIFFERENTIAL - ALGEBRAIC EQUATIONS

LEONID V. KALACHEV
and
ROBERT E. O'MALLEY, JR.
*Department of Applied Mathematics, FS-20*
*University of Washington*
*Seattle, Washington 98195*
*U.S.A.*

ABSTRACT

This paper shows how regularization methods can be used to study the solution structure of boundary value problems for differential - algebraic equations (i.e., constrained differential equations). Attention is primarily restricted to linear problems. The work relies on singular perturbations theory, as developed by Wasow, Harris, and other investigators, and is important for the solution of DAEs which arise in many applications contexts.

## 1. Introduction

Consider the following differential-algebraic system consisting of $m$ scalar ordinary differential equations subject to $0 < n < m$ constraints on the bounded interval $0 \leq t \leq 1$:

$$\begin{cases} \dot{U}_0 = f(U_0, V_0, t) \\ 0 = g(U_0, V_0, t). \end{cases} \tag{1}$$

We must expect the solution of such constrained differential equations to lie on an $m$- or lower-dimensional manifold (cf. Takens (1976) and Rheinboldt (1984)). What boundary values would determine a unique solution is still quite unclear, in contrast to the situation for initial value problems (cf. Brenan et al. (1989), Hairer and Wanner (1991), and Griepentrog et al. (1992)).

The work is supported in part by the National Science Foundation under grant number DMS 9107196.

To study (1), one might introduce the singularly perturbed system

$$\begin{cases} \dot{u} = f(u,v,t) \\ \epsilon\mathcal{R}(u,v,t)\dot{v} = g(u,v,t) \end{cases} \tag{2}$$

for some nonsingular matrix $\mathcal{R}$ and a small positive regularizing parameter $\epsilon$. If, for example, the Jacobian matrix $\frac{\partial}{\partial v}(\mathcal{R}^{-1}(u,v,t)g(u,v,t))$ remains stable, the Tikhonov-Levinson theory guarantees that any initial value problem will have a unique asymptotic solution as $\epsilon \to 0$ with a limiting solution which satisfies the given DAE for $t > 0$ as well as the initial condition $U_0(0) = u(0)$ (cf. Vasil'eva and Butuzov (1990) or O'Malley (1991)). To determine the limiting solution for corresponding boundary value problems requires a knowledge of the structure of the decaying and growing boundary layer solutions for the corresponding stretched systems at $t = 0$ and 1. We note that differential-algebraic systems often arise as limiting solutions to singularly perturbed systems, such as (2), when the fast dynamics is neglected. Then, the parameter $\epsilon$ can naturally represent the size of small inductances or capacitances, low viscosity or compressibility, etc. (cf. Kalachev and O'Malley (1991, 1994)). Regularization methods can be most useful, however, whether or not the parameter introduced has a physical interpretation (cf. Tikhonov and Arsenin (1977) and Morozov (1993)), both for theoretical and computational purposes.

A particularly simple example is the linear index-one DAE

$$\begin{cases} \dot{U}_0 = A(t)U_0 + B(t)V_0 + f(t) \\ 0 = C(t)U_0 + D(t)V_0 + g(t), \end{cases} \tag{3}$$

where $D(t)$ is nonsingular throughout $0 \le t \le 1$, subject to the $m$ boundary conditions

$$R_0U_0(0) + R_1U_0(1) = s. \tag{4}$$

The $m$-dimensional manifold of solutions for this DAE is clearly parameterized by the initial vector $U_0(0)$, which might be determined through the boundary condition. We can regularize the problem simply by introducing the singularly perturbed system

$$\begin{cases} \dot{u} = A(t)u + B(t)v + f(t) \\ -\epsilon D(t)\dot{v} = C(t)u + D(t)v + g(t) \end{cases} \tag{5}$$

subject to the $m + n$ boundary conditions

$$\begin{cases} R_0u(0) + R_1u(1) = s \\ v(0) = q. \end{cases} \tag{6}$$

We note that the regularization (5) is identical to using the classical matrix pencil of Kronecker when the coefficients are constant matrices (cf. Gantmacher (1960)). The Tikhonov-Levinson theory implies that the two-point problem (5)-(6) has a unique solution of the form

$$\begin{cases} u(t,\epsilon) = U_0(t) + O(\epsilon) \\ v(t,\epsilon) = V_0(t) + e^{-t/\epsilon}(v(0) - V_0(0)) + O(\epsilon), \end{cases} \tag{7}$$

where $\begin{pmatrix} U_0 \\ V_0 \end{pmatrix}$ satisfies (3)-(4), provided the matrix

$$R_0 + R_1\Psi(1) \quad \text{is nonsingular,}$$

where $\Psi(t)$ is the fundamental matrix defined by

$$\dot{\Psi} = (A - BD^{-1}C)\Psi, \ \Psi(0) = I_m.$$

We note that the fast variable $v$ in (5) will feature on initial jump as $\epsilon \to 0$ if $v(0) \neq V_0(0) = D^{-1}(0)(C(0)U_0(0) + g(0))$, and that the DAE's solution is obtained using the system (5) with $\epsilon = 0$ and the boundary conditions (6) with the initial condition for $v$ canceled. Such cancellation laws for singularly perturbed linear differential equations were obtained in Wasow (1941) and Harris (1960). Their generalization to fully nonlinear problems has not been achievable.

## 2. An Index-Two Problem

A more challenging DAE is provided by the linear "pure index two" or "Hessenberg" system:

$$\begin{cases} \dot{U}_0 = A(t)U_0 + B(t)V_0 + f(t) \\ 0 = C(t)U_0 + g(t) \end{cases} \tag{8}$$

when the $n \times n$ matrix

$$C(t)B(t) \quad \text{is nonsingular} \tag{9}$$

throughout the interval $0 \le t \le 1$. We can then eliminate $V_0$ from the system since differentiating the constraint with respect to $t$ implies that $C(AU_0 + BV_0 + f) + \dot{C}U_0 + \dot{g} = 0$, so

$$V_0(t) = -(CB)^{-1}[(CA + \dot{C})U_0 + Cf + \dot{g}]. \tag{10}$$

Thus, we are left with the DAE

$$\begin{cases} \dot{U}_0 = \mathcal{A}(t)U_0 + \mathbf{f}(t) \\ 0 = C(t)U_0 + g(t) \end{cases} \tag{11}$$

where

$$\mathcal{A}(t) \equiv [I - B(CB)^{-1}C]A - B(CB)^{-1}\dot{C}$$

and

$$\mathbf{f}(t) \equiv [I - B(CB)^{-1}C]f - B(CB)^{-1}\dot{g}.$$

Since (9) implies that $C$ has full rank $n$, it is convenient to introduce the matrix

$$K(t) \equiv C'(CC')^{-1}C \tag{12}$$

(where the prime denotes transposition) which projects onto the range of $C$. The structure of the solution space for either the DAE (8) or (11) is now clear, since $K(t)U_0 = -C'(CC')^{-1}g$ is completely determined through the constraint and only its orthogonal complement needs to be obtained through the differential system. The $(m-n)$-dimensional manifold of solutions will clearly be parameterized through the components of the initial vector $(I_m - K(0))U_0(0)$.

To regularize (11), we will instead first introduce a vector Lagrange multiplier $\mu$ into the differential equation and we will also use it as a scaled slack variable in the constraint to define the regularized system

$$\begin{cases} \dot{u} = \mathcal{A}(t)u + \mathbf{f}(t) + C'(t)\mu \\ \epsilon S(t)\mu = C(t)u + g(t) \end{cases} \tag{13}$$

for some nonsingular matrix $S(t)$. We note that such multipliers have important physical meanings in certain mechanical applications. Also note that for $\mu = 0$, we obtain the DAE (1). The relation of such regularizations to corresponding stabilization methods should be observed (cf. Ascher et al. (1993)). Alternatively, eliminating $\mu$ provides us the singularly perturbed linear system

$$\epsilon\dot{u} = C'S^{-1}(Cu + g) + \epsilon(\mathcal{A}u + \mathbf{f}). \tag{14}$$

The asymptotic solution of (13), however, is far from straightforward since the limiting system matrix

$$C'S^{-1}C$$

has rank $n < m$ and may possibly also have eigenvalues in both half planes. Nonetheless, we are able to solve (14) asymptotically, subject to the general linear boundary condition that

$$R_0 u(0) + R_1 u(1) = s \tag{15}$$

is prescribed, presuming we use an appropriate scaling matrix $S(t)$ in (13) to decouple the fast and slow motions. The limiting solution within $(0, 1)$ should satisfy the DAE (11).

To proceed, let us now make the kinematic change of variables

$$u(t) = T(t)w(t) \tag{16}$$

where $T(t)$ is a smooth matrix which diagonalizes the projection matrix $K(t)$ of (12), i.e., suppose

$$K(t) \equiv C'(CC')^{-1}C = T(t)\begin{pmatrix} I_n & 0 \\ 0 & 0 \end{pmatrix} T^{-1}(t). \tag{17}$$

Then, $w$ will satisfy the transformed system

$$\epsilon \dot{w} = T^{-1}C'S^{-1}(CTw + g) + \epsilon T^{-1}[(\mathcal{A}T - \dot{T})w + \mathbf{f}]. \tag{18}$$

To simplify (18), we will now select

$$S \equiv CT \begin{pmatrix} \Lambda & 0 \\ 0 & 0 \end{pmatrix} T^{-1}C' \tag{19}$$

where we require that $\Lambda^2(t) = I_n$ in order to make the limiting system matrix

$$T^{-1}C'S^{-1}CT = \begin{bmatrix} \Lambda & 0 \\ 0 & 0 \end{bmatrix}$$

block diagonal, but leaving $\Lambda$ otherwise free. This follows since

$$S^{-1} = (CC')^{-1}CT \begin{bmatrix} \Lambda & 0 \\ 0 & 0 \end{bmatrix} T^{-1}C'(CC')^{-1}.$$

Other choices for $S$ may also be tried (cf., e.g., Kalachev and O'Malley (1994)). Decomposing

$$w = \begin{pmatrix} y \\ z \end{pmatrix} \tag{20}$$

finally yields the decoupled fast-slow system

$$\begin{cases} \epsilon\dot{y} = \Lambda(y + [T^{-1}C'(CC')^{-1}g]_1) + \epsilon[T^{-1}((\mathcal{A}T - \dot{T})\begin{pmatrix} y \\ z \end{pmatrix} + \mathbf{f})]_1 \\ \dot{z} = [T^{-1}((\mathcal{A}T - \dot{T})\begin{pmatrix} y \\ z \end{pmatrix} + \mathbf{f})]_2 \end{cases} \tag{21}$$

where the subscripts 1 and 2 refer to the first $n$ and last $m-n$ components of the indicated vector. Because $\Lambda(t)$ remains nonsingular, the Tikhonov-Levinson theory shows that the bounded asymptotic solutions of the singularly perturbed system (21) must have the form

$$\begin{cases} y(t,\epsilon) = Y_0(t) + e^{\Lambda(0)t/\epsilon}\mathcal{K}(0)\alpha(0) + e^{-\Lambda(1)(1-t)/\epsilon}\mathcal{L}(1)\beta(0) + O(\epsilon) \\ z(t,\epsilon) = Z_0(t) + O(\epsilon) \end{cases} \tag{22}$$

(cf. Harris (1973)). Here the matrix $\mathcal{K}(0)$ is a projection onto the (say, $k$-dimensional) stable eigenspace of $\Lambda(0)$ and $\mathcal{L}(1)$ projects onto the (($n-k$)-dimensional) unstable eigenspace of $\Lambda(1)$. Moreover, the outer limit $\begin{pmatrix} Y_0(t) \\ Z_0(t) \end{pmatrix}$ will satisfy the limiting system, i.e.

$$\begin{cases} Y_0(t) = -[T^{-1}C'(CC')^{-1}g]_1 \\ \text{and} \\ \dot{Z}_0 = [T^{-1}(\mathcal{A}T - \dot{T})\begin{pmatrix} (-T^{-1}C'(CC')^{-1}g)_1 \\ Z_0 \end{pmatrix} + \mathbf{f}]_2. \end{cases} \tag{23}$$

We observe that the differential system for $Z_0$ correspondings to the "essential underlying ordinary differential equation" of Ascher and Petzold (1992). More completely, variation of parameters implies the general solution

$$Z_0(t) = \Phi(t)[Z_0(0) + \int_0^t \Phi^{-1}(T^{-1}(\mathcal{A}T - \dot{T})\begin{pmatrix} (-T^{-1}C'(CC')^{-1}g)_1 \\ 0 \end{pmatrix} + \mathbf{f})_2 ds \tag{24}$$

(parameterized by $Z_0(0)$) where $\Phi(t)$ is the fundamental matrix defined by

$$\dot{\Phi} = [T^{-1}(\mathcal{A}T - \dot{T})\begin{pmatrix} 0 \\ \Phi \end{pmatrix}]_2, \quad \Phi(0) = I_{m-n}.$$

It is easy to check that

$$U_0(t) = T(t)\begin{pmatrix} Y_0(t) \\ Z_0(t) \end{pmatrix} \tag{25}$$

explicitly provides the $(m-n)$-dimensional solution manifold for the DAE (11) and the original pure index-two DAE (8).

We note that the $m$ boundary conditions (15) can be re-expressed as

$$R_0T(0)\begin{pmatrix} y(0,\epsilon) \\ z(0,\epsilon)\end{pmatrix} + R_1T(1)\begin{pmatrix} y(1,\epsilon) \\ z(1,\epsilon)\end{pmatrix} = s. \tag{26}$$

Moreover, (22)–(24) imply that

$$\begin{cases} y(0,\epsilon) = -[T^{-1}(0)C'(0)(C(0)C'(0))^{-1}g(0)]_1 + \mathcal{K}(0)\alpha(0) + O(\epsilon), \\ z(0,\epsilon) = Z_0(0) + O(\epsilon), \\ y(1,\epsilon) = -[T^{-1}(1)C'(1)(C(1)C'(1))^{-1}g(1)]_1 + \mathcal{L}(1)\beta(0) + O(\epsilon), \\ \text{and} \\ z(1,\epsilon) = \Phi(1)[Z_0(0) + \int_0^1 \Phi^{-1}(T^{-1}((AT-\dot{T})\begin{pmatrix}(-T^{-1}C'(CC')^{-1}g)_1 \\ 0\end{pmatrix} \\ \qquad + \mathbf{f}))_2 ds + O(\epsilon). \end{cases} \tag{27}$$

Thus, (26) can be rewritten as a linear system

$$\Delta_\Lambda \begin{pmatrix} \gamma \\ Z_0(0)\end{pmatrix} = \widetilde{s} \tag{28}$$

of $m$ equations where $\gamma$ is an $n$-vector denoting the $k$ possibly nonzero components of the projected $n$-vector $\mathcal{K}(0)\alpha(0)$ and the $n-k$ possibly nonzero components of $\mathcal{L}(1)\beta(0)$. If the matrix $\Lambda(t)$ in (19) is selected so that $\Delta_\Lambda$ is invertible, the corresponding regularized problem (14)–(15) will have a unique asymptotic solution for $\epsilon$ sufficiently small specified by $Z_0(0)$ and $\gamma$. It will feature endpoint boundary layers (so the limiting solution $\begin{pmatrix} Y_0 \\ Z_0\end{pmatrix}$ will be inconsistent with the prescribed boundary conditions) unless $\gamma = 0$. We note that for inconsistent boundary conditions different solutions $U_0$ of the DAE (14) would be obtained through (25) (i.e. as outer limits of the regularized boundary value problem for any given boundary vector $s$) if different choices of $\Lambda(t)$ provided invertible coefficient matrices $\Delta_\Lambda$ in (28).

## 3. An Example

Consider the index-two problem (8) with

$$A = \begin{pmatrix} 0 & 1 & 1 \\ 0 & 1 & 0 \\ 1 & 0 & 0 \end{pmatrix}, \quad B = \begin{pmatrix} 0 & 1 \\ 1 & 0 \\ 0 & 1 \end{pmatrix}, \quad f = \begin{pmatrix} 1 \\ 0 \\ 0 \end{pmatrix},$$

$$C = \begin{pmatrix} 1 & 1 & 0 \\ 1 & 0 & 1 \end{pmatrix}, \quad \text{and} \quad g = \begin{pmatrix} 1 \\ 0 \end{pmatrix}.$$

Let us introduce the regularization (13) with $S$ defined by (19) with

$$T = \begin{pmatrix} 1 & 1 & -1 \\ 0 & 1 & 1 \\ 1 & 0 & 1 \end{pmatrix} \quad \text{and} \quad \Lambda = \begin{pmatrix} \lambda_1 & 0 \\ 0 & \lambda_2 \end{pmatrix}.$$

Moreover, let us impose the three boundary conditions (15) for the resulting singularly perturbed system with

$$R_0 = \begin{pmatrix} 1 & 0 & 0 \\ 0 & 0 & 0 \\ 0 & 0 & 0 \end{pmatrix}, \quad R_1 = \begin{pmatrix} 0 & 0 & 0 \\ 0 & 1 & 0 \\ 0 & 0 & 1 \end{pmatrix}, \quad \text{and} \quad s = \begin{pmatrix} s_1 \\ s_2 \\ s_3 \end{pmatrix}$$

arbitrary. Then $\begin{pmatrix} y_1 \\ y_2 \\ z \end{pmatrix} \equiv T^{-1}u$ will satisfy the singularly perturbed boundary value problem

$$\begin{cases} \epsilon \dot{y}_1 = \lambda_1(y_1 - \frac{1}{3}), & y_1(0) + y_2(0) - z(0) = s_1 \\ \epsilon \dot{y}_2 = \lambda_2(y_2 + \frac{2}{3}), & y_2(1) + z(1) = s_2 \\ \dot{z} = -\frac{3}{2}z - \frac{1}{2}, & y_1(1) + z(1) = s_3. \end{cases} \tag{29}$$

Note that the limiting DAE (without boundary conditions) will have the general solution

$$\begin{cases} Y_{10} = \frac{1}{3}, \\ Y_{20} = -\frac{2}{3}, \\ Z_0(t) = -\frac{1}{3} + e^{-\frac{3}{2}t}(Z_0(0) + \frac{1}{3}). \end{cases} \tag{30}$$

*(a)*. Taking

$$\Lambda = \begin{pmatrix} -1 & 0 \\ 0 & 1 \end{pmatrix},$$

the singularly perturbed system (29) has bounded asymptotic solutions of the form

$$\begin{cases} y_1(t,\epsilon) = \frac{1}{3} + e^{-t/\epsilon}(y_1(0,\epsilon) - \frac{1}{3}), \\ y_2(t,\epsilon) = -\frac{2}{3} + e^{-(1-t)/\epsilon}(y_2(1,\epsilon) + \frac{2}{3}), \\ z(t,\epsilon) = -\frac{1}{3} + e^{-\frac{3}{2}t}(z(0,\epsilon) + \frac{1}{3}) = -\frac{1}{3} + e^{\frac{3}{2}(1-t)}(z(1,\epsilon) + \frac{1}{3}). \end{cases}$$

Neglecting only asymptotically exponentially small terms, the boundary conditions of (29) imply that we must have

$$\begin{cases} y_1(0,\epsilon) - \frac{2}{3} - z(0,\epsilon) \sim s_1, \\ y_2(1,\epsilon) - \frac{1}{3} + e^{-3/2}(z(0,\epsilon) + \frac{1}{3}) \sim s_2, \\ \frac{1}{3} + (-\frac{1}{3} + e^{-3/2}(z(0,\epsilon) + \frac{1}{3})) \sim s_3. \end{cases}$$

Thus, the unique corresponding asymptotic solution is

$$\begin{cases} y_1(t,\epsilon) \sim \frac{1}{3} + e^{-t/\epsilon}(s_1 + s_3 e^{3/2}) \\ y_2(t,\epsilon) \sim -\frac{2}{3} + e^{-(1-t)/\epsilon}(1 + s_2 - s_3) \\ z(t,\epsilon) \sim -\frac{1}{3} + e^{\frac{3}{2}(1-t)} s_3. \end{cases} \tag{31}$$

All solutions of the DAE are obtained as limiting solutions within $(0,1)$ as $s_3$ varies. To solve the DAE, we need retain only the terminal condition $Y_{10}(1) + Z_0(1) = s_3$. The initial condition will be inconsistent with the DAE unless $s_1 = -s_3 e^{3/2}$ and the first terminal condition will be inconsistent unless $s_2 = s_3 - 1$. In that special case, no boundary layers are necessary.

*(b).* If we instead take

$$\Lambda = \begin{pmatrix} 1 & 0 \\ 0 & -1, \end{pmatrix}$$

the regularized problem will have the unique asymptotic solution

$$\begin{cases} y_1(t,\epsilon) \sim \frac{1}{3} + e^{-(1-t)/\epsilon}[s_3 - s_2 - 1], \\ y_2(t,\epsilon) \sim -\frac{2}{3} + e^{-t/\epsilon}[s_1 + (s_2+1)e^{3/2}], \\ z(t,\epsilon) \sim -\frac{1}{3} + e^{\frac{3}{2}(1-t)}(s_2 + 1). \end{cases} \tag{32}$$

Now all solutions of the DAE are obtained as $s_2$ varies. To solve the DAE, we use only the terminal condition $Y_{20}(1) + Z_0(1) = s_2$. The remaining boundary conditions are now inconsistent unless $s_1 = -(s_2+1)e^{3/2}$ and $s_3 = s_2 + 1$.

*(c).* Finally, for

$$\Lambda = \begin{pmatrix} 1 & 0 \\ 0 & 1 \end{pmatrix},$$

the regularized problem has the unique terminal layer solution

$$\begin{cases} y_1(t,\epsilon) \sim \frac{1}{3} + e^{-(1-t)/\epsilon}[s_3 + e^{-3/2} s_1], \\ y_2(t,\epsilon) \sim -\frac{2}{3} + e^{-(1-t)/\epsilon}[s_2 + 1 + e^{-\frac{3}{2}} s_1], \\ z(t,\epsilon) \sim -\frac{1}{3} - e^{-\frac{3t}{2}} s_1. \end{cases} \tag{33}$$

Now all solutions of the DAE are obtained by letting $s_1$ vary and $\epsilon \to 0$. The consistency conditions (corresponding to trivial boundary layers) are the same as for the other two regularizations.

Because the solution space of this DAE is one-dimensional, while the given boundary conditions consist of one initial condition and two terminal conditions, one must expect that any appropriate regularized problem would need a terminal boundary layer. If one tried to regularize using the stable matrix

$$\Lambda = \begin{pmatrix} -1 & 0 \\ 0 & -1 \end{pmatrix},$$

all bounded asymptotic solutions of the singularly perturbed system would be of the form

$$\begin{cases} y_1(t,\epsilon) = \frac{1}{3} + e^{-t/\epsilon}(y_1(0,\epsilon) - \frac{1}{3}), \\ y_2(t,\epsilon) = -\frac{2}{3} + e^{-t/\epsilon}(y_2(0,\epsilon) + \frac{2}{3}), \\ z(t,\epsilon) = -\frac{1}{3} + e^{\frac{3}{2}(1-t)}(z(1,\epsilon) + \frac{1}{3}). \end{cases}$$

To satisfy the boundary conditions asymptotically, we'd need

$$\begin{cases} y_1(0,\epsilon) + y_2(0,\epsilon) - z(0,\epsilon) \sim s_1 \\ -\frac{2}{3} + z(1,\epsilon) \sim s_2 \\ \frac{1}{3} + z(1,\epsilon) \sim s_3. \end{cases}$$

This system is, however, self-contradictory (corresponding to a singular matrix $\Delta_\Lambda$ in (28)) unless $s_3 = s_2 + 1$, indicating that the proposed regularization is inappropriate. The complication with this proposed regularization occurs because there aren't two initial conditions to cancel in order to define the reduced problem satisfied by the solution of the given DAE.

## References


1. U.M. Ascher, H. Chin, and S. Reich (1993), "Stabilization of DAEs and invariant manifolds," preprint.

2. U.M. Ascher and L.R. Petzold (1992), "Projected collocation for higher-order higher index differential-algebraic equations," *J. Comp. Appl. Math.* 43, 243–259.

3. K.E. Brenan, S.L. Campbell, and L.R. Petzold (1989), *Numerical Solution of Initial Value Problems in Differential-Algebraic Equations*, North-Holland, Amsterdam.

4. E. Griepentrog, M. Hanke, and R. März (1992), *Berlin Seminar on Differential - Algebraic Equations*, Humboldt University, Berlin.

5. E. Hairer and G. Wanner (1991), *Solving Ordinary Differential Equations* II, *Stiff and Differential - Algebraic Problems*, Springer-Verlag, Berlin.

6. W.A. Harris, Jr. (1960), "Singular perturbations of two-point boundary problems for systems of ordinary differential equations," *Arch. Rational Mech. Anal.* 5, 212–225.

7. W.A. Harris, Jr. (1973), "Singularly perturbed boundary value problems revisited," *Lecture Notes in Math.* 312, Springer-Verlag, Berlin, 54–64.

8. L.V. Kalachev and R.E. O'Malley, Jr. (1991), "Regularization of linear differential - algebraic equations", Technical Report, University of Washington.

9. L.V. Kalachev and R.E. O'Malley, Jr. (1994), "Regularization of nonlinear differential - algebraic equations", *SIAM J. Math. Anal.*

10. L.V. Kalachev and R.E. O'Malley, Jr. (1993), "Boundary value problems for differential - algebraic equations", Technical Report, University of Washington.

11. V.A. Morozov (1993), *Regularization Methods for Ill-Posed Problems*, CRC Press, Boca Raton.

12. R.E. O'Malley, Jr. (1991), *Singular Perturbation Methods for Ordinary Differential Equations*, Springer, New York.

13. F. Takens (1976), "Constrained equations: a study of implicit differential equations and their discontinuous solutions", *Lecture Notes in Math.* 525, Springer-Verlag, Berlin, 143–234.

14. A.N. Tikhonov and V.Y. Arsenin (1977), *Methods for Solving Ill-posed Problems*, Wiley, New York.

15. W.C. Rheinholdt (1984). "Differential algebraic systems as differential equations on manifolds", *Math. Comp.* 43, 473–482.

16. A.B. Vasil'eva and V.F. Butuzov (1990), *Asymptotic Methods in Singular Perturbation Theory*, Vysshaya Shkola, Moscow.

17. W. Wasow (1944), "On the asymptotic solution of boundary value problems for ordinary differential equations containing a parameter", *J. Math. and Physics* 23, 173–183.

# Connection Coefficients For A Certain $n^{th}$ Order Differential Equation

T. K. Puttaswamy
Department Of Mathematical Sciences, Ball State University,
Muncie, Indiana - 47306, U. S. A.

## Abstract

This paper is devoted to the solution in the large of the differential equation

$$z^n(\mu^2 - 2\mu z^m + z^{2m})\frac{d^n y}{dz^n} + \sum_{k=0}^{n-1} z^k(a_k + b_k z^m + c_k z^{2m})\frac{d^k y}{dz^k} = 0$$

Here, $m$ is any arbitrary positive integer. The variable $z$ and the constants $\mu, a_k, b_k, c_k$ $(k = 0, 1, 2, \ldots, n-1)$ are complex with $\mu \neq 0$ and $a_{n-1} + b_{n-1}\mu^m + c_{n-1}\mu^{2m} = 0$. We shall also assume that the difference of no two roots of the indicial equation about the regular singular point $z = 0$ is congruent to zero modulo $m$.

The central question in a two point connection problem is to find a linear relationship between two sets of fundamental solutions defined in respective neighborhoods of two different singular points. This has been successfully carried out in this paper between the fundamental solutions at the regular points $z = 0$ and $z = \infty$.

## 1 Nature Of The Present Problem

In order to introduce the investigations of this paper, let us take for consideration the linear homogeneous differential equation of $n^{th}$ order

$$z^n(\mu^2 - 2\mu z^m + z^{2m})\frac{d^n y}{dz^n} + \sum_{k=0}^{n-1} z^k(a_k + b_k z^m + c_k z^{2m})\frac{d^k y}{dz^k} = 0 \qquad \mathbf{(1.1)}$$

Here, $m$ is an arbitrary positive integer. The variable $z$ and the constants, $\mu, a_k, b_k, c_k$ $(k = 0, 1, 2, \ldots, n-1)$ are complex with $\mu \neq 0$ and $a_{n-1} + b_{n-1}\mu^m + c_{n-1}\mu^{2m} = 0$. If $\mu_1, \mu_2, \ldots, \mu_m$ are the roots of $\mu - z^m = 0$, then in the language of Fuch's theory, the differential equation (1.1) will have regular singular points at $z = 0, z = \mu_i$ $(i = 1, 2, \ldots, m)$ and $z = \infty$. In all, (1.1) will have $m + 2$ regular singular points. Since $m$ is an arbitrary positive integer, the differential equation (1.1) will have arbitrary number of regular singular points.

If $\{h\}_k = h(h-1)(h-2)(h-k+1) = \prod_{j=0}^{k-1}(h-j), \quad k \geq 1$
and $\{h\}_0 = 1$, then, the indicial equation about the regular point $z = 0$ is found to be

$$\mu^2\{h\}_n + \sum_{k=0}^{n-1} a_k\{h\}_k = 0 \qquad \textbf{(1.2)}$$

We shall also assume that the roots $h_i (i = 1, 2, \ldots, n)$ of (1.2) are such that the difference of no two of them is congruent to zero modulo $m$. Then, from the existing theory of linear differential equations, (1.1) will have $n$ linearly independent solutions $y_i(z), i = 1, 2, 3, \ldots, n$ about the regular singular point $z = 0$ of the form

$$y_i(z) = z^{h_i} \sum_{k=0}^{\infty} g_i(k) z^{mk}$$

$$\text{with} \qquad g_i(0) = 1 \qquad i = 0, 1, 2, \ldots, n \qquad \textbf{(1.3)}$$

Here, each $g_i(k), k = 1, 2, 3, \ldots$ is a determinate function of $k$ and each of these series solutions in (1.3) will converge in a punctured circle drawn about $z = 0$ and extending up to the nearest of the singular points $z = \mu_i (i = 1, 2, \ldots, m)$. Outside this circle, each becomes divergent. The behavior of $y_i(z) (i = 1, 2, \ldots, n)$ in the neighborhood of $z = \infty$ is not available as far as Fuch's is concerned, even though $z = \infty$ is a regular singular point. However, we know from the established theory of linear differential equations, each of the solutions $y_i(z) (i = 1, 2, \ldots, n)$ about the regular singular point $z = 0$ can be expressed as some linear combination of $n$ linearly independent solutions $\tilde{y}_j(z) (j = 1, 2, \ldots, n)$ about the regular singular point $z = \infty$, having in general, the forms

$$\tilde{y}_j(z) = \frac{1}{z^{l_j}} \sum_{k=0}^{\infty} \frac{\tilde{g}_j(k)}{z^k} \qquad \text{with} \qquad \tilde{y}_j(0) = 1 \qquad \textbf{(1.4)}$$

for $j = 1, 2, \ldots, n$, where $l_j (j = 1, 2, \ldots, n)$ are the indicial exponents about the regular singular point $z = \infty$ and where each of the power series solutions in (1.4) is convergent, where $| z |$ is sufficiently large. The determination of the precise nature of such a linear relationship including the connection coefficients is a problem of great difficulty. These connection coefficients are called Stokes Multipliers. Until a convergent method of computing these connection coefficients is devised, a differential equations can not be considered as solved in the large. It is to the solution of this problem in the large that we address this paper.

## 2 Recurrence Ralations Satisfied by Each $g_i(n)$, $i = 1, 2, \ldots, n$

Substituting $y = \sum_{k=0}^{\infty} g(k) z^{mk+h}$ into (1.1) and equating the coefficients of $z^{mk+h+2}$, we get

$$P_2(k)g(k+2) + P_1(k)g(k+1) + P_0(k)g(k) = 0 \tag{2.1}$$

where

$$P_2(k) = \mu^2\{km+h+2m\}_n + \sum_{i=0}^{n-1} a_i\{km+h+2m\}_i \tag{2.2}$$

$$P_1(k) = -2\mu\{km+h+m\}_n + \sum_{i=0}^{n-1} b_i\{km+h+m\}_i \tag{2.3}$$

$$\text{and} \qquad P_0(k) = \{km+h\}_n + \sum_{i=0}^{n-1} c_i\{km+h\}_i \tag{2.4}$$

We observe that $g(k), k = 1, 2, 3, \ldots$ are completely and correctly determined by (2.1), if we arbitrarily take $g(-1) = 0$, as we do. Furthermore, we make the assumption that when one of $h_i (i = 1, 2, \ldots, n)$ is used for $h$, no root of $P_0(k-2) = 0$ is zero or a negative integer, in which case, we will have $g(-k) = 0$ for $k = 2, 3, \ldots$. Since we wish to regard $k$ as a complex variable, to emphasize this, we replace $k$ by $s$.

Our first objective, then, is to determine that analytic function $u(s)$ of the complex variable $s$, which is a solution of (2.1) and which takes on the initial values $u(0) = 1$ and $u(-1) = 0$.

## 3 Fundamental Solutions Of (2.1)

$$\text{Let} \qquad U_1(s) = u(s) \tag{3.1}$$

$$U_2(s) = u(s+1)$$

Then (2.1) can be put in the matrix form

$$G(s+1) = A(s)G(S) \tag{3.2}$$

where $G(s)$ is a 2x2 matrix which can be decomposed into two column vectors each with two components and $A(s)$ is 2x2 a matrix, representable as a power series in $\frac{1}{s}$ which is convergent in some neighborhood of $s = \infty$. Therefore, there is a finite $s_0$ such that the series

$$A(s) = A_0 + \frac{A_1}{s} + \frac{A_2}{s^2} + \ldots \tag{3.3}$$

converges if $| s |> s_0$,
where

$$A_0 = \begin{pmatrix} 0 & 1 \\ \gamma_0 & \omega_0 \end{pmatrix}$$

and $$A_j = \begin{pmatrix} 0 & 0 \\ \gamma_j & \omega_j \end{pmatrix} j = 1, 2, \ldots$$

with

$$\gamma_0 = -\frac{1}{\mu^2}$$

$$\omega_0 = \frac{2}{\mu}$$

$$\gamma_1 = \frac{1}{\mu^2 m}[\frac{a_{n-1}}{\mu^2} - c_{n-1} + 2nm]$$

$$\omega_1 = -\frac{2}{\mu m}[\frac{a_{n-1}}{\mu^2} + \frac{b_{n-1}}{2\mu} + nm]$$

By using a paper by W. J. A. Culmer and W. A. Harris, Jr[1], we find that the two linearly independent solutions $u_1(s)$ and $u_2(s)$ of (2.1) are expressible by means of known factorial series as follows:

$$u_1(s) = (\frac{1}{\mu})^s s^{r_1}[1 + E_1(s)]$$

$$u_2(s) = (\frac{1}{\mu})^s s^{r_2}[1 + E_2(s)] \qquad \mathbf{(3.4)}$$

where

$$r_1 = \frac{p + \sqrt{p^2 + 4\mu q}}{2}$$

$$r_2 = \frac{p - \sqrt{p^2 + 4\mu q}}{2} \qquad \mathbf{(3.5)}$$

with

$$p = \mu\omega_1 + 1$$

$$q = \mu\gamma_2 + \omega_2 \qquad \mathbf{(3.6)}$$

$u_1(s)$ and $u_2(s)$ are analytic at every point in some right half plane and in this right half plane $E_j(s) \longrightarrow 0$ as $| s | \longrightarrow \infty$ for $j = 1, 2$.

# 4 Main Results and the Main Theorem

We define,

$$f_0(h) = \mu^2\{k\}_n + \sum_{i=0}^{n-1} a_i\{k\}_i \tag{4.1}$$

$$f_1(h) = -2\mu\{k\}_n + \sum_{i=0}^{n-1} b_i\{k\}_i \tag{4.2}$$

Then, we observe, that in view of (1.2), $f_0(h) \equiv 0$.
Moreover, from (2.2) and (2.3) , we have

$$P_2(s) = f_0[m(s+2)+h] \tag{4.3}$$

$$P_1(s) = f_1[m(s+1)+h] \tag{4.4}$$

Again, in view of (1.2), $P_2(-2) = f_0(h) \equiv 0$

So 0 is a root of $P_2(s-2) = \mu^2\{ms+h\}_n + \sum_{i=0}^{n-1} a_i\{ms+h\}_i = 0$

If $\beta_i(i = 1, 2, \ldots, n-1)$ are the other roots of $P_2(s-2) = 0$, we observe that none of the other roots $\beta_i(i = 1, 2, \ldots, n-1)$ can be a positive integer. For, if $\beta_i = N(N = 1, 2, \ldots)$, then $P_2(N-2) = f_0(mN+h) = 0$, which would imply that the difference of two roots of the indicial equation (1.2) about the regular singular point $z = 0$ is congruent to zero modulo $m$, which contradicts our hypothesis.

We further note that if none of $\beta_i(i = 1, 2, \ldots, n-1)$ is a positive integer, then none of the quantities $1 - \beta_i(i = 1, 2, \ldots, n-1)$ can be zero or a negative integer.

**The main theorem is the following:**

**Theorem**

If the indicial exponents $l_j(j = 1, 2, \ldots, n)$ of regular singular $z = \infty$ of the differential equation (1.1) are such that the difference of no two roots of them is congruent to zero modulo $m$, then the solutions $y_i(z)(i = 1, 2, \ldots, n)$ about the regular singular point $z = 0$ as defined in (1.3), when considered for values of $z$ of large modulus and for which $\arg z^m \neq \arg \mu$ may be developed asymptotically in the form

$$y_i(z) \sim \sum_{k=1}^{n} c_k(h_i, l_1, l_2, \ldots, l_n)\tilde{y_k}(z),$$

where $\tilde{y_k}(z), k = 1, 2, \ldots, n$ are the solutions about the regular singular point $z = \infty$, as defined in (1.4)

**(a)** In case, none of the quantities $\frac{h+l_k}{m}, k = 1,2,\ldots,n$ is integer or in case $\frac{h+l_1}{m}$ is an integer $< 2$, the constant $c_1(h_i.l_1,l_2,\ldots,l_n)$ is determined by

$$c_1(h,l_1,l_2,\ldots,l_n) = \frac{e^{\pi i\frac{(h+l_1)}{m}}\Gamma(2-\frac{h+l_1}{m})\prod_{k=1}^{n-1}\Gamma(1-\beta_k)}{\mu^2 k^* \prod_{k=2}^{n}(l_k-l_1)\prod_{k=2}^{n}\Gamma(\frac{(h+l_k}{m}-1)} w(\frac{-h-l_1}{m}) \tag{4.5}$$

with
$$w(s) = v(s+2)f_0(2m-l_1) + v(s+1)f_0(m-l_1)$$
$$k^* = \begin{cases} -\mu\sqrt{p^2+4q}, & \text{if } \; p^2+4q \neq 0 \\ -\mu^3, & \text{if } \; p^2+4q = 0 \end{cases}$$
and
$$v(s) = \begin{vmatrix} u_1(-1) & u_2(-1) \\ u_1(s) & u_2(s) \end{vmatrix}$$

in which $h_i$ is used for $h$, while the remaining constants $c_j(h_i,l_1,l_2,\ldots,l_n), j = 2,3,\ldots,n$ are determined by interchanging $l_1$ and $l_j$.

**(b)** In case, $\frac{h+l_1}{m} = N =$ an integer $\geq 2$, the constant $c_1(h_i,l_1,l_2,\ldots,l_n)$ is determined by

$$c_1(h,l_1,l_2,\ldots,l_n)$$

$$= \frac{\prod_{k=1}^{n-1}\Gamma(1-\beta_k)[f_0(2m-l_1\frac{d}{ds}v(s+2)+f_1(m-l_1)\frac{d}{ds}v(s+1)]_{s=-N}}{\mu^2 k^*\Gamma(N-1)\prod_{k=2}^{n}(l_k-l_1)\prod_{k=2}^{n}\Gamma(\frac{h+l_k}{m}-1)} \tag{4.6}$$

by the use of $h = h_i$, while the constants $c_j(h_i,l_1,l_2,\ldots,l_n), j = 2,3,\ldots,n$ continues to be determined from (4.5) as in (a).

In the above theorem, it must be understood that we have assumed $\mu \neq 0$ in (1.1) and the indicial exponents $h_i(i = 1,2,\ldots,n)$ about the regular singular point $z = 0$ are such that the difference of no two of them is congruent to zero modulo $m$. Moreover, in as much as the series $\tilde{y}_j(z), j = 1,2,\ldots,n$ are known in the present instance to be convergent for the indicated values of $z$, the symbol $\sim$ as used above may be changed to $=$ and the resulting equation may be regarded as those which join the fundamental solutions (1.3) with the fundamental solutions (1.4) upon a suitably constructed Reimann surface having the cut the line drawn from $z^m = \mu$ to $z = \infty$.

Proof of theorem will be given in sections 5 and 6.

# 5 Solutions $y_i(z), i = 1,2,\ldots,n$ About The Regular Singular Point $z = 0$

By rendering precise meaning through certain conventions and by analytic continuation, we find that the two linearly independent solutions $u_1(s)$ and $u_2(s)$ of (2.1) are single valued and analytic throughout the s-plane except for the poles at the points

$$s = \alpha_j - k, k = 0, 1, 2, \ldots, j = 1, 2, \ldots, n \tag{5.1}$$

where $\alpha_j (j = 1, 2, \ldots, n)$ are the roots of $P_o(s) = 0$

Next, we determine the constants $a$ and $b$ such that

$$u(s) = au_1(s) + bu_2(s) \tag{5.2}$$

The constants $a$ and $b$ are evidently to be determined from the initial conditions $u(0) = 1$ and $u(-1) = 0$. However, in order to obtain such a function $u(s)$, we shall assume that none of the poles (5.1) is an integer, the result of removing this restriction, we shall examine at a later point, but it may be observed that, in part this restriction was made at the outset, when we assumed that none of the roots of $P_0(s-2)$ is zero or a negative integer.

If we let

$$D(s) = \begin{vmatrix} u_1(s) & u_2(s) \\ u_1(s+1) & u_2(s+1) \end{vmatrix} \tag{5.3}$$

we find that

$$a = -\frac{u_2(-1)}{D(-1)} \qquad \text{and} \qquad b = \frac{u_1(-1)}{D(-1)} \tag{5.4}$$

Our next objective, then is to find the functional value of the Casorati determinant $D(s)$. By Heyman's theorem, $D(s)$ satisfies

$$\frac{D(s+1)}{D(s)} = \frac{P_0(s)}{P_2(s)} = \frac{\prod_{k=1}^{n}(s-\alpha_k)}{\mu^2 \prod_{k=1}^{n}(s-\beta_k+2)},$$

where $\beta_j, j = 1, 2, \ldots, n$ are the roots of $P_2(s-2) = 0$ Since 0 is a root of $P_2(s-2) = 0$, let $\beta_n$ be this root. Then,

$$D(s) = \frac{k^*(\frac{1}{\mu^2})^s \prod_{k=1}^{n} \Gamma(s-\alpha_k)}{\Gamma(s+2)\prod_{k=1}^{n-1}\Gamma(s-\beta_k+2)} \tag{5.5}$$

So, $$D(-1) = \frac{k^*\mu^2 \prod_{k=1}^{n}\Gamma(-1-\alpha_k)}{\prod_{k=1}^{n-1}\Gamma(1-\beta_k)} \tag{5.6}$$

Since none of the expressions $1 - \beta_j (j = 1, 2, \ldots, n-1)$ can be zero or a negative integer, $D(-1) \neq 0$. Again, we note that the numerator in (5.6) is necessary finite, in as much as we have assumed that $\alpha_j, j = 1, 2, \ldots, n$ are non integers.

So, in view of (5.2) and (5.6),

$$u(s) = \frac{\prod_{k=1}^{n-1}\Gamma(1-\beta_k)}{k^*\mu^2 \prod_{k=1}^{n}\Gamma(-1-\alpha_k)} v(s) \tag{5.7}$$

$$\text{where} \qquad v(s) = \begin{vmatrix} u_1(-1) & u_2(-1) \\ u_1(s) & u_2(s) \end{vmatrix} \tag{5.8}$$

Next, we state the following lemma:

**Lemma:**

If $l_j (j = 1, 2, \ldots, n)$ are the indicial exponents about the regular singular point $z = \infty$ of the differential equation (1.1), then

$$\alpha_j = -(\frac{l_j + h}{m}), j = 1, 2, \ldots, n \tag{5.9}$$

Proof of the lemma will be not given here due to the limitation of space.

It follows that each of the solutions $y_i(z), i = 1, 2, \ldots, n$ of (1.3) about the regular singular point $z = 0$ of the differential equation (1.1) is of the form

$$y(z) = z^h[\frac{-u_2(-1)}{D(-1)} \sum_{k=0}^{\infty} G_1(k)(\frac{z^m}{\mu})^k + \frac{u_1(-1)}{D(-1)} \sum_{k=0}^{\infty} G_2(k)(\frac{z^m}{\mu})^k] \tag{5.10}$$

where

$$G_1(s) = s^{r_1}[1 + E_1(s)]$$
$$G_2(s) = s^{r_2}[1 + E_2(s)]$$

# 6 Asymptotic Behavior of the Solutions $y_i(z)$, $i = 1, 2, \ldots, n$ About the Regular Singular Point $z = 0$.

We apply the modified version of W. B. Ford's theorem [5] to determine the asymptotic development of the functions, defined by the two series in (5.10). Then, we find that the fundamental solution $y(z)$ about the regular singular point $z = 0$ as defined by (1.3), when $| z |$ is large and when $z$ lies along any ray except for which $\arg z^m = \arg \mu$ may be expanded asympotically in the form

$$y(z) \sim -z^h R(z) \tag{6.1}$$

where $R(z)$ represents the sum of the residues of the function

$$\frac{\pi u(s)(-z)^s}{\sin \pi s} \tag{6.2}$$

at the poles (5.1).

Our next objective, then, is to obtain $R(z)$. If $R_{jk}(z)$ is the residue of **(6.2) at** pole $s = \alpha_j - k = -\frac{h+l_j}{m} - k$,
then,

$$R(z) = \sum_{j=1}^{n} \sum_{k=0}^{\infty} R_{jk}(z) \tag{6.3}$$

At this stage, we shall also assume that none of $l_j - l_k (j \neq k, j, k = 1, 2, \ldots, n)$ is congruent to zero modulo $m$. Then the poles (5.1) will all be simple poles of (6.2).

After computing $R_{jk}(z)$, we find from (6.3) that when $| z |$ is large and when $z$ lies along any ray except for which $\arg z^m = \arg \mu$,

$$y(z) \sim \sum_{k=1}^{n} c_k(h, l_1, l_2, \ldots, l_n) \frac{1}{z^{lk}} [1 + \frac{(\;\;)}{z^m} + \frac{(\;\;)}{z^{2m}} + \ldots] \tag{6.4}$$

where $c_1(h, l_1, l_2, \ldots, l_n)$ is given by (4.5) and where $c_k(h, l_1, l_2, \ldots {}_n)$ for $k = 2, 3, \ldots, n$ are obtained from (4.5) by interchanging $l_1$ and $l_k$.

In arriving at this result, we have placed several restrictions. In fact, we have assumed that none of the quantities $\alpha_k = \frac{-h-l_k}{m} (k = 1, 2, \ldots, n)$ and $\alpha_k - \alpha_j = \frac{l_j - l_k}{m} (k \neq j, j, k = 1, 2, \ldots, n)$ is an integer.

Assuming that $\alpha_k - \alpha_j = \frac{l_j - l_k}{m} (k \neq j, j, k = 1, 2, \ldots, n)$ is not an integer, which means that the difference of no two indicial exponents $l_j$ and $l_k (k \neq j, j, k = 1, 2, \ldots, n)$ is congruent to zero modulo $m$, we shall now examine the effects upon (4.5) when one of $\frac{h+l_k}{m}, k = 1, 2, \ldots, n$ is an integer. In this case, it is to be noted that in no case two of these quantities can be integers, for this would imply that the difference of two indicial exponents about $z = \infty$ is conguent to zero modulo $m$, contradicting the hypothesis. Moreover, it will suffice to consider merely the case in which $\frac{h+l_1}{m}$ is an integer in as much as all possible cases may be then covered by merely interchanging those of $l_1$ and $l_j (j = 2, 3, \ldots, n)$.

We shall suppose that $\frac{h+l_1}{m}$ is an integer. Let us divide the possible integral values of $\frac{h+l_1}{m}$ into two classes:

**(a)** $\frac{h+l_1}{m}$ is an integer $< 2$ **(b)** $\frac{h+l_1}{m}$ is an integer $\geq 2$

**Case (a)** $\frac{h+l_1}{m}$ **is an integer** $< 2$

If we now form the differential equation in which $\mu^2$ in (1.1) is changed to $\mu^2 - E$, where $E$ is arbitrarily small and positive, the resulting values of $l_j (j = 1, 2, \ldots, n)$ remains as before, in as much as they depend only on $c_j, j = 0, 1, 2, \ldots, n-1$.

The resulting value of $h$ will be of the form $h - \eta$, where $\eta$ becomes indefinitely small with $E$ and atleast after $E$ has been taken sufficiently small, none of the quantities $\frac{h+l_k-\eta}{m}$ is an integer. The solution $y(z)$ corresponding to (1.3) of the new differential equation will therefore be developable asymptotically in the form obtained by (6.4) replacing $\frac{h}{m}$ by $\frac{h-\eta}{m}$ and the resulting $c_k(h-\eta, l_1, l_2, \ldots, l_n)$ determined from (4.5).

After the expressions thus obtained have been examined, it is noted that when $\frac{h+l_1}{m}$ is an integer $< 2$, each of the following limit exists:

$$\lim_{\eta \longrightarrow 0} \frac{\Gamma(2 - \frac{h+l_k-\eta}{m})}{\prod_{j=1_{j\neq k}}^{n} \Gamma(\frac{h+l_j-\eta}{m} - 1)} \quad \text{for} \quad k = 1, 2, \ldots, n \tag{6.5}$$

Hence we conclude that so long as $\frac{h+l_1}{m}$ is an integer belonging to class (a) mentioned above, the relation (6.4) continues as before.

**Case (b) $\frac{h+l_1}{m}$ is an integer $\geq 2$**

First we consider the case $\frac{h+l_1}{m} = 2$. Proceeding as before, we obtain the new differential equation from (1.1) by replacing $\mu^2$ by $\mu^2 - E$ and hence $h$ by $h - \eta$ in (4.5). We endeavour to examine the limit as $\eta \longrightarrow 0$ of $c_1(h-\eta, l_1, l_2, \ldots, l_n)$ as determined by (4.5). In doing so, there is involved aside from the factor which clearly approaches a limit, the product

$$\Gamma(\frac{\eta}{m})[f_0(h)v(\frac{\eta}{m}) + f_1(h-m)v(\frac{\eta}{m} - 1)] \tag{6.6}$$

Since $f_0(h) \equiv 0$, we have $f_0(h)v(\frac{\eta}{m}) = 0$

Hence, limit of (6.6) as $\eta \longrightarrow 0$

$$= \lim_{\eta \longrightarrow 0} \Gamma(\frac{\eta}{m})[f_1(h-m)v(\frac{\eta}{m} - 1)] \tag{6.7}$$

We observe that as $\eta \longrightarrow 0$ $\Gamma(\frac{\eta}{m})$ becomes infinite to the first order.

Also $\lim_{\eta \longrightarrow 0} f_1(h-m)v(\frac{\eta}{m} - 1) = f_1(h-m)v(-1) = 0$

Thus the limit of (6.7) as $\eta \longrightarrow 0$ is indeterminate. We can however evaluate it upon replacing

$$\Gamma(\frac{\eta}{m}) = \frac{\Gamma(1 + \frac{\eta}{m})}{\frac{\eta}{m}}$$

the result being $mf_1(h-m)[\frac{d}{d\eta}v(\frac{\eta}{m} - 1)]_{\eta=0}$

$$= f_1(h-m)[\frac{d}{ds}v[(s-1)]_{s=0}$$

or as we shall prefer to write it for future purposes

$$[f_0(2m-l_1)\frac{d}{ds}v(s+2)+f_1(m-l_1)\frac{d}{ds}v(s+1)]_{s=\frac{-(h+l_1)}{m}=-2} \tag{6.8}$$

Let us suppose next that $\frac{h+l_1}{m}$ an integer $N>2$. Then, reasoning as in the case $\frac{h+l_1}{m}=2$, we replace $h$ in (4.5) by $h-\eta$ and attempt to obtain the limit of the resulting expressions as $\eta \longrightarrow 0$. No difficulties are encountered except in connection with the product

$$\Gamma(2-N+\frac{\eta}{m})[f_0(2m-l_1)v(2-N+\frac{\eta}{m})+f_1(m-l_1)v(1-N+\frac{\eta}{m})] \tag{6.9}$$

But $$\Gamma(2-N+\frac{\eta}{m})=\frac{(-1)^N\Gamma(1+\frac{\eta}{m})}{(N-2-\frac{\eta}{m})(N-3-\frac{\eta}{m})\dots(1-\frac{\eta}{m})(\frac{\eta}{m})}$$

Since $N>2$,

$$\lim_{\eta\longrightarrow 0} v(2-N+\frac{\eta}{m})=v(2-N)=0$$

and $$\lim_{\eta\longrightarrow 0} v(1-N+\frac{\eta}{m})=v(1-N)=0$$

Hence, we see limit of (6.9) as $\eta \longrightarrow 0$ is

$$\frac{(-1)^N}{\Gamma(N-1)}[f_0(2m-l_1)\frac{d}{ds}v(s+2)+f_1(m-l_1)\frac{d}{ds}v(s+1)]_{s=-N} \tag{6.10}$$

Finally we note that (6.8)may be obtained from (6.10) by putting $N=2$, so that when $\frac{h+l_1}{m}$ is an integer $\geq 2$, we conclude that the relation (6.4) for which $c_1(h,l_1,l_2,\dots,l_n)$ is no longer to be determined from (4.5) but by (4.6), while $c_j(h,l_1,l_2,\dots,l_n)$ for $j=2,3,\dots,n$ are to be determined by interchanging $l_1$ and $l_j$ in (4.5)

Due to the limitation of space, the case when the difference of the indicial exponents about $z=\infty$ is congruent to zero modulo $m$ will not be discussed here and will be published elsewhere as a sequel to this paper. This paper extends and generalizes [3], [4] and [6].

## References

[1] W. J. A. Culmer and W. A. Harris Jr, *'Convergent solutions of Ordinary Linear Homogeneous Difference Equations'*, Pac. Jour. of Math. 13 (1963), 1111 - 1138.

[2] W. B. Ford, *'Asymptotic Developments of Functions Defined by Maclaurin Series'*, Chelsa Publishing Company (1960).

[3] T. K. Puttaswamy, *'Solution in the Large of a Certain $n^{th}$ Order Differential Equation'*, Proceedings of Ramanujan Birth Centenary Year International Symposium on Analysis, Macmillan India Limited (1989), 451 - 465.

[4] T. K. Puttaswamy, *'A connection problem for a $n^{th}$ Order Homogeneous Differential Equation'*, Proceedings of the International Conference on Theory and Applications of Differential Equations, Ohio University Press (1989), 317 - 321.

[5] T. K. Puttaswamy, *'A Generalization of a Theorem of W. B. Ford'*, Proceedings of the International Conference on New Trends in Geometry Function Theory and Applications, World Scientific (1991), 90-95.

[6] T. K. Puttaswamy, *'Stokes Multiplier for a Certain $n^{th}$ Order Differential Equation'*, to appear in the Proceedings of the Third International Colloquim on Differential Equations held at Plovdiv, Bulgaria during August 18-22, 1992.

[7] T. K. Puttaswamy, *'Stokes Multiplier for a Certain Third Order Differential Equation'*, to appear in the Special Volume of the Journal of Ramanujan Mathematical Society, dedicated to the memory of late Professor M. Venkataraman.

# SYNTHESIZING OPTIMAL CONTROLS FOR NONLINEAR SYSTEMS WITH NONQUADRATIC COST CRITERIA

by

David L. Russell

Department of Mathematics
Virginia Polytechnic Institute and State University
Blacksburg, VA 24061

Abstract

We describe the use of system decoupling techniques, particularly related to the stable manifold theorem, to obtain optimal controls for nonlinear systems and nonquadratic cost criteria in nonlinear synthesized form.

1. Foreword As the subject material of this talk differs somewhat from that of the majority of talks in this International Symposium, some orientation may benefit the listener/reader. The author very happily acknowledges his particular debt to two of the honorees of this symposium, Professors Harris and Sibuya. Both played a decisive part assisting the writer in the development of his Phd thesis [ 14 ] at the University of Minnesota in 1964. Subsequently the author turned his main research effort toward the then developing theories of Optimal Control and Control of Distributed Parameter Systems. While this area has a "flavor" quite different from either the classical or modern Asymptotic Theory of ordinary differential equations, many of the techniques learned for the Asymptotic Theory, perhaps not surprizingly, have turned out to be crucial in Control investigations. That is particularly true of the topic presented here, which makes essential use of the stable manifold theorem [ 4 ], [ 5 ], itself a statement about asymptotic properties of ordinary differential equations, and quite similar in its proof to many of the diagonalization results which were beginning to play an increasingly strong role in Asymptotic Theory during the 1960's. Thus the author is delighted to acknowledge an even stronger debt to the honorees as techniques originally learned under their supervision continue to be useful and essential in the development of his current research program.

The full length paper on which this talk is based, authored by the present writer and his PhD student, Xiaohong Zhang, will appear in [ 17 ]. The emphasis in this short presentation, however, is somewhat different than in that work.

2. Linear Quadratic Optimal Control Theory This theory, now almost "over familiar", was developed by Kalman and Bucy [ 7 ], [ 8 ] in the 1960's and subsequently became, at least until 1980, when the $H^\infty$ theory [ 6 ] was introduced, a favored mathematical tool in design of linear control systems. It concerns a linear control system

$$\dot{x} = A(t)\, x + B(t)\, u, \quad x \in R^n,\ u \in R^m , \tag{2.01}$$

wherein x is the state variable and u is the control, or input, variable. For a given (positive definite here, for simplicity) symmetric weighting matrices

$$\begin{bmatrix} W(t) & R(t) \\ R(t)^* & U(t) \end{bmatrix}, \ (n+m) \times (n+m), \quad V,\ n \times n,$$

one poses the optimal control problem, given an initial state $x_0$ for (2.01), of finding a control $\hat{u}(t) \in L^2[t_0,t_1]$ minimizing the cost

$$\int_{t_0}^{t_1} \begin{bmatrix} x(t)^* & u(t)^* \end{bmatrix} \begin{bmatrix} W(t) & R(t) \\ R(t)^* & U(t) \end{bmatrix} \begin{bmatrix} x(t) \\ u(t) \end{bmatrix} dt + x(t_1)^* V x(t_1) , \tag{2.02}$$

it being understood that x(t), with $x(t_0) = x_0$, and u(t) together satisfy (2.01). It turns out that the optimal control $\hat{u}(t)$, with resulting optimal trajectory $\hat{x}(t)$, is unique and is obtained in the following way [ 1 ]. One first solves the matrix Riccati differential equation, with given terminal data,

$$\dot{Q}(t)+A(t)^*Q(t)+Q(t)A(t)+W(t)-\hat{K}(t)^*U(t)\hat{K}(t) = 0, \ Q(t_1) = V,$$

wherein

$$\hat{K}(t) = -\, U(t)^{-1}\left[B(t)^*Q(t)+R(t)^*\right] ;$$

it turns out that the solution $Q(t)$ is defined, symmetric and positive definite over the whole interval $[t_0,t_1]$. With $Q(t)$ thus determined, one constructs the solution $\hat{x}(t)$ of

$$\dot{\hat{x}}(t) = \left[A(t)+B(t)\hat{K}(t)\right] \hat{x}(t),\ \hat{x}(t_0) = x_0, \tag{2.03}$$

over $[t_0,t_1]$. Then

$$\hat{u}(t) \equiv \hat{K}(t)\ \hat{x}(t),\ t \in [t_0,t_1], \tag{2.04}$$

so that (2.03) is, in fact, the same as

$$\dot{\hat{x}}(t) = A(t)\ \hat{x}(t) + B(t)\ \hat{u}(t)\ .$$

For the constant coefficient case wherein all matrices in (2.01) and (2.02) are constant, the corresponding problem on the interval $[0,\infty)$ involves the unique symmetric positive definite solution $Q$ of the matrix quadratic equations

$$A^*Q + QA + W + \hat{K}^*U\ \hat{K}\ ,\ \hat{K} = -U^{-1}\left[B^*Q + R^*\right] \tag{2.05}$$

in the same framework, so that

$$\hat{u}(t) \equiv \hat{K}\ \hat{x}(t),\ \dot{\hat{x}}(t) \equiv \left[A + B\hat{K}\right] \hat{x}(t) \equiv A\ \hat{x}(t) + B\hat{u}(t),\ t \in [0,\infty).$$

In this case, which is generally the most pertinent for practical control design, it turns out that the "closed loop", or "synthesized" system matrix $A + B\hat{K}$ is a stability matrix, i.e., all of its eigenvalues have negative real parts. The manner in which solutions of the closed loop system

$$\dot{\hat{x}}(t) = \left[A + B\hat{K}\right] \hat{x}(t)$$

approach 0 as $t \to \infty$ is strongly influenced by the choice of the matrices $W$, $R$ and $U$, allowing various properties of the closed loop system to be realized by appropriate choice of these matrices.

A quite analogous theory [ 2 ], [ 3 ] exists for discrete linear systems

$$x_{k+1} = A_k x_k + B_k u_k, \ k = K_0, \ldots, K_1 ,$$

$$x_{k+1} = A\, x_k + B\, u_k, \ k = 0,1, \ldots, \quad \text{(constant coeffient case)}$$

with cost sums analogous to the integral in (2.02). The formulae for the optimal control $\hat{u}_k$ in this case are similar but rather more complicated; more on this in §4.

3. <u>NLNQ Problems</u> The acronym NLNQ stands for <u>N</u>on<u>L</u>inear, <u>N</u>on<u>Q</u>uadratic; to the extent that the theory involves only small, perturbational, departures from the linear theory of the preceding section, it might well also be interpreted as meaning <u>N</u>early <u>L</u>inear, <u>N</u>early <u>Q</u>uadratic. This theory began with the 1967 University of Minnesota PhD thesis of D. L. Lukes, the main results being subsequently published in [ 11 ], [ 12 ]. In [ 4 ],[ 17 ] we show that the finite interval time varying coefficient case, not explicitly studied by Lukes but also not very difficult, can also be developed along these lines but, for the present, we consider only autonomous systems. The theory, then, concerns a system of differential equations

$$\dot{x} = F(x,u) = A\,x + B\,u + f(x,u)\ , \ x(0) = x_0 \text{ given,} \qquad (3.01)$$

or a system of recursion equations

$$x_{k+1} = F(x_k,u_k) = A\,x_k + B\,u_k + f(x_k,u_k)\ , \ x_0 \text{ given,} \qquad (3.02)$$

with f, assumed continuously differentiable, and its first order partial derivatives with respect to the components of x and u all vanishing at (0,0), so that

$$f(x,u) = o\left(\|x\|+\|u\|\right), \ \|x\|, \ \|u\| \to 0 .$$

In each case it is desired to find an optimal control, $\hat{u}(t)$ or $\hat{u}_k$, minimizing, as appropriate,

$$\int_0^\infty G(x(t),u(t)\ dt\ , \tag{3.03}$$

or

$$\sum_{k=1}^{\infty} G(x_k,u_k)\ , \tag{3.04}$$

with G(x,u) twice continuously differentiable and

$$G(x,u) = \begin{bmatrix} x^* & u^* \end{bmatrix} \begin{bmatrix} W & R \\ R^* & U \end{bmatrix} \begin{bmatrix} x \\ u \end{bmatrix} + g(x,u)\ , \tag{3.05}$$

$$g(x,u) = o\left(\|x\|^2+\|u\|^2\right),\ \|x\|,\ \|u\| \to \infty\ .$$

A quick example in the differential equations case would consist of the pendulum system

$$\begin{bmatrix} \dot{x}_1 \\ \dot{x}_2 \end{bmatrix} = \begin{bmatrix} x_2 \\ -\sin x_1 + (\cos x_1)\ u \end{bmatrix} = \begin{bmatrix} 0 & 1 \\ -1 & 0 \end{bmatrix} \begin{bmatrix} x_1 \\ x_2 \end{bmatrix} + \begin{bmatrix} 0 \\ 1 \end{bmatrix} u + \cdots$$

and the cost functional, corresponding to system energy plus a quadratic control cost,

$$\int_0^\infty \left[ \frac{1}{2} x_2(t)^2 + \left[1 - \cos x_1(t)\right] + \frac{1}{2} u(t)^2\right] dt\ ,$$

for which the leading quadratic form involves the matrix

$$\begin{bmatrix} W & R \\ R^* & U \end{bmatrix} = \frac{1}{2} \begin{bmatrix} 1 & 0 & 0 \\ 0 & 1 & 0 \\ 0 & 0 & 1 \end{bmatrix} .$$

The differential equations case was treated by D. L. Lukes in his thesis. He showed that there is a neighborhood, $\mathfrak{N}$, of the origin in $R^{m+n}$ and a continuously differentiable "feedback" function

$$\hat{\mathbb{K}}(x) = \hat{K}\ x + \hat{\mathfrak{K}}(x),\ \hat{\mathfrak{K}}(x) = o(\|x\|),\ \|x\| \to \infty, \tag{3.06}$$

$\hat{K}$ being the optimal feedback matrix for the truncated linear quadratic problem, as described in the preceding section, such that the (locally) unique optimal control $\hat{u}(t)$ is synthesized by

$$\hat{u}(t) = \hat{\mathbb{K}}(\hat{x}(t)) \qquad (3.07)$$

and the synthesized system

$$\dot{\hat{x}}(t) = F(\hat{x}(t),\hat{\mathbb{K}}(\hat{x}(t))) \equiv F_{\hat{\mathbb{K}}}(\hat{x}(t)) \qquad (3.08)$$

has the origin as an asymptotically stable critical point.

We will give a very brief sketch of the proof of this result. From the Pontryagin Maximum Principle [ 10 ] we obtain the necessary conditions to the effect that for each $t \in [0,\infty)$ the optimal control value $\hat{u}(t)$ solves the optimization problem

$$\max_u \left\{-\lambda^*(t)F(\hat{x}(t),u) - G(\hat{x}(t),u)\right\}$$

leading to

$$\lambda^*(t)\frac{\partial F}{\partial u}(\hat{x}(t),\hat{u}(t)) + \frac{\partial G}{\partial u}(\hat{x}(t),\hat{u}(t)) = 0 \ , \qquad (3.09)$$

where $\lambda(t)$ satisfies the differential equation

$$\dot{\lambda}^*(t) = -\lambda(t)^*\frac{\partial F}{\partial x}(\hat{x}(t),\hat{u}(t)) - \frac{\partial G}{\partial x}(\hat{x}(t),\hat{u}(t)) \qquad (3.10)$$

with

$$\lim_{t\to\infty} \lambda(t) = 0 \ . \qquad (3.11)$$

Taking account of the forms (3.01), (3.02) of F and G (3.09) can be solved near (0,0) to give

$$\hat{u}(t) = \hat{L}(\hat{x}(t),\lambda(t)) = -U^{-1}\left[R^*\hat{x}(t) + B^*\lambda(t)\right] + \hat{\ell}(\hat{x}(t),\lambda(t)) \ . \qquad (3.12)$$

Then writing $\lambda(t)$ in the form (cf.(2.05))

$$\lambda(t) = 2Q\,\hat{x}(t) + \eta(t)\ , \tag{3.13}$$

the coupled $\hat{x}$, $\eta$ equations may be seen to take the linearly block triangulated form

$$\begin{bmatrix} \dot{\hat{x}}(t) \\ \dot{\eta}(t) \end{bmatrix} = \begin{bmatrix} A+B\hat{K} & -\frac{1}{2}BU^{-1}B^{*} \\ 0 & -(A+B\hat{K})^{*} \end{bmatrix} \begin{bmatrix} \hat{x}(t) \\ \eta(t) \end{bmatrix} + \begin{bmatrix} f_1(\hat{x}(t),\eta(t)) \\ f_2(\hat{x}(t),\eta(t)) \end{bmatrix} .$$

The fact that $A+B\hat{K}$ has all eigenvalues with real parts negative while $-(A+B\hat{K})^{*}$ has all eigenvalues with positive real parts enables one at this point to invoke the stable manifold theorem [ 4 ],[ 5 ] to show that in some neighborhood of the origin in $R^{2n}$ we have a unique continuously differentiable relationship

$$\eta(t) = \hat{\eta}(\hat{x}(t))$$

which, substituted into (3.12), (3.13), yields

$$\hat{u}(t) = \hat{L}(\hat{x}(t), 2Q\,\hat{x}(t) + \hat{\eta}(\hat{x}(t)) =$$

$$= -U^{-1}\left[R^{*} + B^{*}Q\right]\hat{x}(t) + \hat{\ell}(\hat{x}(t), 2Q\,\hat{x}(t) + \hat{\eta}(\hat{x}(t)) = \hat{K}\,\hat{x}(t) + \hat{\kappa}(\hat{x}(t))\ ,$$

thereby establishing the relationship (3.06) as claimed. While this argument is entirely local, it turns out that the nonlinear feedback relationship must, in fact, extend over the range of all values of x which can be controlled to the origin in such a way as to keep the cost (3.03) finite.

An analogous argument applies in the case of the discrete system (3.02) with the cost sum (3.04). We will have more to say about this in the next section. However, to offer an analog of the approach we use in that section, we note that there is another way in which the results described above for the continuous time case can be obtained.

The Hamilton - Jacobi - Bellman framework for solution of the optimal control problem discussed here involves the so-called *optimal value function* $J(x_0)$ whose value at $x_0$ is the least achievable value for the cost (3.03), given that the control, solution pair x(t), u(t) is required to satisfy (3.01) and

$$\lim_{t\to\infty} x(t) = 0 .$$

The dynamic programming formalism characterizes the optimal control as the (locally unique) solution of the equation

$$\frac{\partial J}{\partial x}(\hat{x}(t)) \frac{\partial F}{\partial u}(\hat{x}(t),\hat{u}) + \frac{\partial G}{\partial u}(\hat{x}(t),\hat{u}) = 0 ,$$

which yields $\hat{u}$ in the partially synthesized form

$$\hat{u}(t) = \hat{u}(\hat{x}(t),\frac{\partial J}{\partial x}(\hat{x}(t))).$$

It is then shown that for all x in some neighborhood of the origin in $R^n$ the optimal value function J(x) satisfies the Hamilton - Jacobi partial differential equation

$$\frac{\partial J}{\partial x}(x) \frac{\partial F}{\partial x}(x,\hat{u}(x,\frac{\partial J}{\partial x}(x))) + \frac{\partial G}{\partial x}(x,\hat{u}(x,\frac{\partial J}{\partial x}(x))) = 0 . \tag{3.14}$$

This can be seen to be a scalar, first order nonlinear hyperbolic partial differential equation for J(x). The characteristic curves turn out to be optimal trajectories, i.e., the solutions of

$$\dot{\hat{x}}(t) = F(\hat{x}(t),\hat{u}(\hat{x}(t),\frac{\partial J}{\partial x}(\hat{x}(t))) , \tag{3.15}$$

which then turns out to be the synthesized optimal system (3.08). Unfortunately, there are substantial difficulties in establishing, even locally, the existence and uniqueness of a continuously differentiable solution J(x) of (3.14), due in large part to the fact that the data J(0) = 0 for J is given only at the point x = 0 which is the common limit of the curves $\hat{x}(t)$ as $t \to \infty$. In fact, the easiest way to establish the existence and uniqueness of J(x) is to carry out the procedure described earlier, showing the solution $\lambda(t)$ of the adjoint equation (3.10) to have the form (3.13); then one sets

$$\frac{\partial J}{\partial x}(x) = 2Q\, x + \hat{\eta}(x) .$$

4. <u>The Discrete NLNQ Theory</u> As indicated in the preceding §3 we are now concerned with a system of recursion equations with a prescribed initial state:

$$x_{k+1} = F(x_k,u_k) = A\, x_k + B\, u_k + f(x_k,u_k) \;,\; x_0 \text{ given}, \qquad (4.01)$$

with the pair A, B controllable and with f, assumed continuously differentiable, and its first order partial derivatives with respect to the components of x and u all vanishing at (0,0), so that

$$f(x,u) = o\left[\|x\|+\|u\|\right],\; \|x\|,\; \|u\| \to 0 \;.$$

A control sequence $\{\hat{u}_k\} \in \ell^2$ is sought minimizing the cost sum

$$\sum_{k=1}^{\infty} G(x_k,u_k) \;, \qquad (4.02)$$

with G(x,u) twice continuously differentiable and, as in the foregoing discussion of the continuous time problem,

$$G(x,u) = \begin{bmatrix} x^* & u^* \end{bmatrix}\begin{bmatrix} W & R \\ R^* & U \end{bmatrix}\begin{bmatrix} x \\ u \end{bmatrix} + g(x,u) \;, \qquad (4.03)$$

$$g(x,u) = o\left[\|x\|^2+\|u\|^2\right],\; \|x\|,\; \|u\| \to \infty \;.$$

The state vector lies in $R^n$; ordinarily we would also require that the control u should lie in a finite dimensional space $R^m$. However, the work carried out in [ 17 ], which was particularly concerned with NLNQ theory in the case of periodic differential equations, makes use of an infinite dimensional control through replacement of the T-periodic ordinary differential equation

$$\dot{x}(t) = \Phi(t,x(t),u(t)),\; \Phi(t+T,x,u) \equiv \Phi(t,x,u) \equiv C(t)x+D(t)u+\varphi(t,x,u),$$

by its integrated form

$$x_{k+1} \equiv x((k+1)T) = \Psi(T,0)x(kT) + \int_0^T \Psi(T,s)\left[D(s)u(s)+\varphi(s,x(s),u(s))\right]ds$$

$$\equiv A\ x_k + Bu_k + f(x_k,u_k), \quad Bu \equiv \int_0^T \Psi(T,s)D(s)u(s)ds, \ u \in C[0,T] \ .$$

It turns out that this introduces no real problem, the discrete NLNQ theory to be described here experiences no problem with infinite dimensional control vectors u as long as the associated Jacobians of F and G with respect to u are bounded.

With the control problem posed as indicated, we are interested in the question as to whether the optimal control sequence $\{\hat{u}_k\}$ can be written in synthesized form

$$\hat{u}_k = \hat{\mathbb{K}}(\hat{x}_k), \tag{4.04}$$

the optimal solution sequence $\hat{x}_k$ satisfying the synthesized system

$$\hat{x}_{k+1} = F(\hat{x}_k, \hat{\mathbb{K}}(\hat{x}_k)) \equiv F_{\hat{\mathbb{K}}}(\hat{x}_k) \ . \tag{4.05}$$

Here the discrete version of the Maximum Principle [ 3 ] yields a coupled state - adjoint system, with boundary requirements as indicated, taking the form

$$x_{k+1} = A\ x_k + B\ u_k + f(x_k,u_k), \quad x_0 \text{ given}, \quad \lim_{k\to\infty} x_k = 0 \ , \tag{4.06}$$

$$\lambda_{k+1}{}^{*}A - \lambda_k{}^{*} + 2x_k{}^{*}W + 2u_k{}^{*}U + h(\lambda_{k+1},x_k,u_k), \quad \lim_{k\to\infty} \lambda_k = 0, \tag{4.07}$$

and the characterization of the optimal control:

$$\lambda_{k+1}{}^{*}\frac{\partial F}{\partial u}(\hat{x}_k,\hat{u}_k) + \frac{\partial G}{\partial u}(\hat{x}_k,\hat{u}_k) = 0 \ . \tag{4.08}$$

The last equation can be solved, locally, to give

$$\hat{u}_k = -\frac{1}{2}\ U^{-1}\left[B^{*}\lambda_{k+1} + R^{*}\hat{x}_k\right] + \hat{\ell}(\hat{x}_k,\lambda_{k+1}) \ . \tag{4.09}$$

Here, again, the linear - quadratic theory is well known (cf.[ 3 ]); the optimal control in that case is given in the form of linear state feedback with a feedback matrix $\hat{K}$ and the optimal cost is $x_0^* P x_0$, where P, $\hat{K}$ are the unique solution pair of

$$P = W + R\hat{K} + \hat{K}^*R^* + \hat{K}^*U\hat{K} + (A+B\hat{K})^*P(A+B\hat{K}) \ , \qquad (4.10)$$

$$\hat{K} = -\left[U+B^*PB\right]^{-1}\left[R^*+B^*PA\right] \ . \qquad (4.11)$$

To treat the discrete NLNQ problem we set

$$\lambda_k = 2P\,x_k + \eta_k \ , \quad \hat{A} = A + B\hat{K} \ , \qquad (4.12)$$

(parallelling the differential equations case, $\hat{A}$ is a discrete stability matrix, i.e., all of its eigenvalues lie in the interior of the unit circle in the complex plane) and using (4.06),(4.07),(4.10), (4.11) we obtain a modified system, linearly decoupled, of the form

$$\hat{x}_{k+1} = \hat{A}\,\hat{x}_k + C\,\eta_k + f_1(\hat{x}_k,\eta_k) \ , \qquad (4.13)$$

$$\hat{A}^*\eta_{k+1} = \eta_k + f_2(x_k,\eta_k) \ . \qquad (4.14)$$

As in the differential equations case, we seek a functional relationship

$$\eta_k = \hat{\eta}(\hat{x}_k) \qquad (4.15)$$

such that if $\hat{x}_k$ satisfies (4.13) with $\eta_k$ replaced by $\hat{\eta}(\hat{x}_k)$, then $\eta_k$, as given by (4.15), will identically satisfy (4.14). This amounts, as in the corresponding differential equations case, to obtaining an equation describing the stable manifold of the discrete system (4.13), (4.14). It is possible to do this by means of a discrete analog of the integral equation used in [ 4 ] for the continuous time case, or one can cite general results of this sort as in [ 9 ],[ 13 ],[ 16 ]. However, we choose here to describe a method which adapts easily to finite element type numerical approximation.

Making the indicated substitutions one arrives at

$$\hat{A}^* \hat{\eta}\left[\hat{A}\ \hat{x}_k + C\ \hat{\eta}(\hat{x}_k) + f_1(\hat{x}_k,\hat{\eta}(\hat{x}_k))\right] = \hat{\eta}(\hat{x}_k) + f_2(\hat{x}_k,\hat{\eta}(\hat{x}_k)) .$$

Considering all initial states and corresponding optimal solutions, we see that $\eta(x)$ should satisfy the functional equation

$$\hat{\eta}(x) = \hat{A}^* \hat{\eta}\left[\hat{A}\ x + C\ \hat{\eta}(x) + f_1(x,\hat{\eta}(x))\right] - f_2(x,\hat{\eta}(x)) , \quad (4.16)$$

which we may with some justification call the *fundamental equation of discrete NLNQ theory.* It is an alternate characterization of the stable manifold for the partially stable discrete dynamical system (4.13),(4.14).

We will give a very brief sketch of the local existence and uniqueness of a solution $\hat{\eta}(x)$ of (4.16). We let $C_n^1(\mathfrak{X})$, where $\mathfrak{X}$ is a region in $R^n$, be the Banach space of continuously differentiable n-vector functions defined on $\mathfrak{X}$ with the norm

$$\|p\|_1 = \|p(0)\|_{R^n} + \sup_{x\in\mathfrak{X}} \left\| \frac{\partial p}{\partial x} \right\|_{R^{n,n}}$$

and let $C_n^{1,\varepsilon}$, $\varepsilon > 0$, be the closed subset of $C_n^1$ containing functions p such that $p(0) = 0$ and

$$\|p\|_1 = \sup_{x\in\mathfrak{X}} \left\| \frac{\partial p}{\partial x} \right\|_{R^{n,n}} \leqq \varepsilon .$$

We define the map $\mathcal{P}:C_n^{1,\varepsilon} \to C_n^1$ by

$$\left[\mathcal{P}(\hat{\eta})\right](x) = \hat{A}^* \hat{\eta}\left[\hat{A}\ x + C\ \hat{\eta}(x) + f_1(x,\hat{\eta}(x))\right] - f_2(x,\hat{\eta}(x)) ; \quad (4.17)$$

if we can show that, in fact, $\mathcal{P}:C_n^{1,\varepsilon} \to C_n^{1,\varepsilon}$ and is a contraction with respect to $\| \ \|_1$. The contraction property follows almost immediately from the n-dimensional version of the chain rule:

$$\frac{\partial}{\partial x}\left[\mathcal{P}(\hat{\eta})\right](x) = \hat{A}^*\frac{\partial\hat{\eta}}{\partial x}\left[\hat{A}x+C\hat{\eta}(x)+\cdots\right]\left[\hat{A}+C\frac{\partial\eta}{\partial x}(x)+\cdots\right] = \hat{A}^*\frac{\partial\hat{\eta}}{\partial x}\left[\hat{A}x+\cdots\right]\hat{A} +\cdots .$$

Since $\hat{A}$ is a discrete stability matrix, a linear change of variable, which we need not detail here, allows us to assume $\|\hat{A}\| < 1$. Then the map (4.17) can be seen to map $C_n^{1,\varepsilon}$ into itself and to be a contraction with respect to $\| \ \|_1$, provided $\mathfrak{X}$ is confined to a suitably small neighborhood of $x = 0$ and is invariant under the map

$$y = \hat{A}x + C\eta(x) + f_1(x,\eta(x)) \tag{4.18}$$

for $\eta \in C_n^{1,\varepsilon}$, $\varepsilon$ sufficiently small.

Thus it remains to identify small domains $\mathfrak{X}$ which are invariant under (4.18) when $\|\eta\|_1$ is small. We do this with the aid of the discrete Liapounov matrix equation

$$\mathbb{P} = \hat{A}^*\mathbb{P}\,\hat{A} + I\ ,$$

known [ 1 ],[ 3 ] to have a unique symmetric positive definite solution $\mathbb{P}$ just in case $\hat{A}$ is a discrete stability matrix, as is true in this instance. For small $r > 0$ we let $\mathfrak{X} = \mathfrak{X}_r$ be the ellipsoid

$$\mathfrak{X}_r = \left\{x \in R^n \middle| \ x^*\mathbb{P}\,x < r\right\}.$$

Then (cf.(4.18)) we have

$$y^*\mathbb{P}\,y = \left[\hat{A}\,x + C\,\eta(x) + f_1(x,\eta(x))\right]^*\mathbb{P}\left[\hat{A}\,x + C\,\eta(x) + f_1(x,\eta(x))\right]$$

$$= x^*\left[\mathbb{P} - I\right]x + \cdots < \ x^*\mathbb{P}\,x$$

if $\eta \in C_n^{1,\varepsilon}$ and $r$ and $\varepsilon$ are sufficiently small. The omitted details for application of the contraction fixed point theorem are routine. The unique fixed point is then the desired function $\hat{\eta}(x)$.

Numerical approximation of $\hat{\eta}(x)$ is carried out by replacing that function by a finite element approximant $\tilde{\eta}(x)$, determined by its values at a finite number of nodes, and then using the recursive scheme

$$\tilde{\eta}_{k+1}(x) = \hat{A}^* \tilde{\eta}_k\left[\hat{A}\ x + C\ \tilde{\eta}_k(x) + f_1(x,\tilde{\eta}_k(x))\right] - f_2(x,\tilde{\eta}_k(x))$$

to compute the values of $\tilde{\eta}_{k+1}(x)$ at the nodal points x; note that the finite element functions determined by the values of $\tilde{\eta}_k$ at the nodal points are needed in order to evaluate

$$\tilde{\eta}_k\left[\hat{A}\ x + C\ \tilde{\eta}_k(x) + f_1(x,\tilde{\eta}_k(x))\right] \tag{4.19}$$

and the region $\tilde{\mathfrak{X}}$ in which the approximation $\tilde{\eta}$ is to be constructed needs to be invariant with respect to the argument,

$$\hat{A}\ x + C\ \tilde{\eta}_k(x) + f_1(x,\tilde{\eta}_k(x)),$$

of $\tilde{\eta}_k$ in (4.19). This last requirement is a major restriction in potential use of this method for approximation of $\hat{\eta}(x)$ and hence of the optimal feedback law $\hat{u}_k = \mathbb{K}(\hat{x}_k)$.

## References

[ 1 ] Anderson, B.D.O., and J.B. Moore: *Linear Optimal Control,* Prentice-Hall, Inc., Englewood Cliffs, NJ, 1971

[ 2 ] Aoki, M. *Optimal Control and System Theory in Dynamic Economic Analysis,* North-Holland Pub. Co., New York, Oxford, Amsterdam, 1976

[ 3 ] Chow, G. *Analysis and Control of Dynamic Economic Systems,* Wiley Series in Probability, John Wiley & Sons, Inc., New York, London, Sydney, Toronto, 1975

[ 4 ] Coddington, E.A., and N. Levinson: *Theory of Ordinary Differential Equations,* McGraw-Hill Book Co., New York, 1955

[ 5 ] Cronin, J. *Differential Equations: Introduction and Qualitative Theory,* Vol. 54, Pure and Applied Mathematics, Marcel Dekker, Inc., New York, 1980

[ 6 ] Francis, B. A.: *"A Course in $H^\infty$ Control Theory"*, Vol. 88 of *Lecture Notes in Control and Information Sciences*, Springer-Verlag, Berlin, Heidelberg, New York, 1987.

[ 7 ] Kalman, R.E.: *Contributions to the theory of optimal control*, Bol. Soc. Mat. Mexicana, 5 (1960), pp. 102 - 119

[ 8 ] Kalman, R.E., and R.S. Bucy: *New results in linear prediction and filtering theory*, J. Basic Engr. (Trans. ASME, Ser. D) 83D (1961), pp. 95 - 100

[ 9 ] Kelley, A.: *The stable, center-stable, center, center unstable and unstable manifolds*, appendix in *Transversal Mappings and Flows*, R. Abraham and J. Robbin, Benjamin, New York, 1967

[ 10 ] Lee, E.B., and L.W. Markus: *Foundations of Optimal Control Theory*, John Wiley & Sons, Inc., 1967

[ 11 ] Lukes, D.L.: *Optimal regulation of nonlinear dynamical systems*, SIAM J. Control 7 (1969), pp. 75 - 100

[ 12 ] Lukes, D.L.: *Stabilizability and optimal control*, Funkcialaj Ekvacioj, 11 (1968), pp. 39 - 50

[ 13 ] Nemytskii, V. V., and V. V. Stepanov: *Qualitative Theory of Differential Equations*, Princeton University Press, Princeton, N. J., 1960

[ 14 ] Russell, D. L.: *Analytic simplification of linear systems with a combined transition - regular singular point*, Thesis, University of Minnesota, Minneapolis, August, 1964

[ 15 ] Russell, D. L., and X. Zhang: *Extending linear - quadratic optimal control laws to nonlinear systems and / or non-quadratic cost criteria*, to appear in the L.W. Markus Festschrift volume, N. Everitt, E.B. Lee, Eds., Marcel Dekker, Inc., New York

[ 16 ] Smale, S.: *Stable manifolds for differential equations and diffeomorphisms*, Annali Scuola Norm. Pisa, Ser. 3, 17 (1963), Fasc. 1-2.

[ 17 ] Zhang,X.: Thesis, Department of Mathematics, Virginia Polytechnic Institute and State University, 1993

# Solution In The Large Of A Certain Third Order Differential Equation Of Langer

T. K. Puttaswamy
Department Of Mathematical Sciences, Ball State University,
Muncie, Indiana - 47306, U. S. A.

T.V. Sastry
Department Of Mathematics. Bradley University
Peoria, Illinois 61625, U. S. A.

## Abstract

The goal of this paper is to solve in the large the differential equation of third order

$$\frac{d^3y}{dt^3} + t\frac{dy}{dt} + cy = 0$$

where the variable t and the constant c are complex.

The solution in the large is essentially a two point connection problem. The central question in a two point connection problem is to find a linear relationship between two sets of fundamental solutions defined in the respective neighborhoods of two distinct points. This has been successfully carried in this paper between the fundamental solutions at the ordinary point $t = 0$ and the irregular singular point $t = \infty$.

## 1 Introduction

R. E. Langer examined the behavior of the solutions of the differential equation

$$y''' + \lambda^2 zy' + 3\mu\lambda^2 y = 0 \qquad \textbf{(1.1)}$$

where the variable $z$ is complex, $\lambda$ is a complex parameter of large absolute value and the constant $\mu$ is arbitrary, real or complex. The equation (1.1) has associated with it, the so called characteristic or auxillary equation

$$\rho^3 + \rho z = 0 \qquad \textbf{(1.2)}$$

Hence, at the ordinary point $z = 0$, all the roots of the auxillary equation (1.2) are coincident and $z = 0$ is a multiple turning point of (1.1). Langer used the method of 'related' equations to solve his problem. The constuction of such a related equation must be based upon a suitable differential equation whose solutions are completely known and which manifeste the same configuration of the characteristic roots as the given equation.

If we make a change of the independent variable by the transformation

$$t = \lambda^{\frac{2}{3}} z \tag{1.3}$$

then the equation (1.1) becomes

$$\frac{d^3y}{dt^3} + t\frac{dy}{dt} + cy = 0 \tag{1.4}$$

with $c = 3\mu$.

The goal of this paper is to solve in the large the differential equation (1.4) where the variable $t$ and the constant $c$ are complex. In the language of Fuch's theory, $t = 0$ is an ordinary point and $t = \infty$ is an irregular singular point of (1.4). Solution in the large is essentially a two point connection problem. The central question in a two point connection problem is to find a linear relationship between two sets of fundamental solutions defined in the respective neighborhoods of two distinct points. This has been successfully carried out in this paper between the fundamental solutions at the ordinary point $t = 0$ and the irregular singular point at $t = \infty$.

# 2 Principle Solutions at the Ordinary Point $t = 0$

Let $v$ be the differential operator $t\frac{d}{dt}$. Then, it follows that $t^n\frac{d^n}{dt^n} = v(v-1)(v-2)\ldots(v-n+1)$

Now, multiplying (1.4) by $t^3$, (1.4) can be put in the form

$$f(v) + t^3 g(v)y = 0 \tag{2.1}$$

where

$$f(v) = v(v-1)(v-2) \tag{2.2}$$

$$\text{and} \qquad g(v) = v + c \tag{2.3}$$

The indicial equation of (1.4) for the ordinary point $t = 0$ is

$$f(v) = v(v-1)(v-2) = 0 \tag{2.4}$$

Hence, the indicial exponents for the ordinary point $t = 0$ are

$$\lambda_j = j - 1, j = 1, 2, 3 \tag{2.5}$$

Then, from the established theory of linear differential equations, the differential (1.4) will have three linearly independent solutions $y_j(t), j = 1, 2, 3$ of the form

$$y_j(t) = t^{\lambda_j} \sum_{k=0}^{\infty} c_{jk} t^{3k} \tag{2.6}$$

where $\lambda_j$ is given by (2.5)

To obtain the solutions corresponding to $\lambda_j (j = 1, 2, 3)$, substitute (2.6) into (1.4).

Using the power shift $\quad f(v)t^a = f(a)t^a \quad$ (2.7)

we obtain the two terms recurrence relation

$$f(\lambda_j + 3k)c_{jk} + g(\lambda_j + 3(k-1))c_{j,k-1} = 0, \qquad k \geq 1 \tag{2.8}$$

where $f(v)$ and $g(v)$ are given by (2.2) and (2.3) respectively.

If we choose $\quad c_{j0} = \dfrac{1}{(j-1)!} \quad$ (2.9)

iteration yields

$$c_{jk} = \frac{(-3)^k \prod_{r=0}^{k-1} (\frac{j+c-1}{3} + r)}{(j + 3k - 1)!}, \qquad k \geq 1 \tag{2.10}$$

and

$$y_j(t) = \frac{t^{j-1}}{(j-1)!} + \sum_{k=1}^{\infty} \frac{(-3)^k t^{3k+j-1}}{(j+3k-1)!} \prod_{r=0}^{k-1} (\frac{j+c-1}{3} + r) \tag{2.11}$$

for $j = 1, 2, 3$, which converges in the open plane of the extended complex z-plane.

# 3 Formal Solutions $\tilde{y}_j(t), j = 1, 2, 3$ about the Irregular Singular Point $t = \infty$

The point $t = \infty$, as already noted, is an irregular singular point of our differential equation (1.4). G. D. Birkoff [1] has completely discussed the existence, form and range of the formal solutions of the homogeneous linear differential equation

$$\frac{d^n y}{dz^n} + a_1(z)\frac{d^{n-1}y}{dz^{n-1}} + \ldots + a_n(z)y = 0 \tag{3.1}$$

where the coefficients $a_r(z)$ are developable ($|z| < R$, where $R$ is sufficiently large) in convergent series

$$a_r(z) = z^{rk}[a_{r,0} + \frac{a_{r,1}}{z} + \frac{a_{r,2}}{z^2} + \ldots] \tag{3.2}$$

$r = 1, 2, 3, \ldots, n$, $k$ being zero or a positive integer. Applying Birkoff's results, we see that the equation (1.4) has the characteristic equation

$$\rho^3 = 0 \tag{3.3}$$

and so has all its roots equal to zero, with which the formal solutions become exponential in character as follows:

In the differential equation (1.4), let us transform the independent variable $t$ to $w$ by placing

$$t = w^{2/3} \tag{3.4}$$

The resulting differential equation becomes

$$\frac{d^3y}{dw^3} + \frac{1}{w}\frac{d^2y}{dw^2} + \frac{1}{9}\frac{dy}{dw}(4 - \frac{1}{w^2}) + \frac{8c}{27w}y = 0 \tag{3.5}$$

For the differential equation (3.5), the characteristic equation is found to be

$$\rho^3 + \frac{4}{9} = 0 \tag{3.6}$$

whose roots are $$\rho_1 = \frac{2}{3}i, \rho_2 = -\frac{2}{3}i, \rho_3 = 0 \tag{3.7}$$

Hence, the differential equation (3.5) will have 3 linearly independent formal solutions $\tilde{y}_j(w), j = 1, 2, 3$, for which when $|w|$ is sufficiently large,

$$\tilde{y}_j(w) \sim s_j(w), \qquad j = 1, 2, 3 \tag{3.8}$$

where $s_j(w)$ are the divergent series given by

$$s_j(w) = e^{\rho_j w} \omega^{m_j} \sum_{k=0}^{\infty} \frac{l_{j,k}}{\omega^k} \tag{3.9}$$

Here, $m_j$ and $l_{j,k}$ for $j = 1, 2, 3$ and $k = 0, 1, 2, \ldots$ are constants, whose values may be determined through the usual method of undetermined coefficients, by substituting (3.9) into the differential equation (3.5). By doing so, we find that

$$m_1 = m_2 = \frac{c}{3} - \frac{1}{2} \tag{3.10}$$

$$m_3 = -\frac{2}{3}c \tag{3.11}$$

$$l_{j,0} = 1 \qquad \text{for} \qquad j = 1,2,3 \tag{3.12}$$

$$l_{2,k} = (-1)^k l_{1,k} \tag{3.13}$$

$$\text{and} \qquad l_{3,k} = 0 \qquad \text{for} \qquad k = 1,3,5\ldots \tag{3.14}$$

Thus,

$$s_1(w) = e^{\frac{2}{3}iw} w^{\frac{c}{3}-\frac{1}{2}} \sum_{k=0}^{\infty} \frac{a_k}{w_k} \tag{3.15}$$

$$s_2(w) = e^{-\frac{2}{3}iw} w^{\frac{c}{3}-\frac{1}{2}} \sum_{k=0}^{\infty} \frac{a_k}{(-w)^k} \tag{3.16}$$

$$s_3(w) = w^{-\frac{2}{3}c} \sum_{k=0}^{\infty} \frac{b_k}{w^{2k}} \tag{3.17}$$

$$\text{where} \qquad a_k = l_{1,k} \qquad \text{and} \qquad b_k = l_{3,2k} \tag{3.18}$$

with $a_0 = b_0 = 1$

However, it is convenient at this point to introduce into discussion the variable $z$ which is given by

$$z = \frac{1}{3}iw = \frac{1}{3}it^{3/2} \tag{3.19}$$

Then, from (3.15), (3.16) and (3.17),

$$s_1(w) = e^{2z} z^{\frac{c}{3}-\frac{1}{2}} \sum_{k=0}^{\infty} \frac{a_k}{(-3zi)^k} \tag{3.20}$$

$$s_2(w) = e^{-2z} z^{\frac{c}{3}-\frac{1}{2}} \sum_{k=0}^{\infty} \frac{a_k}{(3zi)^k} \tag{3.21}$$

$$s_3(w) = z^{-\frac{2}{3}} \sum_{k=0}^{\infty} \frac{b_k}{(3zi)^k} \tag{3.22}$$

For each integral index $q$, the inequality

$$(q-1)\pi \leq \arg z \leq q\pi \tag{3.23}$$

defines a half plane and thus by (3.19) defines a sector

$$(\frac{2}{3}q - 1)\pi \leq \arg t \leq (\frac{2}{3}q - \frac{1}{3})\pi \tag{3.24}$$

of the $t$ - plane. Any three of these sectors with consecutive indicies will cover the whole $t$-plane.

Thus, the differential equation (1.4) has a set of formal solutions $\tilde{y}_{j,q}, j = 1,2,3$ which for $t$ in the sector (3.24) are asympotically represented as follows:

$$\tilde{y}_{1,q}(t) \sim e^{2\sigma z} z^{\frac{c}{3}-\frac{1}{2}} \sum_{k=0}^{\infty} \frac{a_k}{(-3i\sigma z)^k} \qquad \textbf{(3.25)}$$

$$\tilde{y}_{2,q}(t) \sim e^{-2\sigma z} z^{\frac{c}{3}-\frac{1}{2}} \sum_{k=0}^{\infty} \frac{a_k}{(3i\sigma z)^k} \qquad \textbf{(3.26)}$$

$$\tilde{y}_{3,q}(t) \sim z^{-\frac{2}{3}c} \sum_{k=0}^{\infty} \frac{b_k}{(3zi)^k} \qquad \textbf{(3.27)}$$

with $\sigma = (-1)^q$ and $a_0 = b_0 = 1$ **(3.28)**

# 4 Main Results and Main Theorems

If $c$ is zero or a negative integer, the value $\frac{j+c-1}{3}$ is likewise zero or a negative integer for one of the values $j = 1, 2, 3$. We will show in section 5 that the series for the respective solutions $y_j(t), j = 1, 2, 3$ as given by (2.11) as well as $\tilde{y}_{3,q}$ as given by (3.27) are then terminating and these solutions are accordingly polynomials in $t$.

The main two theorems of this paper are the following:

## Theorem 1

If $c$ is zero or a negative integer, in which case $\frac{j+c-1}{3} = -p$, where $p$ is zero or a positive integer for one of the indicies $j = 1, 2, 3$, the formal solution $\tilde{y}_{3,q}(t)$ as given by (3.27) about the irregular singular point $t = \infty$ as well as the set of the three linearly independent solutions $y_j(t), j = 1, 2, 3$ ae given by (2.11) about the ordinary point $t = 0$ of the differential equation (1.4) are polynomials in $t$ and

$$\tilde{y}_{3,q}(t) = \frac{(3p + j - 1)!}{3^p} \frac{1}{p!} y_j(z) \qquad \textbf{(4.1)}$$

In as much as the right handside of (4.1) does not depend on the integral index $q$, we observe that the solutions $\tilde{y}_{3,q}(t)$ are the same for all integral index $q$.

## Theorem 2

If $\frac{j+c-1}{3}$ is neither zero nor a negative integer, the three linearly independent solutions $y_j(t), j = 1, 2, 3$ as given by (2.11) of the differential equation (1.4) for values of $t$ of large modulus in the sector (3.24) for any integral index $q$ can be developed asymptotically as follows:

$$y_j(t) \sim c_{j1}^q \tilde{y}_{1,q} + c_{j2}^q \tilde{y}_{2,q} + c_{j3}^q \tilde{y}_{3,q} \qquad \textbf{(4.2)}$$

Here, $\tilde{y}_{j,q}, j = 1,2,3$ as given by (3.25), (3.26) and (3.27) are the formal solutions of (1.4) about the irregular singular point $t = \infty$ with

$$c^q_{j1} = A_j A_0(-2\sigma)^\theta \tag{4.3}$$

$$c^q_{j2} = A_j A_0(2\sigma)^\theta \tag{4.4}$$

$$c^q_{j3} = A_j(-27)^{-(\frac{j+c-1}{3})} \tag{4.5}$$

$$A_j = \frac{(-3i)^{2/3(j-1)}}{\Gamma(\frac{j+c-1}{3})} \tag{4.6}$$

$$A_0 = \sqrt{\frac{\pi}{2}}2^{\frac{2j-c+1}{3}}3^{\frac{1}{2}-j} \tag{4.7}$$

$$\theta = \frac{c+\frac{1}{2}-2j}{3} \tag{4.8}$$

$$\sigma = (-1)^q \tag{4.9}$$

The proofs of theorem 1 and 2 will be given in series 5 and 6 respectively.

# 5 Polynomial Solutions

$$\text{Let} \qquad \frac{j+c-1}{3} = -p \tag{5.1}$$

where $p$ is zero or a positive integer. Then $y_j(t), j = 1,2,3$ as given by (2.11) can be written as

$$y_j(t) = \frac{t^{j-1}}{(j-1)!} + \sum_{k=1}^{\infty} \frac{(-3)^k t^{3k+j-1}}{(3k+j-1)!} \prod_{r=0}^{k-1}(r-p) \tag{5.2}$$

$$= \frac{t^{j-1}}{(j-1)!} + \sum_{k=1}^{\infty}(3)^k t^{3k+j-1} \frac{1}{(3k+j-1)!} \prod_{r=0}^{k-1}(p-r)$$

$$y_j(t) = \frac{t^{j-1}}{(j-1)!} + \sum_{k=1}^{p} \frac{3^k t^{3k+j-1}}{(3k+j-1)!} \frac{p!}{(p-k)!}$$

$$= \sum_{k=0}^{P} \frac{3^k t^{3k+j-1}}{(3k+j-1)!} \frac{p!}{(p-k)!} \tag{5.3}$$

Again, from (3.27), we see that

$$\tilde{y}_{3,q} \sim t^{-c} \sum_{k=0}^{\infty} \frac{b_k}{t^{3k}} \qquad \text{with} \quad b_0 = 1 \tag{5.4}$$

Substituting (5.4) into the differential equation (1.4) and equating the coefficients of $t^{-3k-c}$, we get the recurrence relation

$$b_k = (-1)^3\{3(k-1)c\}\{3(k-1)c+1\}\dots\{3k+c-1\}b_{k-1} \qquad \text{for } k \geq 1 \tag{5.5}$$

Iteration yields,

$$b_k = \frac{(-1)^{3k}c(c+1)(c+2)\dots(3k+c-1)}{k!3^k} \qquad \text{for} \quad k \geq 1 \tag{5.6}$$

From (5.1), this becomes

$$b_k = \frac{(-1)^{3k}(1-j-3p)(2-j-3p)\dots(3k-j-p)}{k!3^k} \qquad \text{for} \quad k \geq 1 \tag{5.7}$$

It is seen from (5.7) that every coefficient $b_k$ is zero if $k > p$

Hence, the series $\tilde{y}_{3,q}(t)$ as given in (5.4) as also terminating, so that $\tilde{y}_{3,q}$ is also a polynomial in $t$

From (5.1) and (5.4), $\tilde{y}_{3,q}(t)$ can be written as

$$\tilde{y}_{3,q}(t) = t^{3p+j-1} + \sum_{k=1}^{p} t^{3(p-k)+j-1}\left[\frac{(3p+j-1)(3p+j-2)\dots(3p+j+3k)}{k!3^k}\right] \tag{5.8}$$

$$\text{Set} \quad p-k=m$$

Then,

$$\tilde{y}_{3,q}(t) = t^{3p+j-1} + \sum_{m=0}^{p-1} t^{3m+j-1}\frac{(3p+j-1)(3p+j-2)\dots(3m+j)}{(p-m)!3^{p-m}}$$

$$= \sum_{m=0}^{p} t^{3m+j-1}\frac{(3p+j-1)!}{(3m+j-1!)}\frac{3^m}{3^p(p-m)!}$$

$$= \frac{(3p+j-1)!}{3^p}\sum_{m=0}^{p}\frac{t^{3m+j-1}3^m}{(3m+j-1)!(p-m)!} \tag{5.9}$$

$$= \frac{(3p+j-1)!}{3^p}\frac{1}{p!}y_j(t) \tag{5.10}$$

# 6 Asympotic behavior of $y_j(t), j = 1,2,3$ in the neighborhood of $t = \infty$

We assume that $\frac{j+c-1}{3} \neq -p$, where $p$ is zero or a positive integer.

The solutions $y_j(t), j = 1,2,3$ as given by (2.11) can be written as

$$y_j(t) = \frac{t^{j-1}}{(j-1)!} + \sum_{k=1}^{\infty}\frac{(-3)^k t^{3k+j-1}}{\Gamma(3k+j)}\frac{\Gamma(\frac{j+c-1}{3}+k)}{\Gamma(\frac{j+c-1}{3})}$$

$$= \frac{t^{j-1}}{\Gamma(\frac{j+c-1}{3})} \sum_{k=0}^{\infty} \frac{(-3t^3)^k \Gamma(\frac{j+c-1}{3}+k)}{\Gamma(3k+j)} \tag{6.1}$$

In terms of $z$ as given by (3.19),

$$y_j(t) = A_j z^{\frac{2}{3}(j-1)} F(27z^2) \tag{6.2}$$

$$\text{where} \qquad A_j = \frac{(-3i)^{\frac{2}{3}(j-1)}}{\Gamma(\frac{j+c-1}{3})} \tag{6.3}$$

$$F(z) = \sum_{k=0}^{\infty} \frac{f(k)}{\Gamma(k+1)} z^k \tag{6.4}$$

$$\text{and} \qquad f(s) = \frac{\Gamma(s+\frac{j+c-1}{3})|\Gamma(s+1)}{\Gamma(j+3s)} \tag{6.5}$$

We apply E. M. Wright's paper [3] on generalized hypergeometric function to find the asymptotic development of the function $F(z)$ in (6.4). According to this paper, if $|\arg(-z)| \leq \pi$, $F(z)$ as given by (6.4) can be written as

$$F(z) \sim I(Z_1(z)) + I(Z_2(z)) + J(-Z) \tag{6.6}$$

where

$$I(z) = A_0 Z^{\theta} e^{Z} \sum_{m=0}^{\infty} \frac{c_m}{Z_m} \qquad \text{with } c_0 = 1 \tag{6.7}$$

$$A_0 = \frac{\sqrt{\pi}}{2} 2^{\frac{2j-c+1}{3}} 3^{\frac{1}{2}-j} \tag{6.8}$$

$$\theta = \frac{c+\frac{1}{2}-2j}{3} \tag{6.9}$$

$$Z_1(z) = \frac{2}{3\sqrt{3}}(-z)^{\frac{1}{2}} i \tag{6.10}$$

$$Z_2(z) = \frac{-2}{3\sqrt{3}}(-z)^{\frac{1}{2}} i \tag{6.11}$$

and $J(z)$ is the sum of the residues of the function

$$G(s,z) = \Gamma(-s) f(s) z^s \tag{6.12}$$

at the poles of $f(s)$ as given by (6.5)

$$Z_1(27z^2) = Z_1(27z^2 e^{-2qi\pi}) = -2z\sigma \tag{6.13}$$

$$Z_2(27z^2) = Z_2(27z^2 e^{-2qi\pi}) = 2z\sigma \tag{6.14}$$

where $\sigma = (-1)^q$

$$I(Z_1(27z^2)) = A_0(-2\sigma)^\theta z^\theta e^{-2z\sigma} \sum_{m=0}^{\infty} \frac{c_m}{(-2z\sigma)^m} \quad \textbf{(6.15)}$$

$$I(Z_2(27z^2)) = A_0(2\sigma)^\theta z^\theta e^{2z\sigma} \sum_{m=0}^{\infty} \frac{c_m}{(2z\sigma)^m} \quad \textbf{(6.16)}$$

$$\text{with} \qquad c_0 = 1$$

Hence from (6.2),

$$y_j(z) \sim A_j z^{\frac{2}{3}(j-1)}[I(Z_1(27z^2) + I(Z_2(27z^2)) + J(-27z^2)]$$

$$\text{So,} \qquad y_j(z) \sim A_j A_0(-2\sigma)^\theta e^{-2z\sigma} z^{\frac{c}{3}-\frac{1}{2}} \sum_{m=0}^{\infty} \frac{c_m}{(2z\sigma)^m}$$

$$+A_j A_0(2\sigma)^\theta e^{2z\sigma} z^{\frac{c}{3}-\frac{1}{2}} \sum_{m=0}^{\infty} \frac{c_m}{(2z\sigma)^m} + A_j z^{\frac{2}{3}(j-1)} J(-27z^2)$$

$$\text{Thus,} y_j(z) \sim c_{j1}^q \tilde{y}_{1,q}(t) + c_{j2}^q \tilde{y}_{2,q} + A_j z^{2/3(j-1)} J(-27z^2) \quad \textbf{(6.17)}$$

where $c_{j1}^q$ and $c_{j2}^q$ are given by (4.3) and (4.4) respectively

Our next problem, then, is to find $J(z)$, the sum of the residue of the function $G(s,z)$ as given by (6.12).
Since the factor $\frac{\Gamma(s+1)}{\Gamma(j+3s)}$ is analytic, $f(s)$ has poles only where $\Gamma(s + \frac{j+c-1}{3})$ has poles. Those poles are the points

$$s = s_l = -l - (\frac{j+c-1}{3}), \quad \textbf{(6.18)}$$

$l = 0, 1, 2, \ldots$
So, if $J_l(z)$ is the residue of the function $G(s,z)$ as given by (6.12) at the pole $s = s_l = -l - (\frac{j+c-1}{3})$,

$$J(z) = \sum_{l=0}^{\infty} J_l(z)$$

We next proceed to find $J_l(z)$, Here two cases arise:

(a) $\frac{j+c-1}{3}$ is not an integer

(b) $c$ is not a positive integer

we will have two alternate evaluations of $J_l(z)$ corresponding to the two cases cited above.

**Case (a)**

$$G(s,z) = \Gamma(-s)f(s)z^s = \frac{\Gamma(-s)\Gamma(s+\frac{j+c-1}{3})\Gamma(s+1)}{\Gamma(3s+j)}z^s \tag{6.19}$$

$$\text{Since} \qquad \Gamma(z) = \frac{\pi}{\Gamma(1-z)\sin \pi z} \tag{6.20}$$

We have $\Gamma(-s)\Gamma(s+1) = \frac{\pi}{\sin \pi s}$

$$\text{and} \qquad \Gamma(s+\frac{j+c-1}{3}) = \frac{\pi}{\Gamma(1-s-\frac{j+c-1}{3})\sin(-s+\frac{j+c-1}{3})}$$

$$\text{Hence} \qquad G(s,z) = \frac{\pi^2}{\sin \pi s}\frac{z^s}{\Gamma(1-s-\frac{j+c-1}{3})\Gamma(3s+j)\sin \pi(-s+\frac{j+c-1}{3})} \tag{6.21}$$

$$\text{Now} \qquad J_l(z) = \lim_{s\to s_l}(s-s_l)G(s,z)$$

$$= \frac{z^{s_l}\pi^2}{\sin \pi s_l}\frac{1}{\Gamma(1-s_l-\frac{j+c-1}{3})}\frac{1}{\Gamma(3s_l+j)}\lim_{s\to s_l}\frac{s-s_l}{\sin \pi(s+\frac{j+c-1}{3})}$$

$$= \frac{z^{s_l}\pi^2}{\sin \pi s_l}\frac{1}{\Gamma(1+l)}\frac{1}{\Gamma(-3l+c-1)}\frac{1}{\pi\cos(s_l+\frac{j+c-1}{3})}$$

$$= \frac{z^{s_l}\pi}{\Gamma(1+l)\Gamma(-3l+c-1)\sin \pi(\frac{j+c-1}{3})} \tag{6.22}$$

**Case (b)**

From (6.20), we have

$$\frac{\pi}{\Gamma(3s+j)} = \Gamma(1-3s-j)\sin \pi(3s+j)$$

So from (6.21),

$$G(s,z) = \frac{\pi}{\sin \pi s}\frac{z^s\Gamma(1-3s-j)\sin \pi(3s+j)}{\Gamma(1-s-\frac{j+c-1}{3})\sin \pi(s+\frac{j+c-1}{3})}$$

$$= \pi\frac{\sin 3s\pi}{\sin \pi s}\frac{z^s\Gamma(1-3s-j)}{\Gamma(1-s-\frac{j+c-1}{3})}\frac{(-1)^j}{\sin \pi(s+\frac{j+c-1}{3})}$$

$$= \pi[1+2\cos 2s\pi]\frac{z^s\Gamma(1-3s-j)}{\Gamma(1-s-\frac{j+c-1}{3})}\frac{(-1)^j}{\sin \pi(s+\frac{j+c-1}{3})}$$

$$J_l(z) = \lim_{s\to s_l}(s-s_l)G(s,z)$$

$$= z^{s_l}(-1)^j \frac{\pi\Gamma(3l+c)}{\Gamma(1+l)}[1+2\cos\pi(\frac{j+c-1}{3}) \lim_{s\to s_l} \frac{s-s_l}{\sin\pi(s+\frac{j+c-1}{3})}]$$

$$= z^{s_l}(-1)^{j+l} \frac{\Gamma(3l+c)}{\Gamma(1+l)}[1+2\cos\pi(\frac{j+c-1}{3})]$$

We observe that in every case, atleast one of these two evaluations (6.22) on (6.23) for $J_l(z)$ is valid, because when $c$ is zero or a negative integer, $\frac{j+c-1}{3}$ is likewise zero or a negative integer and we have precluded the non positive integral values of $\frac{j+c-1}{3}$.

Thus

$$J(z) = \sum_{l=0}^{\infty}(z)^{s_l} R_l$$

where $R_l$ is either (6.22) or (6.23)

So, $\qquad J(-27z^2) = (-27z^2)^{-\frac{j+c-1}{3}} \sum_{l=0}^{\infty} z^{-2l} R_l$

Finally, from (6.17),

$$y_j(t) \sim c_{j1}^q \tilde{y}_{1,q}(t) + c_{j2}^q \tilde{y}_{2,q} + c_{j3}^q \tilde{y}_{3,q}$$

where $c_{j3}^q = A_j(-27)^{\frac{j+c-1}{3}}$ as given in (4.5)

This completes the proof of the theorem 2.

## References


[1] G.D. Birkoff, *'Collected Mathematical papers'*, Dover, 1968.

[2] R. E. Langer, *'The Solutions of the differential equation $v''' + \lambda z v' + 3\mu\lambda^2 v = 0$'*, Duke Math. Jour (22), 1955a, pp 525-542.

[3] E. M. Wright, *'The asymptotic expansion of the generalized hypergeometric function'*, Proceedings of London Mathematical Society, Second Series, Vol. 46 (1940), pp 389-408.

# A MAJORANT METHOD FOR DIFFERENTIAL EQUATIONS WITH A SINGULAR PARAMETER

REINHARD SCHÄFKE
*Fachbereich Mathematik, Universität Essen,*
*45117 Essen, Germany*

The majorant method is not new. It has been used extensively in the theory of ordinary differential equations, e.g. to show convergence of power series solutions at a regular point or a regular singular point. Recently, Gérard [3], Gérard and Lutz [4] and Gérard, Lutz und Schäfke [5] have used the majorant method for a variety of convergence problems for power and factorial series.

In this lecture, I would like to present four applications of the majorant method to *singularly perturbed* differential equations. Here the existence of formal power series solutions in the parameter is well known, but the growth of the coefficients was unknown or difficult to determine. As the coefficients are analytic functions of the independent variable and the recursion determining them contains derivatives, the use of appropriate norms - which have to be compatible with differentiation - on spaces of analytic functions is necessary if we want to use the majorant method. I will give three examples of such norms.

## 1. The Gevrey property for a singularly perturbed problem with invertible leading matrix

Here we want to give a different proof of the main theorem of Sibuya's article [15].

**Theorem 1:** *Consider a system of ordinary differential equations*

$$\varepsilon^{\sigma}\frac{dy}{dx} = F(x,\varepsilon,y) \equiv f(x,\varepsilon) + A(x,\varepsilon)y + \sum_{|\mathbf{p}|\geq 2} f_{\mathbf{p}}(x,\varepsilon)y^{\mathbf{p}} \quad , \textit{ where} \tag{1}$$

1. *$\sigma$ is a positive integer; $y, f$ and $f_p$ are n-dimensional vectors; $A$ is an $n \times n$-matrix; $\mathbf{p} = (p_1, ...p_n)$ and the $p_j$ are nonnegative integers; $|\mathbf{p}| = p_1 + ... + p_n$; $y^{\mathbf{p}} = y_1^{p_1}y_2^{p_2}...y_n^{p_n}$ and $y_j$ is the j-th entry of the column vector $y$,*
2. *$F(x,\varepsilon,y)$ is analytic in $|x| < r_0$, $|\varepsilon| < r_1$, $|y| < r_2$, where $r_0, r_1, r_2$ are positive numbers,*
3. *$f(0,0) = 0$, $A(0,0)$ is invertible.*

*Then the system admits a unique formal solution $y = \sum_{m\geq 0} a_m(x)\varepsilon^m$, where the coefficients $a_m(x)$ are holomorphic and bounded for $|x| < r_3$ for some $r_3$ such that $0 < r_3 < r_0$. Furthermore this formal series is Gevrey of order $1/\sigma$ uniformly on $|x| < r_3$ if we choose $r_3$ sufficiently small.*

Here Gevrey ... means that there are two nonnegative numbers $C_0, C_1$ such that $|a_m(x)| \leq C_0(m!)^{1/\sigma}{C_1}^m$ for all $|x| < r_3$ and $m = 0, 1, ....$

First observe that we can find $y(x,0)$ using the implicit function theorem and hence can reduce the theorem to the case where $f(x,0)=0$ and only $y$ with $y(x,0)=0$ has to be found. Observe that $r_0, r_1, r_2$ might change. Reducing $r_0$ further, if necessary, we can find the coefficients of $y$ from the differential equation by comparing coefficients of $\varepsilon^k$ on both sides. To make this more obvious we rewrite the differential equation

$$y = -A(x,0)^{-1}\left[f(x,\varepsilon) + (A(x,\varepsilon)-A(x,0))y - \varepsilon^\sigma \frac{dy}{dx} + \sum_{|\mathbf{p}|\geq 2} f_{\mathbf{p}}(x,\varepsilon)y^{\mathbf{p}}\right] \tag{2}$$

Observe that if the coefficients of $\varepsilon, ..., \varepsilon^l$ in $y$ are already known then the coefficient of $\varepsilon^{l+1}$ in $y$ is just the coefficient of $\varepsilon^{l+1}$ of the right hand side of (2) and can thus be computed from the preceding ones. By reducing $r_0, r_1, r_2$ a little further we can assume that $f, f_{\mathbf{p}}, A, A(x,0)^{-1}$ and $\sum_{|\mathbf{p}|\geq 2} f_{\mathbf{p}}(x,\varepsilon)y^{\mathbf{p}}$ are bounded for $|x| < r_0, |\varepsilon| < r_1, |y| < r_2$. (For vectors we use the maximum norm, for matrices a compatible matrix norm.)

Now we introduce a scale of norms for functions analytic in $|x| < r = r_0$ by

$$\|a\|_l = \sup_{|x|<r} |a(x)|\,(r-|x|)^l \ .$$

Here we need the important theorem (cf. W. Walter [16])
**Nagumo's Lemma:** *If $a$ is analytic in $|x| < r$ then*

$$\|a'\|_{l+1} \leq e(l+1)\,\|a\|_l \ .$$

The proof is not difficult and only uses Cauchy's theorem.

We want to apply a majorant argument to (2), so we define majorant series.
**Definition:** *We say that $y(x,\varepsilon) = \sum_{l=0}^{\infty} y_l(x)\varepsilon^l$ is majorized by $g(z) = \sum_{l=0}^{\infty} g_l z^l$ or in symbols*

$$y(x,\varepsilon) \ll g(z) \text{ if } \|y_l\|_l \leq g_l \text{ for } l=0,1,\ldots.$$

The Cauchy product of two series and Nagumo's Lemma show
**Lemma:** *Assume that $y(x,\varepsilon) \ll g(z)$ and $\tilde{y}(x,\varepsilon) \ll \tilde{g}(z)$. Then*

$$y(x,\varepsilon)\tilde{y}(x,\varepsilon) \ll g(z)\tilde{g}(z) \quad \text{and} \quad \varepsilon\frac{dy}{dx}(x,\varepsilon) \ll ez\frac{d}{dz}(zg(z)) \ .$$

The above remark on how $y$ can be computed implies with the lemma that $y(x,\varepsilon) \ll g(z)$ where $g(z)$ is the solution with $g(0)=0$ of

$$g = M\left[\phi(z) + \alpha(z)g + e\,z^\sigma \frac{d}{dz}(zg(z)) + \sum_{l\geq 2}\phi_l(z)g^l\right] \ . \tag{3}$$

Here $M = \sup|A(x,0)^{-1}|$, $\phi(z)$ is a convergent(!) majorant of $f(x,\varepsilon)$, $\alpha(z)$ is a convergent(!) majorant of $A(x,\varepsilon) - A(x,0)$ and $\phi_l(z)$ are convergent majorants of $\sum_{|\mathbf{p}|=l} f_{\mathbf{p}}(x,\varepsilon)$. Observe that the series $\sum_{l\geq 2} \phi_l(z) g^l$ is convergent if $|g|$ is small and that $\phi(0) = \alpha(0) = 0$.

Now it is known that (3) has a unique power series solution $g(z)$ which is Gevrey of order $1/\sigma$. Since our solution $y$ of (1) has $g(z)$ as majorant we see immediately that $y$ is also Gevrey of order $1/\sigma$ uniformly near $x = 0$.

Now what can be done with such a Gevrey formal solution of (1) towards properties of *actual* solutions? Here we assume $\sigma = 1$ for simplicity. Then using the formal solution of theorem 1 we can define

$$h(x,t) := \sum_{m=1}^{\infty} a_m(x) \frac{t^{m-1}}{(m-1)!}$$

for complex $t$ with $|t| < R := 1/C_1$. Furthermore we can construct the following *incomplete Laplace transform*

$$\tilde{y}(x,\varepsilon) := \int_0^{\rho} e^{-t/\varepsilon} h(x,t)\,dt \tag{4}$$

where $0 < \rho < R$. Then we have

$$\tilde{y}(x,\varepsilon) \sim \sum_{m=1}^{\infty} a_m(x)\varepsilon^m$$

for $\mathrm{Re}(1/\varepsilon) \to \infty$ and thus

$$\tilde{d}(x,\varepsilon) := \varepsilon \frac{d\tilde{y}}{dx}(x,\varepsilon) - f(x,\varepsilon) - A(x,\varepsilon)\tilde{y}(x,\varepsilon) - \sum_{|\mathbf{p}|\geq 2} f_{\mathbf{p}}(x,\varepsilon)\tilde{y}^{\mathbf{p}} \sim 0 + 0\varepsilon + 0\varepsilon^2 + \dots\ .$$

Even better, because $\tilde{d}$ is also the Laplace transform of some function with a power series expansion in $|t| < R$, we have an estimate

$$\tilde{d}(x,\varepsilon) = \mathcal{O}(e^{-\rho/\varepsilon})$$

for $\mathrm{Re}(1/\varepsilon) \to \infty$, uniformly for $|x| \leq \tilde{r}$ where $\tilde{r} < r$ is arbitrary.

Now $y = \tilde{y} + v$ transforms (1) into

$$\varepsilon \frac{dv}{dx} = \tilde{d}(x,\varepsilon) + \tilde{A}(x,\varepsilon)v + \sum_{|\mathbf{p}|\geq 2} \tilde{f}_{\mathbf{p}}(x,\varepsilon)v^{\mathbf{p}} \tag{5}$$

where the coefficient functions are defined for $\mathrm{Re}(1/\varepsilon)$ sufficiently large and $|x| \leq \tilde{r}$ and furthermore $A(x,0) = \tilde{A}(x,0)$. By the well known method described in Wasow's book [18] it can be shown that (5) has an actual solution $v = v(x,\varepsilon)$ which is *exponentially small* in some given sector for $\varepsilon$, i.e. $|\arg\varepsilon - \varphi| \leq \varepsilon_0$, $|\varepsilon| \leq \varepsilon_0$ and some neighborhood of $x = 0$, i.e. $|x| \leq \varepsilon_0$ where $\varepsilon_0$ is sufficiently small.

Thus we have constucted an *actual solution* $y := \tilde{y} + v$ in the above $(x, \varepsilon)$–domain which has the formal solution of theorem 1 as its asymptotic series. Moreover, this construction shows that $y$ also has Gevrey–1 character for $\varepsilon \to 0$ in the following sense. For every $\varepsilon_1 < \varepsilon_0$ there are $C_3$ and $K$ such that

$$\left| y(x,\varepsilon) - \sum_{m=1}^{M-1} a_m(x)\varepsilon^m \right| \leq K\, M!\, C_3^M$$

for every $M$ and $(x, \varepsilon)$ with $|\arg \varepsilon - \varphi| \leq \varepsilon_1$, $|\varepsilon| \leq \varepsilon_1$ and $|x| \leq \varepsilon_1$.

Recently, the above results have been generalized by M. Canalis–Durand, J.P. Ramis, Y. Sibuya and R.S. [2] to a problem with *singular* leading matrix. Consider

$$\varepsilon^\sigma \frac{dy}{dx} = f(x,\varepsilon) + A(x,\varepsilon)y + B(x,\varepsilon)a + \sum_{|\mathbf{p}|+|\mathbf{q}|\geq 2} f_{\mathbf{pq}}(x,\varepsilon) y^{\mathbf{p}} a^{\mathbf{q}} \,, \tag{6}$$

where $f$, $A$, $B$, $f_{\mathbf{pq}}$ satisfy similar conditions to theorem 1, 1., 2., 3. . Furthermore we assume $f(x, 0) = 0$, but only that

$$\det(A(x,0)) = x^m d_m + \ldots \quad \text{with } d_m \neq 0 \,.$$

Hence without a special choice of the additional parameter vector $a$ there could be no solutions of (6). Instead of the invertibility of $A(0,0)$ we now assume that the mapping

$$\Phi : \begin{cases} \mathcal{H}(D_{r_0})^n \times \mathbb{C}^m & \to \quad \mathcal{H}(D_{r_0})^n \\ (y,a) & \mapsto \quad A(x,0)y + B(x,0)a \end{cases}$$

is bijective. ($\mathcal{H}(D_r)$ means the space of all complex valued functions which are analytic in the open disk $D_r$ around 0 with radius $r$.) In the case $m = 1$ and especially for $n = 2$ such problems have been studied by M. Canalis–Durand (e.g. [1]).

Here we have the following

**Theorem 2**: *There is a unique formal series $\hat{a} = \hat{a}(\varepsilon) = \sum_{j=1}^\infty c_j \varepsilon^j$ such that (6) with this choice of the parameter vector has a (also uniquely determined) formal solution $\hat{y} = \hat{y}(x, \varepsilon) = \sum_{j=1}^\infty b_j(x)\varepsilon^j$ where $b_j(x)$ are analytic functions. Moreover, $\hat{a}$ and $\hat{y}$ are Gevrey $1/\sigma$ (cf. definition below theorem 1).*

The proof is a modification of the proof of theorem 1. We rewrite (6) in a way similar to (2):

$$\begin{aligned} &A(x,0)y + B(x,0)a = \\ &\quad -\left[ f(x,\varepsilon) + \tilde{A}(x,\varepsilon)y + \tilde{B}(x,\varepsilon)a - \varepsilon^\sigma \frac{dy}{dx} + \sum_{|\mathbf{p}|+|\mathbf{q}|\geq 2} f_{\mathbf{pq}}(x,\varepsilon) y^{\mathbf{p}} a^{\mathbf{q}} \right] \end{aligned} \tag{7}$$

The condition on the above mapping $\Phi$ guarantees that (7) has a unique formal power series solution $(\hat{y}, \hat{a})$. The only real difference in the proof of theorem 2 concerns the inverse of the mapping $\Phi$. Let us denote the components of $\Phi^{-1}(u)$ by $P_1u$ and $P_2u$. Then we have to show that $P_1$ and $P_2$ are compatible with the Nagumo norms $\|...\|_l$. Unfortunately, for the original Nagumo norms this does *not* seem to be the case; but the following modifications satisfy our needs. Put

$$d(x) := \begin{cases} r - |x| & \text{if } \rho \leq |x| < r \\ r - \rho & \text{if } |x| \leq \rho \end{cases}$$

Then define

$$\|u\|_l := \sup_{|x|<r} |u(x)|\, d(x)^l \ .$$

With this norm, Nagumo's lemma still holds, but there also exists a $K > 0$ such that $\|P_j u\|_l \leq K\,\|u\|_l$ for all $u$ and $l$. The rest of the proof of theorem 2 is as in the proof of theorem 1.

The existence of an *actual* solution of (6) with $\hat{y}$, $\hat{a}$ as its asymptotic series which has a Gevrey–$1/\sigma$ character for $\varepsilon \to 0$ can also be proved.

## 2. Uniform simplification in a full neighborhood of a turning point

This problem, which was extensively discussed by Y. Sibuya [12] and earlier in special cases by W. Wasow [17] and R.Y. Lee [10], deals with an ordinary differential equation of second order with a singular parameter $\varepsilon$

$$\varepsilon^2 y'' = (x^m + \varepsilon b(x,\varepsilon))y \tag{8}$$

where $b(x,\varepsilon)$ is analytic near (0,0). Here $x = 0$ is – in general – a turning point and the problem consists in simplifying the equation by a linear transformation as much as possible *in a full neighborhood* $|x| < r$ of the turning point $x = 0$.

The linear transformation we want is of the form

$$z = A(x,\varepsilon)y + B(x,\varepsilon)\varepsilon y' \tag{9}$$

and we want that it simplifies (8) to

$$\varepsilon^2 z'' = (x^m + \varepsilon c(x,\varepsilon))z, \quad \text{where } c = \sum_{\nu=0}^{m-2} c_\nu(\varepsilon)x^\nu \tag{10}$$

is now a *polynomial* in $x$ of degree at most $m-2$. (In the case $m = 1$ this means $c \equiv 0$) We cannot expect to find such a transformation and such a polynomial $c$ where $A$, $B$ are analytic near $x = 0, \varepsilon = 0$. The best we can expect is functions $A, B, c_\nu$ which are analytic in $|x| < r$ and in a sector $S_\varepsilon$ given by $|\varepsilon| < \delta$, $|\arg\varepsilon - \phi| < \delta$ and have a Gevrey–1 asymptotic expansion for $\varepsilon \to 0$ in $S_\varepsilon$ uniformly for $|x| < r$.

We rewrite the equations for $A, B$ and the $c_\nu$ in a form similar to (2), (7) (omitting the arguments $(x,\varepsilon)$ for brevity)

$$\begin{aligned} 2x^m B' + mx^{m-1}B + (b - \textstyle\sum_{\nu=0}^{m-2} c_\nu(\varepsilon)x^\nu)A &= -\varepsilon A'' - 2\varepsilon B'b - \varepsilon Bb' \\ 2A' + (b - \textstyle\sum_{\nu=0}^{m-2} c_\nu(\varepsilon)x^\nu)B &= -\varepsilon B'' \end{aligned} \tag{11}$$

and first seek a formal solution $\hat{A}, \hat{B}, \hat{c}_\nu$ (also satisfying $\hat{A}(x,0)^2 - x^m \hat{B}(x,0)^2 \equiv 1$). The existence of a formal solution was already treated by R. Hanson and D. Russel [7], [8]. By rearranging their proof in a way similar to that of theorem 2 it can be shown that the formal series $\hat{A}, \hat{B}, \hat{c}_\nu$ are indeed also Gevrey–1. The appearance of a *second* derivative on the right hand side is not a problem because of the first order ode on the left hand side. This implies that for the computation of some coefficient in the recursion we have to solve such an ode and this compensates one derivative.

Now in the very same way as after theorem 2, one can prove the existence of *actual* solutions $A$, $B$, $c_\nu$ to (11) which have the above $\hat{A}$, $\hat{B}$, $\hat{c}_\nu$ as their Gevrey–1 asymptotic expansions as we wanted. I should remark, however, that Sibuya's work does not require that we start with an equation (8) where $b$ is analytic in $\varepsilon$; he only needs that $b$ has an asymptotic expansion for $\varepsilon \longrightarrow 0$. On the other hand, Sibuya needs additional restrictions on $b$ in the case $m > 2$ which are not necessary here. More important, as stated in [14], Sibuya's proof yields a uniform asymptotic expansion for $A$, $B$ in *any* closed subdisk of the disk $|x| < R$ where $b$ in (8) is analytic. Our proof – in the present form – only yields a uniform asymptotic expansion close to $x = 0$, but this might be overcome by choosing a more appropriate norm.

## 3. Separation of fast and slow variables

The third example is taken from the theory of averaging. Neishtadt [11] considers the following system with a "fast" variable $\varphi$ called *phase* and several slow variables $x_1, ..., x_n$ depending upon a real variable $t$

$$\dot{x} = \varepsilon f(x, \varphi, \varepsilon)\ , \quad \dot{\varphi} = \omega(x) + \varepsilon g(x, \varphi, \varepsilon)\ , \tag{12}$$

where $x \in \mathbb{R}^n$, $\omega, f, g$ are defined and real analytic for $x$ in some convex domain $D$, all real $\varphi$ and small $\varepsilon$ and where $f$ and $g$ are $2\pi$–periodic in $\varphi$. We assume that $\omega(x) \geq c > 0$ throughout $D$. The problem consists in eliminating $\varphi$ from the right hand sides of (12) by a nonlinear transformation

$$x = y + \varepsilon U(y, \psi, \varepsilon)\ , \quad \varphi = \psi + \varepsilon V(y, \psi, \varepsilon) \tag{13}$$

where $U$ and $V$ are $2\pi$–periodic in $\psi$. The simplified system is wanted in the form

$$\dot{y} = \varepsilon F(y, \varepsilon)\ , \quad \dot{\psi} = \omega(y) + \varepsilon \Omega(y, \varepsilon) \tag{14}$$

i.e. the slow variables now satisfy an "averaged" equation independent of the fast phase. As an inessential normalisation of the transformation, we impose the condition that $U(y, 0, \varepsilon) \equiv 0$ and $V(y, 0, \varepsilon) \equiv 0$.

Unfortunately, it is not always possible to find an analytic simplification (13). Neishtadt has (not rigourously) proved that there exists a *formal* (w.r.t. $\varepsilon$) transformation (13) that leads to a simplified formal system (14), that the formal power series in $\varepsilon$ that occur are Gevrey–1 and that there exists an actual transformation of the form (13) defined in $D$ and some $\varepsilon$–sector which reduces (12) to (14) *except for exponentially small error terms.* We want to prove this result by our method of majorants.

First we write down the equations determining $U$, $V$, $F$, $\Omega$ and then, as before, rewrite them so that we can see the existence of a formal power series solution easily. These transformation equations are ( the variables $(y, \psi, \varepsilon)$ are omitted for brevity)

$$\begin{aligned} \omega\frac{\partial U}{\partial \psi} + F &= f(y+\varepsilon U, \psi + \varepsilon V, \varepsilon) - \varepsilon\Omega\frac{\partial U}{\partial \psi} - \varepsilon F\frac{\partial U}{\partial y} \\ \omega\frac{\partial V}{\partial \psi} + \Omega &= g(y+\varepsilon U, \psi + \varepsilon V, \varepsilon) - \varepsilon\Omega\frac{\partial V}{\partial \psi} - \varepsilon F\frac{\partial V}{\partial y} \end{aligned} \tag{15}$$

If we expand here in powers of $\varepsilon$ we see that all our coefficients are determined recursively and the equations for the $m$-th coefficient have the form

$$\omega\frac{\partial U_m}{\partial \psi} + F_m = R_m\ , \qquad \omega\frac{\partial V_m}{\partial \psi} + \Omega_m = S_m$$

where $R_m$ and $S_m$ denote terms only depending upon $U_j, V_j, F_j, \Omega_j$ with $j < m$ which can be considered as known. Since both equations are analogous we consider only the first one. As $F_m$ is wanted to be independent of $\psi$ and $U_m$ is wanted to be $2\pi$–periodic in $\psi$ we can find $F_m$ by integration over the period.

$$F_m(y) = \langle R_m \rangle := \tfrac{1}{2\pi}\int_0^{2\pi} R_m(y,\psi)\,d\psi, \quad \omega(y)U_m(y,\psi) = \int_0^{\psi} R_m(y,\theta)d\theta - \psi F_m(y) \tag{16}$$

Here we use the following modification of the Nagumo norms. For functions $C = C(y,\psi)$ defined and analytic in $\tilde{D}\times S$, where $\tilde{D}$ is a convex neighborhood of $D$ in $\mathbb{C}^n$ and $S$ is a neighborhood of $\mathbb{R}$ in $\mathbb{C}$ we put

$$\|C\|_l := \sup\left\{ \left(|C(y,\psi)| + \left|\tfrac{\partial C}{\partial \psi}(y,\psi)\right|\right) \operatorname{dist}(y, \partial\tilde{D})^l \ :\ \psi \in S,\, y \in \tilde{D} \right\}\ .$$

Then these norms are compatible with differentiation $\partial/\partial y$ as before and, as can be seen from (16), the mappings $R_m \mapsto F_m$ and $R_m \mapsto U_m$ are compatible with the norms, too. Again, we can find a majorant differential equation with a singular variable for (15) which is of Poincaré rank 1 and has a Gevrey–1 formal solution. This proves the Gevrey–1 character of the formal solutions of (15). The incomplete Laplace transform, constructed in the same way as near (4), then yields the wanted analytic transformation reducing (12) to (14) except for exponentially small error terms. Thus Neishtadt's theorem is proved completely and correctly.

## 4. Convergence of formal solutions of meromorphic ODEs containing parameters

The last example does not contain a singular parameter any more, but it fits in this lecture because formal series in a parameter are discussed. With Gérard and Sibuya [6] and Sibuya [13] we consider an equation

$$x^{k+1}y' = f_0(x,\varepsilon) + A(x,\varepsilon)y + \sum_{|\mathbf{p}|\geq 2} f_{\mathbf{p}}(x,\varepsilon)y^{\mathbf{p}} \tag{17}$$

with the usual assumptions of analyticity near $x = 0, \varepsilon = 0$ and uniform convergence of the series near $y = 0$. We assume also that $k$ is a nonnegative integer.

From the works mentioned above, the following theorem is known, proved by several different methods. *If $A(0,0)$ is invertible and (17) has a formal solution $y(x,\varepsilon) = \sum_{|\mathbf{q}|\geq 0} a_{\mathbf{q}}(x)\varepsilon^{\mathbf{q}}$ where each $a_{\mathbf{q}}(x)$ is analytic in $x = 0$ then this formal series is convergent.* In this section, we want to give a proof of a generalisation of the above result based on the majorant method. We prove

**Theorem 3**: *Assume that the differential equation $x^{k+1}y' = A(x,0)y$ has no non-trivial solution $y$ which is analytic near $x = 0$. Then every formal solution $y(x,\varepsilon) = \sum_{|\mathbf{q}|\geq 0} a_{\mathbf{q}}(x)\varepsilon^{\mathbf{q}}$ of (17), where each $a_{\mathbf{q}}(x)$ is analytic near $x = 0$, is convergent.*

First we again reduce to the case $f_0(x,0) \equiv 0$ by a shift $y \to y + a_0(x)$. Then we rewrite the equation so that we see the recursion relation determining the $a_{\mathbf{q}}$ easily.

$$x^{k+1}y' - A(x,0)y = f_0(x,\varepsilon) + (A(x,\varepsilon) - A(x,0))y + \sum_{|\mathbf{p}|\geq 2} f_{\mathbf{p}}(x,\varepsilon)y^{\mathbf{p}} \qquad (18)$$

As $y(x,0) \equiv 0$, the coefficient of some power $\varepsilon^{\mathbf{q}}$ on the right hand side only depends upon $a_{\mathbf{l}}(x)$ with $|\mathbf{l}| < |\mathbf{q}|$. Because of our assumption, this implies that there is at most one formal solution $y(x,\varepsilon) = \sum_{|\mathbf{q}|\geq 1} a_{\mathbf{q}}(x)\varepsilon^{\mathbf{q}}$ where each $a_{\mathbf{q}}(x)$ is analytic in $x = 0$. Furthermore, we conclude that *all* $a_{\mathbf{q}}(x)$ are analytic in the same disk $|x| < r$ where all the coefficients of the right hand side of (17) are analytic.

In this example, we use old fashioned norms on the space of functions analytic in $|x| < r$. For an arbitrary $\rho$, $0 < \rho < r$, we define

$$\|c\| := \sum_{l=0}^{\infty} |c_l|\, \rho^l$$

if $c(x)$ is analytic in $|x| < r$. The most important point now is

**Lemma**: *Under the assumption of the theorem, there is a number $K > 0$ with the following property. Whenever $a(x)$ and $b(x)$ are analytic in $|x| < r$ and satisfy*

$$x^{k+1}a'(x) + A(x,0)a(x) = b(x) \qquad (|x| < r)$$

*then their norms satisfy $\|a\| \leq K\, \|b\|$ .*

I cannot give the proof of the lemma here, I just want to mention that the main idea is to reduce the problem to some finite dimensional space of polynomials $b(x)$ by means of an argument similar to that in Harris, Sibuya, Weinberg [9].

The rest of the proof is similar to section 1. We let $g(x,\varepsilon) = \sum g_{\mathbf{l}}(x)\varepsilon^{\mathbf{l}} \ll \sum h_m z^m$ denote the set of conditions

$$\sum_{|\mathbf{l}|=m} |g_{\mathbf{l}}| \leq h_m \qquad (m = 1, 2, 3, ...)\ .$$

Then by the lemma and the relation (18), the formal series $y(x,\varepsilon)$ is majorized by the unique solution $u(z) = \sum_{m=1}^{\infty} u_m z^m$ of the majorant equation

$$u = K\left[\hat{f}(z) + \hat{A}(z)u + \sum_{p=2}^{\infty} \hat{f}_p(z)u^z\right]$$

constructed in the already familiar way. Since $\hat{f}(0) = 0$ and $\hat{A}(0) = 0$ this solution $u(z)$ is convergent by the implicit function theorem and therefore also $y(x, \varepsilon)$ must be convergent. Thus our theorem is proved.

**References:**


[1] M. Canalis–Durand, Solution formelle Gevrey d'une équation singulièrement perturbée: le cas multidimensionnel, preprint 1992.

[2] M. Canalis–Durand, J.P. Ramis, Y. Sibuya, R. Schäfke, Gevrey solutions of singularly perturbed differential and difference equations, in preparation.

[3] R. Gérard, Une classe d'équations différentielles nonlinéaires à singularité régulière, Funk. Ekvac. **29** (1986), 55–76.

[4] R. Gérard and D.A. Lutz, Convergent factorial series solutions of singular operator equations, Analysis **10** (1990), 99–145.

[5] R. Gérard, D.A. Lutz, and R. Schäfke, Analytic Reduction of Some Nonlinear Difference Equations, preprint 1993.

[6] R. Gérard, Y. Sibuya, Etude de certains systèmes de Pfaff avec singularités, C. R. Acad. Sc. Paris 284 (1977), 57–60.

[7] R.J. Hanson, implification of second order systems of ordinary differential equations with a turning point, SIAM J. Math. Anal. **16** (1968), 1059–1080.

[8] R.J. Hanson, D.L. Russell, Classification and reduction of second order systems at a turning point,J. Math. and Phys. **46** (1967), 74–92.

[9] W. Harris, Jr., Y. Sibuya, L. Weinberg, Holomorphic solutions of linear differential systems at singular points, Arch. Rational Mech. Anal., **35** (1969), 245–248.

[10] R.Y. Lee, On uniform simplification of linear differential in a full neighborhood of a turning point, J. Math. Anal. Appl. **27** (1969), 501–510.

[11] A.I. Neishtadt, The separation of motions in systems with rapidly rotating phase, J. Appl. Math. Mech., **48** (1984), 134–139.

[12] Y. Sibuya, Uniform simplification in a full neighborhood of a transition point, Memoirs AMS **149**, American Mathematical Society, Providence, RI, 1974.

[13] Y. Sibuya, Convergence of power series solutions of a linear Pfaffian system at an irregular singular point, Keio Engineering Reports, **31** (1978), 79–86.

[14] Y. Sibuya, A theorem concerning uniform simplification at a transition point and the problem of resonance, SIAM J. Math. Anal. **12** (1981), 653–668.

[15] Y. Sibuya, Gevrey property of formal solutions in a parameter, Asymptotic and computational analysis (ed. R. Wong), **124** (1990), 393–401.

[16] W. Walter, An elementary proof of the Cauchy–Kovalewsky theorem, Amer. Math. Monthly (1985), 115–125.

[17] W. Wasow, Asymptotic expansions for ordinary differential equations, New York, 1965.

[18] W. Wasow, Linear turning point theory, New York 1985.

# ON THE DOUBLE CONFLUENT HEUN EQUATION

DIETER SCHMIDT
Universität Essen, FB 6 - Mathematik,
D-45117 Essen, Germany

and

GERHARD WOLF
Universität Essen, FB 6 - Mathematik,
D-45117 Essen, Germany

ABSTRACT

In this report we give a short survey of the global theory of the Double Confluent Heun Equation.

## 0. Introduction

The Heun Equation is (a canonical form of) the general second-order Fuchsian differential equation with four singular points being located at 0, 1, $a \in \mathbf{C} \setminus \{0, 1\}$ and $\infty$. Confluence of the singularities $a$ and $\infty$ yields the Confluent Heun Equation (CHE); further confluence of the singularities 0 and 1 yields then the Double Confluent Heun Equation (DCHE). We consider here the following *'symmetric canonical form'* of the DCHE

$$D^2y + \alpha\left(z + \frac{1}{z}\right) Dy + \left((\beta_1 + \tfrac{1}{2})\alpha z + (\frac{\alpha^2}{2} - \gamma) + (\beta_{-1} - \tfrac{1}{2})\frac{\alpha}{z}\right) y = 0, \quad (1)$$

where $D$ denotes $z\,\frac{d}{dz}$. The independent variable $z$ is regarded as a complex variable, the quantities $\alpha, \beta_{-1}, \beta_1$ and $\gamma$ are arbitrary complex parameters, and the solutions $y$ are to be analytic functions with respect to the variable $z$ as well as to the parameters.

(1) is a meromorphic ODE, the only singular points are 0 and $\infty$. Since all solutions can be continued analytically within the domain $\mathbf{C}^* := \mathbf{C} \setminus \{0\}$, the resulting global solutions are - in general - not single valued on $\mathbf{C}^*$ but on the Riemannian surface of the logarithm over $\mathbf{C}^*$, from now on denoted by $\Omega$. In view of this, all solutions of (1) will be regarded throughout as 'global solutions', being defined and holomorphic on $\Omega$.

In the nondegenerate case $\alpha \neq 0$, the singular points 0 and $\infty$ are both irregular singularities of rank one. Thus, there exist uniquely-defined fundamental sets of solutions about each singularity, possessing a characteristic asymptotic behaviour within special sectors. These sectors are bounded by the so-called Stokes rays, depending on the argument of $\alpha$. This suggest to let also $\alpha$ vary in $\Omega$ rather than in $\mathbf{C}^*$.

The main problems of a global analysis of the DCHE are the so-called *'connection relations'* between the various asymptotic solutions. Essentially, one has to distinguish two main types of connection problems. The *'lateral connection problem'* deals with the relations between the various asymptotic solutions in different sectors at the same singularity, and is associated with the concept of *'Stokes phenomenon'*. The *'central connection problem'*, on the other hand, deals with the relations between the various asymptotic solutions at different singularities. In the case of the DCHE both problems are very closely related.

In [4] we have established a general theory of the DCHE with regard to the global behaviour of the solutions as analytic functions of the independent variable $z$, as well as the parameters $\alpha, \beta_{-1}, \beta_1, \gamma$. The intention of the present paper is to report on the main features of this theory.

## 1. General Features

### *1.1. Transformations; Standard Forms*

We begin by considering the (class of) general second-order linear ODEs with two singularities in the closed complex plane, both at most irregular singularities of rank 1, which we assume to be situated at 0 and $\infty$. Such an ODE can be written in the form

$$D^2y + a(z)Dy + b(z)y = 0, \quad D := z\,\frac{d}{dz} \tag{2}$$

with

$$a(z) = \sum_{j=-1}^{1} a_j\, z^j\,, \quad b(z) = \sum_{j=-2}^{2} b_j\, z^j\,. \tag{3}$$

We call this equation *'General Double-Confluent Heun Equation'* (gDCHE).

The only transformations of the dependent variable leaving the class of gDCHEs invariant are

$$y(z) = z^{\zeta_0}\,\exp(\zeta_1\, z - \zeta_{-1}\, z^{-1})\,\tilde{y}(z). \tag{4}$$

For every fixed gDCHE (2), there is exactly one transformation of this type yielding the following *'normal form'* of the gDCHE

$$D^2y + B(z)y = 0 \tag{5}$$

with

$$B(z) := b(z) - \left(\frac{a(z)}{2}\right)^2 - \frac{1}{2}\,Da\,(z) =: \sum_{j=-2}^{2} B_j z^j. \tag{6}$$

$B$ and thus the $B_j\,(j = -2, \ldots, 2)$ are invariants of the gDCHE (2) by transformations of type (4). On the other hand, for every fixed gDCHE (2), there are 4 transformations of type (4) yielding the following *'canonical form'* of the **gDCHE**

$$D^2y + (\alpha_1 z + \frac{\alpha_{-1}}{z})Dy + \left((B_1 + \frac{\alpha_1}{2})z + (B_0 + \frac{\alpha_1\alpha_{-1}}{2}) + (B_{-1} - \frac{\alpha_{-1}}{2})\frac{1}{z}\right) y = 0, \tag{7}$$

with $\alpha_j{}^2 = -4\, B_{2j}, \quad (j = -1, 1)$.

The actual type of the singularities of a gDCHE can be read off from the corresponding normal or canonical forms. In particular, in the *'nondegenerate case'* $B_2\, B_{-2} \neq 0$, which is equivalent to $\alpha_1\, \alpha_{-1} \neq 0$, both 0 and $\infty$ are irregular singularities of rank 1.

The only transformations of the independent variable leaving the class of gDCHEs invariant are of the form

$$z = \tau\, \tilde{z}^k, \quad y(z) = \tilde{y}(\tilde{z}), \quad \tau \in \mathbf{C}^*, \quad k \in \{-1, 1\}. \tag{8}$$

Moreover, these transformations also leave the class of normal forms and also the class of canonical forms of the gDCHE invariant. In particular, in the nondegenerate case $\alpha_1\, \alpha_{-1} \neq 0$, we find that there exist 4 transformations of type (8) reducing the canonical form of the gDCHE (7) to the DCHE (1).

Summarizing the preceding considerations we obtain

PROPOSITION: *In the the nondegenerate case* $B_2\, B_{-2} \neq 0$, *the* gDCHE (2) *is reducible to the* DCHE (1) *by applying a product of suitable transformations of type* (4) *and* (8) .

PROPOSITION: *There are 16 transformations, namely*

$$y(z) = \exp(\zeta_1\, z - \zeta_{-1}\, / z)\; \tilde{y}(\tilde{z}), \qquad z = \tau\, \tilde{z}^k$$

*with*

$$\zeta_j = \frac{\alpha}{2}(\varepsilon_j - 1), \;\; \varepsilon_j \in \{-1, 1\}, \;\; (j = -1, 1); \;\; \tau^2 = \varepsilon_1\, \varepsilon_{-1}; \;\; k \in \{-1, 1\},$$

*which leave the class of* DCHEs *invariant.*

### *1.2. The Transformation Group of the DCHE*

Considering now the DCHE (1), we introduce the parameter domain

$$\Lambda := \{(\alpha, \beta, \gamma) \,|\, \alpha \in \Omega, \;\; \beta := (\beta_{-1}, \beta_1) \in \mathbf{C}^2, \;\; \gamma \in \mathbf{C}\}.$$

Then, all solutions will principally be holomorphic functions of the variable $(\alpha, \beta, \gamma; z) \in \Lambda \times \Omega$.

The following transformations, defined for arbitrary functions $y : \Lambda \times \Omega \to \mathbf{C}$, will play a fundamental role in our investigations

$$\begin{aligned}
(t_1 y)(\alpha,\beta,\gamma;z) &:= e^{-\alpha z}\, y(e^{i\pi/2}\alpha, \beta_{-1}, -\beta_1, \gamma; e^{i\pi/2} z),\\
(t_{-1} y)(\alpha,\beta,\gamma;z) &:= e^{\alpha/z}\, y(e^{i\pi/2}\alpha, -\beta_{-1}, \beta_1, \gamma; e^{-i\pi/2} z),\\
(t_0 y)(\alpha,\beta,\gamma;z) &:= e^{-\alpha z+\alpha/z}\, y(\alpha, \beta_1, \beta_{-1}, \gamma; 1/z),\\
(m\, y)(\alpha,\beta,\gamma;z) &:= y(\alpha,\beta,\gamma; e^{2\pi i} z),\\
(s\, y)(\alpha,\beta,\gamma;z) &:= y(e^{2\pi i}\alpha,\beta,\gamma;z).
\end{aligned} \tag{9}$$

One also needs the parameter-substitutions induced by the transformations $t_j$ and $s$, which we denote by $\tau_j$ and $\sigma$, respectively.

By the considerations in section *1.1* we obtain

PROPOSITION:
(i) *The transformations $t_0$ and $t_1$ generate an infinite group, denoted by $\mathcal{T}$, containing in particular $t_{-1}, m$ and $s$. The following relations are valid:*

$$\begin{aligned}
&{t_0}^2 = id; \quad t_1 \circ t_{-1} = t_{-1} \circ t_1, \quad t_1 \circ t_0 = t_0 \circ t_{-1},\\
&m = (t_1 \circ t_{-1}^{-1})^2 = {t_1}^2 \circ t_{-1}^{-2}, \quad s = (t_1 \circ t_{-1})^2 = {t_1}^2 \circ t_{-1}^{2}.
\end{aligned}$$

(ii) *For each $(\alpha,\beta,\gamma) \in \Lambda$, let $y(\alpha,\beta,\gamma;.) : \Omega \to \mathbf{C}$ be a solution of the corresponding* DCHE (1). *Then, for each $t \in \mathcal{T}$ and each $(\alpha,\beta,\gamma) \in \Lambda$, the function*

$$(t\, y)(\alpha,\beta,\gamma;.) : \Omega \to \mathbf{C}$$

*is also a solution of the corresponding* DCHE.

### *1.3. The Wronskian; A General Connection Formula*

Let $y_1, y_2$ be arbitrary solutions of the DCHE (1). Then their Wronskian

$$w[y_1,y_2] := y_1\, Dy_2 - y_2\, Dy_1,$$

has the representation

$$w[y_1,y_2](z) =: [y_1,y_2]\; e^{-\alpha z+\alpha/z}, \quad (z \in \mathbf{C}^*),$$

where $[y_1,y_2] \in \mathbf{C}$ is a constant. The so-defined brackets $[.,.]$ are an alternating bilinear functional on the solution-space of the DCHE with

$$[y_1,y_2] \neq 0 \iff y_1,\ y_2 \textit{ fundamental set of solutions.}$$

For any further solution $y$ of the DCHE we have the *'general connection formula'*

$$[y_1,y_2]\; y(z) = [y,y_2]\; y_1(z) + [y_1,y]\; y_2(z), \quad (z \in \Omega). \tag{10}$$

## 2. The Analytic Behaviour of the Solutions

### *2.1. The Basic Asymptotic Solution*

For $(\alpha,\beta,\gamma)\in\Lambda$ the DCHE (1) has a unique solution $\Psi(\alpha,\beta,\gamma;.):\Omega\to\mathbf{C}$ with the asymptotic representation

$$\Psi(\alpha,\beta,\gamma;z)\sim(\alpha\,z)^{-\beta_1-\frac{1}{2}}\sum_{n=0}^{\infty}\psi_n\,(\alpha\,z)^{-n} \tag{11}$$

for $z\to\infty$ within the sector $-\frac{\pi}{2}<\arg(\alpha\,z)<\frac{5\pi}{2}$. $\Psi$ is a holomorphic function of $(\alpha,\beta,\gamma;z)\in\Lambda\times\Omega$. The coefficients $\psi_n=\psi_n(\alpha,\beta,\gamma)$ are polynomials in $(\alpha^2,\beta,\gamma)$, uniquely given by

$$\begin{aligned}(n+1)\,\psi_{n+1} \quad &- \quad ((n+\beta_1+\tfrac{1}{2})^2-\gamma+\alpha^2/2)\;\psi_n \quad + \\ &+ \quad \alpha^2\,(n+\beta_1-\beta_{-1})\,\psi_{n-1}=0\,, \quad (n\in\mathbf{N}), \\ &\psi_0:=1,\quad \psi_{-1}:=0.\end{aligned} \tag{12}$$

We consider the transformation behaviour of $\Psi$ by $t\in\mathcal{T}$. The asymptotic behaviour of $t_1\Psi$ is

$$(t_1\Psi)(\alpha,\beta,\gamma;z)=e^{-\alpha z}\,(e^{i\pi}\,\alpha z)^{\beta_1-\frac{1}{2}}\,(1+\mathcal{O}(1/z)) \tag{13}$$

for $z\to\infty$ in $-\frac{3\pi}{2}<\arg(\alpha z)<\frac{3\pi}{2}$. Inserting this and the asymptotic behaviour of $\Psi$ into $w[\Psi,t_1\Psi]$ furnishes

$$[\Psi,t_1\Psi]=i\,e^{i\pi\beta_1}. \tag{14}$$

In particular we find that $\Psi$ and $t_1\Psi$ constitute a fundamental set. By doing the same with $t_1^2\Psi$, we obtain

$$[t_1^2\Psi,t_1\Psi]=-i\,e^{-i\pi\beta_1}. \tag{15}$$

This yields the *'first basic connection relation'* for $\Psi$

$$t_1^2\Psi=-e^{-2\pi i\beta_1}\,(\Psi-q\,t_1\Psi)\,, \tag{16}$$

where

$$q(\alpha,\beta,\gamma):=i\,e^{i\pi\beta_1}\,[t_1^2\Psi,\Psi]. \tag{17}$$

We denote $q$ *'basic connection coefficient'* of the DCHE. It is a holomorphic function of $(\alpha,\beta,\gamma)\in\Lambda$.

By the asymptotic behaviour of $t_{-1}\Psi$ and $\Psi$ follows, as above,

$$t_{-1}\Psi=\Psi. \tag{18}$$

Applying $t_{-1}$ on the first basic connection relation gives then

$$\tau_{-1}q=q. \tag{19}$$

Thus, in particular, $q$ is a holomorphic function of $(\alpha^2,\beta,\gamma)\in\mathbf{C}^*\times\mathbf{C}^3$. By later consideration follows that $q$ is even holomorphic on $\mathbf{C}^4$.

*2.2. Fundamental Sets of Asymptotic Solutions at ∞ and at 0*

At first, we introduce a fundamental set of solutions with characteristic asymptotic behaviour at $\infty$:

$$\begin{aligned} y_{\infty 1}(\alpha,\beta,\gamma;z) &:= \Psi(\alpha,\beta,\gamma;z), \\ y_{\infty 2}(\alpha,\beta,\gamma;z) &:= i\,e^{-i\pi\beta_1}\,e^{-\alpha z}\,\Psi(e^{i\pi/2}\,\alpha,\beta_{-1},-\beta_1,\gamma;e^{i\pi/2}\,z). \end{aligned} \tag{20}$$

These functions constitute a fundamental solution $Y_\infty := (y_{\infty 1}, y_{\infty 2})$ for the DCHE (1) with the asymptotic behaviour for $z \to \infty$

$$\begin{aligned} y_{\infty 1}(z) &= (\alpha z)^{-\beta_1-\frac{1}{2}}\,(1+\mathcal{O}(1/z)) \quad && in \;\; -\tfrac{\pi}{2} < \arg(\alpha\,z) < \tfrac{5\pi}{2}, \\ y_{\infty 2}(z) &= e^{-\alpha z}\,(\alpha\,z)^{\beta_1-\frac{1}{2}}\,(1+\mathcal{O}(1/z)) \quad && in \;\; -\tfrac{3\pi}{2} < \arg(\alpha z) < \tfrac{3\pi}{2}. \end{aligned} \tag{21}$$

We have

$$[y_{\infty 1}, y_{\infty 2}] = -1. \tag{22}$$

The following transformation-formulas are valid:

$$t_{-1}\,Y_\infty = Y_\infty, \quad t_1 Y_\infty =: Y_\infty \circ Q, \tag{23}$$

where

$$Q =: \begin{pmatrix} 0 & -i\,e^{-i\pi\beta_1} \\ -i\,e^{i\pi\beta_1} & q \end{pmatrix}. \tag{24}$$

Furtheron, we introduce a fundamental set of solutions with characteristic asymptotic behaviour at 0:

$$\begin{aligned} y_{01}(\alpha,\beta,\gamma;z) &:= e^{-\alpha z+\alpha/z}\,\Psi(\alpha,\beta_1,\beta_{-1},\gamma;1/z), \\ y_{02}(\alpha,\beta,\gamma;z) &:= i\,e^{-i\pi\beta_{-1}}\,e^{-\alpha z}\,\Psi(e^{i\pi/2}\alpha,\beta_1,-\beta_{-1},\gamma;e^{i\pi/2}/z). \end{aligned} \tag{25}$$

These functions constitute a fundamental solution $Y_0 := (y_{01}, y_{02}) = t_0\,Y_\infty$ for the DCHE (1) with asymptotic behaviour for $z \to 0$

$$\begin{aligned} y_{01}(z) &= e^{\alpha/z}\,(\alpha/z)^{-\beta_{-1}-\frac{1}{2}}\,(1+\mathcal{O}(z)) \quad && in \;\; -\tfrac{\pi}{2} < \arg(\alpha/z) < \tfrac{5\pi}{2}, \\ y_{02}(z) &= (\alpha/z)^{\beta_{-1}-\frac{1}{2}}\,(1+\mathcal{O}(z)) \quad && in \;\; -\tfrac{3\pi}{2} < \arg(\alpha/z) < \tfrac{3\pi}{2}. \end{aligned} \tag{26}$$

We have

$$[y_{01}, y_{02}] = 1. \tag{27}$$

The following transformation-formulas are valid:

$$t_1\,Y_0 = Y_0, \quad t_{-1}Y_0 = Y_0 \circ \tau_0\,Q. \tag{28}$$

*2.3. On the Central Connection Relation*

Between the fundamental solutions $Y_\infty$ and $Y_0$ introduced in section *2.2*, the following *'central connection relation'* holds

$$Y_0 = t_0\, Y_\infty =: Y_\infty \circ C \tag{29}$$

with *'central connection matrix'*

$$C = \begin{pmatrix} -c^{12} & -c^{22} \\ c^{11} & c^{21} \end{pmatrix}, \quad c^{j\ell} = [y_{0j}, y_{\infty\ell}]. \tag{30}$$

At first, we derive a relation between $Q$ and $C$. By using $t_1 \circ t_0 = t_0 \circ t_{-1}$ we find

$$\begin{aligned} Y_\infty \circ C &= t_0 Y_\infty = t_0\, t_{-1} Y_\infty = t_1\, t_0 Y_\infty \\ &= t_1(Y_\infty \circ C) = t_1 Y_\infty \circ \tau_1 C = Y_\infty \circ Q \circ \tau_1 C, \end{aligned}$$

from which follows

$$C = Q \circ \tau_1 C. \tag{31}$$

Thus, $Q$ can be represented by $C$ through

$$Q = C \circ \tau_1 C^{-1}. \tag{32}$$

Furtheron, we investigate the structure of $C$. With $\det Y_\infty = -1,\ \det Y_0 = 1$ follows

$$\det C = -1. \tag{33}$$

By using ${t_0}^2 = \mathrm{id}$ we find

$$Y_\infty = {t_0}^2 Y_\infty = t_0(Y_\infty \circ C) = Y_0 \circ \tau_0 C,$$

which implies

$$\tau_0 C = C^{-1}. \tag{34}$$

Using $\det C = -1$ yields then

$$\tau_0 c^{11} = c^{11}, \quad \tau_0 c^{22} = c^{22}, \quad \tau_0 c^{12} = c^{21}.$$

Further, looking at the first row of $C = Q \circ \tau_1 C$ gives

$$c^{12} = i\, e^{-i\pi\beta_1}\, \tau_1 c^{11}, \quad c^{22} = i\, e^{-i\pi\beta_1}\, \tau_1 c^{21}.$$

Therefore, we can represent $c^{12}$, $c^{21}$, $c^{22}$ in terms of $c^{11}$.

To achieve a simple structure for the formulas, we introduce

$$c(\alpha, \beta, \gamma) := e^{i\pi(\beta_{-1}+\beta_1)/2}\, [t_0 \Psi, \Psi] \tag{35}$$

and obtain finally

$$C = e^{-i\pi(\beta_{-1}+\beta_1)/2} \begin{pmatrix} -i\,\tau_1 c & \tau_1\tau_{-1}c \\ c & i\,\tau_{-1}c \end{pmatrix}. \tag{36}$$

We call $c$ *'central connection coefficient'* for the DCHE (1). It is a holomorphic function on $\Lambda$ and satisfies $\tau_0 c = c$, which means symmetry in $\beta_{-1}$ and $\beta_1$.

Since $Q$ has a representation in terms of $C$, one can also express $q$ in terms of $c$

$$q = e^{-i\pi\beta_{-1}} \left(c\,\tau_{-1}\tau_1^{\,2}c \;-\; \tau_{-1}c\,\tau_1^{\,2}c\right). \tag{37}$$

By the use of $c$, we can establish a *'second basic connection relation'* for $\Psi$

$$t_0\Psi = -i\,e^{-i\pi(\beta_{-1}+\beta_1)/2} \left(\tau_1 c\;\Psi \;-\; e^{-i\pi\beta_1}\,c\,t_1\Psi\right). \tag{38}$$

*2.4. Laplace Integral for the Basic Asymptotic Solution*

Representing the solutions of the DCHE (1) with Laplace Integrals leads to consider the special CHE

$$t(1-t)v'' - (\beta_1 + (\beta_{-1}+1)(2t-1))\,v' + \left(\alpha^2(t-\tfrac{1}{2}) + \gamma - (\beta_{-1}+\tfrac{1}{2})^2\right) v = 0, \tag{39}$$

having singular points at 0, 1 and $\infty$. 0 and 1 are regular singularities, while $\infty$ is an irregular singularity of rank $\frac{1}{2}$ for $\alpha \neq 0$. Also for the CHE, one needs to introduce only one basic solution, namely the Frobenius solution $\Phi$ at 0 corresponding to the index 0, which is holomorphic with respect to $(\alpha^2, \beta, \gamma; t) \in \mathbf{C}^4 \times (\mathbf{C} \setminus [1,\infty[)$ and can be expanded in a power series

$$\Phi(\alpha^2, \beta, \gamma; t) = \sum_{n=0}^{\infty} \frac{\varphi_n}{\Gamma(1+\beta_{-1}-\beta_1+n)}\, t^n, \quad (|t|<1) \tag{40}$$

with $\varphi_0 = 1$. The $\varphi_n$ are related to the coefficients $\psi_n$ of the asymptotic series of $\Psi$ by

$$\psi_n(\alpha, \beta, \gamma) = \varphi_n(\alpha^2, \beta_1, \beta_{-1}, \gamma), \quad (n \in \mathbf{N}). \tag{41}$$

A detailed study of the CHE (39) yields that the functions

$$\begin{aligned} v_{01}(\alpha^2, \beta, \gamma; t) &:= \Phi(\alpha^2, \beta, \gamma; t), \\ v_{02}(\alpha^2, \beta, \gamma; t) &:= t^{\beta_1-\beta_{-1}}\,\Phi(\alpha^2, \beta_1, \beta_{-1}, \gamma; t), \end{aligned} \tag{42}$$

and

$$\begin{aligned} v_{11}(\alpha^2, \beta, \gamma; t) &:= \Phi(-\alpha^2, \beta_{-1}, -\beta_1, \gamma; 1-t), \\ v_{12}(\alpha^2, \beta, \gamma; t) &:= (1-t)^{-\beta_1-\beta_{-1}}\,\Phi(-\alpha^2, -\beta_1, \beta_{-1}, \gamma; 1-t), \end{aligned} \tag{43}$$

are pairs of Frobenius solutions of the CHE corresponding to the regular singular point 0 and 1, respectively.

Applying the Laplace transform to $v = v_{02}$ with a suitable loop gives

THEOREM: *Let* $0 < \xi < 2\pi$.
*Then, for* $(\alpha, \beta, \gamma; z) \in \Lambda \times \Omega$ *with* $|\arg(\alpha z) - \xi| < \pi/2$

$$\Psi(\alpha, \beta, \gamma; z) = \frac{(\alpha z)^{\frac{1}{2}-\beta_{-1}}}{e^{2\pi i(\beta_1-\beta_{-1})} - 1} \int_{\infty e^{-i\xi}}^{(0+)} e^{-\alpha z t}\, t^{\beta_1-\beta_{-1}}\, \Phi(\alpha^2, \beta_1, \beta_{-1}, \gamma; t)\, dt.$$

*For* $\beta_1 - \beta_{-1} \in \mathbf{Z}$ *one has to take the holomorphic continuation of the expression.*

The Laplace transform achieves a transfer of the first basic connection relation (16) of the DCHE into a connection relation of the CHE (39) between Frobenius solutions at 0 and 1. In particular, it turn out that the basic connection coefficient $q$ coincides with a corresponding coefficient of the connection problem of the CHE at 0 and 1. The latter quantity is well studied, in particular in [3]. Applying these results provides a limit-formula for $q$.

PROPOSITION: *Let* $\psi_n$ , $(n \in \mathbf{N})$, *be the coefficients of the asymptotic series of* $\Psi$. *Then for* $(\alpha, \beta, \gamma) \in \mathbf{C}^4$

$$q(\alpha, \beta, \gamma) = \lim_{n\to\infty} \frac{-2\pi e^{-\pi i \beta_1}\, n!\, \psi_n(\alpha, \beta, \gamma)}{\Gamma(n+1-\beta_{-1}+\beta_1)\,\Gamma(n+\beta_1+\beta_{-1})}, \tag{44}$$

*the convergence being* $\mathcal{O}(n^{-1})$ *as* $n \to \infty$.

## *2.5. Monodromy Behaviour; Characteristic Exponents; Floquet Solutions*

We considers solutions of the DCHE of the form

$$y(z) = z^{\nu}\, h(z) \not\equiv 0, \qquad (z \in \Omega) \tag{45}$$

with $\nu \in \mathbf{C}$ and $h$ holomorphic for $z \in \mathbf{C}^*$. In this situation, $\nu$ is called *'characteristic exponent'* of the DCHE (1) and $y$ *'Floquet solution'* of the DCHE (1) corresponding to the characteristic exponent $\nu$. Expanding $h$ into a Laurent series

$$h(z) = \sum_{n=-\infty}^{\infty} \eta_n\, z^n, \quad (z \in \mathbf{C}^*) \tag{46}$$

and inserting into (1) yields the three-term recurrence relations for the $\eta_n$

$$\begin{aligned} \alpha\,(n+\nu+\beta_{-1}+\tfrac{1}{2})\,\eta_{n+1} + ((n+\nu)^2 - \gamma + \alpha^2/2)\,\eta_n + \\ + \alpha\,(n+\nu+\beta_1-\tfrac{1}{2})\,\eta_{n-1} = 0, \qquad (n \in \mathbf{Z}). \end{aligned} \tag{47}$$

Fundamental for the study of Floquet solutions and characteristic exponents is the *'monodromy operator'* $m$, defined by $(m\,y)(z) = y(e^{2\pi i}\,z)$, $(z \in \Omega)$. When considering

$m$ as an operator on the solution space of the DCHE, a Floquet solution as considered above is just the eigenfunctions of $m$ corresponding to the eigenvalue $e^{2\pi i\nu}$. For a concrete representation of $m$, we introduce

$$m\, Y_\infty =: Y_\infty \circ M_\infty. \tag{48}$$

$M_\infty$ is called *'monodromy matrix'* of the fundamental solution $Y_\infty$. With $m = t_1^2 \circ t_{-1}^{-2}$ we find

$$M_\infty \;=\; Q \circ \tau_1 Q \;=\; \begin{pmatrix} -e^{-2\pi i\beta_1} & -i\, e^{-i\pi\beta_1}\,\tau_1 q \\ -i\, e^{-i\pi\beta_1}\, q & -e^{2\pi i\beta_1} \,+\, q\,\tau_1 q \end{pmatrix}. \tag{49}$$

Since $\nu \in \mathbf{C}$ is a characteristic exponent of the DCHE, if and only if $e^{2\pi i\nu}$ is an eigenvalue of $m$ and thus of $M_\infty$, it follows – by using $\det M_\infty \;=\; 1$ – that the characteristic exponents $\nu$ are determined by

$$\cos 2\pi\nu \;=\; -\cos 2\pi\beta_1 \;+ \tfrac{1}{2}\, q\,\tau_1 q. \tag{50}$$

Then, according to the limit formula for $q$, the characteristic exponents of (1) for fixed parameter values may be computed by using three-term recurrence relations.

We define for arbitrary $\nu \in \mathbf{C}$ and arbitrary solutions $y$ of (1)

$$\mathcal{F}_\nu\, y := m\, y - e^{-2\pi i\nu}\, y, \tag{51}$$

which, again, is a solution of (1). In the case that $\nu$ is a characteristic exponent, $\mathcal{F}_\nu\, y$ is a Floquet solution corresponding to $\nu$.

We introduce for arbitrary $\nu \in \mathbf{C}$ the solution

$$\Psi_\nu := \mathcal{F}_\nu \Psi, \tag{52}$$

which is $\neq 0$ for $q \neq 0$ or $\nu \notin \frac{1}{2}+\beta_1+\mathbf{Z}$. In the case that $\nu$ is a characteristic exponent, $\Psi_\nu$ is a Floquet solution corresponding to $\nu$, called the *'basic Floquet solution'* to $\nu$. It admits the following expansion

$$\Psi_\nu(z) = \sum_{\mu\in\nu+\mathbf{Z}} \eta_\mu\, z^\mu, \quad (z \in \Omega) \tag{53}$$

with uniquely determined coefficients $\eta_\mu = \eta_\mu(\alpha,\beta,\gamma)$, satisfying (47).

Besides $\Psi_\nu$, we introduce for arbitrary $\nu \in \mathbf{C}$ the solution

$$\tilde{\Psi}_\nu := \mathcal{F}_\nu\, y_{01} \;=\; \mathcal{F}_\nu\, t_0\, \Psi \;=\; t_0\, \Psi_\nu. \tag{54}$$

For a characteristic exponent $\nu$ , both functions $\Psi_\nu$ and $\tilde{\Psi}_\nu$ must be linearly dependent. One obtains

$$q\, \tilde{\Psi}_\nu \;=\; -i\, e^{i\pi\beta_1} c_{-\nu}\, \Psi_\nu \tag{55}$$

with

$$c_\nu := \left(\tau_1^{\,2} - e^{-2\pi i\nu}\right) c. \tag{56}$$

Obviously,

$$2\, i \sin(2\pi\nu)\, c = c_\nu - c_{-\nu}. \tag{57}$$

The relations (55), (56) and (57) are essential for determining the central connection coefficient $c$, which means solving the central connection problem for the DCHE (1). In [4] we have derived a representation of $c_\nu$ by an explicit formula, which allows to make effective calculations of this quantity. The methods which we use there are mainly based on the articles [1] and [2].

To close with, we make some few remarks concerning the monodromy behavior of the central connection matrix $C$ with respect to $\alpha$. Applying $\tau_1$ onto $C = Q \circ \tau_1 C$ and using $M_\infty = Q \circ \tau_1 Q$ yields

$$\tau_1^2\, C = M_\infty^{\;-1} \circ C. \tag{58}$$

By applying $\tau_1^{\,2} = \tau_{-1}^{\,2}$ to this equation and using $\tau_{-1}\, M_\infty = M_\infty$ we obtain

$$\sigma\, C = M_\infty^{\;-2} \circ C. \tag{59}$$

This shows that, in general, $C$ as a function of $\alpha$ is not singlevalued on $\mathbf{C}^*$ but has to be considered on $\Omega$.

One can consider (58) for fixed $(\beta, \gamma) \in \mathbf{C}^3$ as a difference equation with respect to $\alpha \in \Omega$, having the 'periodic' coefficient

$$M_\infty^{\;-1}(\alpha e^{\pi i}, \beta, \gamma) = M_\infty^{\;-1}(\alpha, \beta, \gamma).$$

$C(., \beta, \gamma)$ is a holomorphic fundamental solution of (58). Obviously, the characteristic exponents of this difference equation are just the characteristic exponents of the DCHE. For a more detailed study see [4].

## References


[1] M. Kohno, A two point connection problem for nth order single linear ordinary differential equations with an irregular singular point of rank two, *Japan. J. Math.* **40** (1970), 11-62.

[2] F. Naundorf, A connection problem for second order linear differential equations with two irregular singular points. *SIAM J. Math. Anal.*, **7** (1976), 157 - 175.

[3] R. Schäfke, D. Schmidt, The connection problem for general linear ordinary differential equations at two regular singular points with applications in the theory of special functions, *SIAM J. Math.* **11** (1980), 848-862.

[4] Heuns Equations, Theory and Applications, Ed. A. Ronveaux, Part 3, The Double Confluent Heun Equation, D. Schmidt, G. Wolf, Oxford University Press, to appear in 1993.

# ASYMPTOTIC ANALYSIS OF A DIFFERENTIAL-ALGEBRAIC SYSTEM WITH SHOCK STRUCTURE

S. SHAO
Department of Mathematics, Cleveland State University
Cleveland, OH 44115, U.S.A.

Dedicated to Professors H. I. S. on the occasion of this symposium

ABSTRACT

We study the asymptotic behavior of solutions of a differential-algebraic system which is a model of highly nonequilibrium gas flow in the normal shock wave. Using perturbation-numerical method and asymptotic theory, we analyze the limiting behavior of the stable solutions inside the shock and the singular point of the stable solutions. Our results show that the asymptotically stable solutions of the first-order perturbation equations of the system involve a parameter, and they are valid when the values of parameter are in the interval [1,∞] inside the shock.

## 1. Introduction

Consider the first-order system of ordinary differential equations

Eq.(1) $$\frac{d}{dx}(\rho y_1) = 0,$$

Eq.(2) $$\frac{d}{dx}(\rho y_1^2 + y_2) = 0,$$

Eq.(3) $$\frac{d}{dx}(\rho y_1^3 + 2y_1y_2 + 3y_1y_3 + y_4) = 0,$$

Eq.(4) $$\frac{d}{dx}(\rho y_1^3 + 3y_1y_2 + y_5) = -\frac{y_3}{\mu}(y_2 - y_3),$$

$$\frac{d}{dx}\left[\rho y_1^4 + 3y_1^2y_3 + 2y_1y_4 + 5y_1^2y_2 + 2y_1y_5 + f_1(y_1, y_5, y_6, y_7)\right] = \frac{y_3}{\mu}\left[\frac{2}{3}y_4 + 2y_1(y_2 - y_3)\right],$$
Eq.(5)

Eq.(6) $$\frac{d}{dx}[\rho y_1^4 + 6y_1^2y_2 + 4y_1y_4 + f_2(y_1, y_5, y_6, y_7)] = \frac{y_3}{\mu}\left[\frac{1}{2}y_4 - \frac{3}{2}y_5 - 3y_1(y_2 - y_3)\right],$$

together with an algebric equation

Eq.(7) $$y_1f_3(y_1, y_4) = f_4(y_1, y_4, y_6, y_7),$$

and an identity

Eq.(8) $$y_3 = \frac{\rho R}{3}[1 + 5p_1 - 2y_1(1 + p_1) + y_1^2 - y_4],$$

$\rho$, $y_i$ (i = 1, 2,....7), are prescribed at both negative and positive infinity,

we have eight unknowns, $\rho = \rho(x, M)$, $y_i = y_i(x, M)$ (i = 1, 2, ....7) defined on $-\infty \le x \le \infty$ in Eq.(1)-(8), where M is a parameter in $[1, \infty)$, $p_1 = \frac{1}{\gamma M^2}$, $\gamma$, $\mu$ and R are constants, and $f_i$ (i = 1, 2, 3, 4) are smooth given functions ($f_1$, $f_2$, $f_3$ and $f_4$ will be given in Section 2). We note that the system of Eq.(1)-(8) is a first-order differential-algebraic system subject to two sets of boundary conditions prescribed at positive infinity and negative infinity respectively, hence there must be a the shock layer behavior in the solutions of equations (1)-(8) with the given boundary conditions.

The system of Eq.(1)-(8) is a model problem of the Maxwell transfer equations at fourteen moment level for the normal shock wave. The fourteen-moment system is a system of fourteen partial differential equations, which modify the Grad's thirteen-moment system and try to make shock Mach number valid beyond 1.65 for highly nonequilibrium gas flows. (cf. [9], [1], [2]). For the normal shock wave and a steady flow, these fourteen partial differential equations become a system of ordinary differential equations of the form

Eq.(9) $$\frac{d}{dx}(\rho u) = 0,$$

Eq.(10) $$\frac{d}{dx}(\rho u^2 + P) = 0,$$

Eq.(11) $$\frac{d}{dx}(\rho u^3 + 2uP + 3up + S) = 0,$$

Eq.(12) $$\frac{d}{dx}(\rho u^3 + 3uP + S) = -\frac{p}{\mu}(P - p),$$

Eq.(13) $$\frac{d}{dx}\left[\rho u^4 + +3u^2 p + 2uS + 5u^2 P + 2uS + \sum_k Q_{11kk}\right] = \frac{-2p}{3\mu}\left[S + 3u(P-p)\right],$$

Eq.(14) $$\frac{d}{dx}\left[\rho u^4 + 6u^2 P + 4uS + Q_{1111}\right] = \frac{p}{\mu}\left\{\frac{1}{2}S - \frac{3}{2}S_{111} - 3u(P - p)\right\},$$

and one algebraic equation

Eq.(15) $$\left(8T_1 T_2^2 \pi^3\right)^{1/2}\left(1 - \vartheta_1\left[\frac{5}{2} - \frac{T'}{2}\left(\frac{1}{T_1} + \frac{2}{T_2}\right) - \frac{(u'-u)^2}{2T_1}\right] - \vartheta_2\left[\frac{5}{2} - \frac{T''}{2}\left(\frac{1}{T_1} + \frac{2}{T_2}\right) - \frac{(u''-u)^2}{2T_1}\right]\right) = \rho,$$

and an identity $$p = \rho RT,$$

where $\rho$ is the density, P is the pressure, S is the heat flux, $T_i$ (i =1, 2) are temperature components which can be written in terms of velocity u and Mach number M, T', T", u' and u" are real functions, $S_{111}$ is a third-moment unknown, $\sum_k Q_{11kk}$ and $Q_{1111}$ are fourth-moment which can be written in terms of u, T, $\vartheta_1$ and $\vartheta_2$, and $\vartheta_1$, $\vartheta_2$ are two unknown parameters to be determined in the velocity distribution of fourteen-moment system.

The normal schock wave is used to check the accuracy of the fourteen-moment system for the continuum transition regime. Since the structure of a normal shock spans a length somewhere from four to one hundred mean free paths depending on the shock Mack number and gas being study (cf.[5]). Therefore, analyzing the structure of the normal shock wave for the fourteen moment system and studying its asymptotic behavior of the solutions are essential in developing the fourteen moment system. While the Navier-Stokes equations will not accurately reproduce the shock wave structure.

Modivated by such important physical applications in gas dynamics, we present here the asymptotic behavior of the stable solutions of Eq.(1)-(8) as $x \to \pm\infty$, and the asymptotic behavior of the stable solutions as the parameter $M \to 1^+$ (M as the shock Mach number in the physical model. $x \to -\infty$ is corresponding to the upstream limit of the flow and $x \to +\infty$ is corresponding to the downstream limit of the flow, $M \to 1^+$ is concerned the lower Mach number near the shock). From the mathematical point of view, using perturbation-numerical method and asymptotic theory, we analyze the limiting behavior of the stable solutions inside the shock and the sigular point of the stable solutions as $M \to 1^+$. Our results will help the engineers in study the shock wave structure in the continuum transition regime for highly nonequilibrium gas flows.

The outline of the paper is as follows. In Section 2 we present the asymptotic solutions of system (1.1) as $x \to \pm\infty$. Shock layer behavior as $M \to 1^+$ is studied in Section 3. Section 4 is our conclusions.

## 2. Perturbed the System at Both Boundaries.

We first introduce a new spatial variable $\eta$ defined by

$$\eta = \int \frac{y_3}{\mu} d\xi, \quad \text{where } \xi = Ax, \text{ A is a fixed constant.}$$

Using identity (8) and normalizing the variables and integrating both sides of Eq.(1)-(3), yield

Eq.(16) $$\rho y_1 = 1,$$

Eq.(17) $$y_1 + y_2 = p_1 + 1,$$

Eq.(18) $$y_1^2 + 2y_1y_2 + 3y_1y_3 + y_4 = 1 + 5p_1,$$

Eq.(19) $$\frac{d}{d\eta}[\,3(1+p_1)y_1 - 2y_1^2 + y_5\,] = -(y_2 - y_3),$$

Eq.(20) $$\frac{d}{d\eta}[\,(1+5p_1)y_1 + 3(1+p_1)y_1^2 - y_1^3 + 2y_1y_5 + y_1y_4 + f_1] = -\frac{2}{3}y_4 - 2y_1(y_2 - y_3),$$

Eq.(21) $$\frac{d}{d\eta}[\,2(1+p_1)y_1^2 - \frac{5}{3}y_1^3 + \frac{4}{3}y_1y_5 + y_1y_4 + \frac{1}{3}f_2] = -\frac{1}{2}y_5 + \frac{1}{6}y_4 - y_1(y_2 - y_3),$$

Eq.(22) $$y_1 f_3(y_1, y_4) = f_4(y_1, y_4, y_6, y_7),$$

where $p_1 = \frac{1}{\gamma M^2}$, $\gamma = \frac{5}{3}$, $T_1 = y_1(1+p_1-y_1)$, $T_1+2T_2 = [\,1+5p_1 - 2y_1(1+p_1) + y_1^2 - y_4]$,

$$f_1(y_1, y_5, y_6, y_7) = F_1 + 2F_2,\ f_2(y_1, y_5, y_6, y_7) = F_1,$$

$$F_1 = 3\left(\frac{T_1^2}{y_1}\right) + 3\left(\frac{y_6}{y_1}\right)\left\{p_1^2 + 2p_1(1-y_1)^2 + \frac{(1-y_1)^4}{3} + T_1^2 - 2p_1T_1 - 2T_1(1-y_1)^2\right\}$$
$$+ 3\left(\frac{y_7}{y_1}\right)\left\{T^{**2} + 2T^{**}[y_1(+\infty,M)-y_1]^2 + \frac{[y_1(+\infty,M)-y_1]^4}{3} + T_1^2 - 2T^{**}T_1\right.$$
$$\left. - 2T_1[y_1(+\infty,M)-y_1]^2\right\},$$

$$F_2 = \frac{T_1T_2}{y_1} + \left(\frac{y_6}{y_1}\right)\left\{p_1^2 + T_1T_2 - p_1T_1 - p_1T_2 + p_1(1-y_1)^2 + T_2(1-y_1)^2\right\}$$
$$+ \left(\frac{y_7}{y_1}\right)\left\{T^{**2} + T_1T_2 - T_1T^{**} - T_2T^{**} + T^{**}[y_1(+\infty,M)-y_1]^2 - T_2[y_1(+\infty,M)-y_1]^2\right\},$$

$$f_3(y_1, y_4) = \left(8T_1T_2^2\pi^3\right)^{1/2},\quad f_4(y_1, y_4, y_6, y_7) = \Big(1 - y_6[\,\frac{5}{2} - \frac{p_1}{2}\left(\frac{1}{T_1} + \frac{2}{T_2}\right) - \frac{[y_1(-\infty,M)-y_1]^2}{2T_1}]$$
$$- y_7[\,\frac{5}{2} - \frac{T^{**}}{2}\left(\frac{1}{T_1} + \frac{2}{T_2}\right) - \frac{[y_1(+\infty,M)-y_1]^2}{2T_1}]\Big)^{-1},$$

$$T^{**} = T^b\Big(1+\frac{2\gamma(M^2-1)}{(\gamma+1)}\Big)\Big(\frac{2+(\gamma-1)M^2}{(\gamma+1)M^2}\Big) = \frac{p_1(5M^2-1)(3+M^2)}{16M^2} \qquad \text{as } \eta\to-\infty,$$

$$T^{**} = T^b/\Big\{\Big(1+\frac{2\gamma(M^2-1)}{(\gamma+1)}\Big)\Big(\frac{2+(\gamma-1)M^2}{(\gamma+1)M^2}\Big)\Big\} = \frac{(3-p_1)(1+5p_1)M^2}{(5M^2-1)(3+M^2)} \qquad \text{as } \eta\to+\infty.$$

where $T^b$ will be given later in (d), the boundary conditions will be transformed later. We note that here are three first-order differential equations Eq.(19)-(21) in the system. the rest are four algebraic equations. The algebraic equations (16)-(18) can be used to further simplify the above system to a system consisting three differential equations and one algebraic equation with four unknowns $y_1$, $y_5$, $y_6$ and $y_7$. We now define four new unknowns $z_i = z_i(\eta, M)$ $(i = 1, 2, 3, 4)$ such that

$$z_1 = y_1,\ z_2 = y_5,\ z_3 = y_6,\ z_4 = y_7, \tag{a}$$

with the boundary conditions

$$z_i^- = z_i(-\infty, M), \qquad z_i^+ = z_i(+\infty, M), \qquad i = 1, 2, 3, 4, \tag{b}$$

where

$$z_1(-\infty,M) = 1\ ,\ z_2(-\infty,M) = 0,\ z_3(-\infty,M) = 1-\Big(\frac{1}{2p_1\pi}\Big)^{3/2},\ z_4(-\infty,M) = 0,$$

$$z_1(+\infty,M) = \frac{1+5p_1}{4},\ z_2(+\infty,M) = 0,$$

$$z_3(+\infty,M) = 0,\ z_4(+\infty,M) = 1-\Big(\frac{32}{(3-p_1)^3(1+5p_1)\pi^3}\Big)^{3/2}.$$

Since we are interested in the asymptotic behavior of the solutions of Eq.(16)-(22) as $\eta \to \pm\infty$, before we rewrite the system (16)-(22) in terms of new variables $z_i = z_i(\eta, M)$ $(i = 1, 2, 3, 4)$, we define the solution $z_i$ by

$$z_i = z_i(\eta, M) = \begin{cases} Z_i^- & \text{for } \eta\to-\infty \\ & \\ Z_i^+ & \text{for } \eta\to+\infty, \end{cases} \quad + O(\varepsilon^{n+1}), \tag{c}$$

where $\displaystyle Z_i^- = z_i^- + \sum_{k=1}^{n} z_{ik}^-(\eta,M)\varepsilon^k, \qquad Z_i^+ = z_i^+ + \sum_{k=1}^{n} z_{ik}^+(\eta,M)\varepsilon^k.$

and $0 < \varepsilon \ll 1$. We now substitute (a), (b) and (c) into Eq.(19)-(21) and use equations Eq.(16)-(18) to simplify the system (16)-(22), then we obtain four first-order perturbation equations (order of $O(\varepsilon)$) with four variables $z_{i1} = z_{i1}(\eta, M)$ $(i = 1, 2, 3, 4)$,

Eq.(23) $\displaystyle [3(1+p_1) - 4z^b]\frac{dz_{11}}{d\eta} + \frac{dz_{21}}{d\eta} = \Big[\frac{4}{3} - \frac{(1+5p_1)}{3(z^b)^2}\Big]z_{11} - \frac{5}{9z^b}z_{21} - \frac{q_1}{3z^b}z^*,$

$$[(1+5p_1)+6(1+p_1)z^b-9\{(z^b)^2+z^b\}+\frac{T^b(1+p_1)}{z^b}-2(\frac{T^b}{z^b})^2]\frac{dz_{11}}{d\eta}+[\frac{11z^b}{3}-\frac{5T^b}{3z^b}]\frac{dz_{21}}{d\eta}$$

Eq.(24) $$+[q_2-\frac{q_1T^b}{z^b}+q_1z^b]\frac{dz^*}{d\eta}=2(\frac{4}{3}-\frac{(1+5p_1)}{3(z^b)^2})z_{11}-\frac{20}{9}z_{21}-\frac{4q_1}{3}z^*,$$

$$[4(1+p_1)z^b-5z^b+(1+p_1-z^b)(1+p_1-3z^b)]\frac{dz_{11}}{d\eta}+[\frac{4z^b}{3}]\frac{dz_{21}}{d\eta}+q_3\frac{dz^*}{d\eta}$$

Eq.(25) $$=(\frac{4}{3}-\frac{(1+5p_1)}{3(z^b)^2})z_{11}-\frac{7}{9}z_{21}-\frac{q_1}{6}z^*,$$

$$z_{11}(8T_1T_2^2\pi^3)^{1/2}(1-z^{**}-(z_{31}^-)[\frac{5}{2}-\frac{p_1}{2}(\frac{1}{T_1}+\frac{2}{T_2})-\frac{[z_1^--z_1]^2}{2T_1}]$$

Eq.(26) $$-(z_{41}^+)[\frac{5}{2}-\frac{T^{**}}{2}(\frac{1}{T_1}+\frac{2}{T_2})-\frac{[z_1^+-z_1]^2}{2T_1}]) = 1,$$

where $p_1=\frac{1}{\gamma M^2}$, $\gamma=\frac{5}{3}$, $T_1=p_1+(1-p_1)z_{11}$, $T_1+2T_2=[3p_1-2p_1z_{11}-\frac{5}{3}z_{21}-q_1z_{41}]$,

$$\begin{cases} z^b = z_1^- = 1 \\ T^b = p_1 \\ z^* = z_{41} \\ z^{**} = z_3^- \end{cases} \text{if } \eta\to-\infty, \qquad \begin{cases} z^b = z_1^+ = (1+5p_1)/4 \\ T^b = (3+p_1)(1+5p_1)/16 \\ z^* = z_{31} \\ z^{**} = z_4^+ \end{cases} \text{if } \eta\to+\infty, \qquad \text{(d)}$$

and $q_1=\frac{2\omega^3}{3y_1(-\infty,M)}$ as $\eta\to-\infty$, $q_1=\frac{-2\omega^3}{3y_1(+\infty,M)}$ as $\eta\to+\infty$, $\omega=\frac{-3}{4}(\frac{1}{M^2}-1)$,

$$q_2=\frac{[5T^{**2}+8T^{**}\omega^2+\omega^4+5(T^b)^2-10T^{**}T^b-8T^b\omega^2]}{z^b}.$$

$$q_3=\frac{[T^{**2}+(T^b)^2+2T^{**}\omega^2-2T^{**}T^b-2T^b\omega^2+\omega^4/3]}{z^b}.$$

We note that $z_{i1}=z_{i1}^-(\eta, M)$ ($i=1, 2, 4$) at $\eta=-\infty$ and $z_{i1}=z_{i1}^+(\eta, M)$ ($i=1, 2, 3$) at $\eta=+\infty$. The equations (23), (24) and (25) are homogeneous first-order ordinary differential equations which are decoupled from algebraic equation (26). Therefore, the solutions of Eq.(23)-(25) have the form

$$z_{i1}^-=z_{i1}^-(\eta, M)=c_{i1}v_{i1}e^{\lambda_1\eta}+c_{i2}v_{i2}e^{\lambda_2\eta}+c_{i3}v_{i3}e^{\lambda_3\eta}, \quad i=1, 2, 4,$$
$$z_{i1}^+=z_{i1}^+(\eta, M)=d_{i1}w_{i1}e^{\zeta_1\eta}+d_{i2}w_{i2}e^{\zeta_2\eta}+d_{i3}w_{i3}e^{\zeta_3\eta}, \quad i=1, 2, 3,$$

where for each i, $\lambda_i=\lambda_i(M)$ and $v_{ji}=v_{ji}(M)$ ($j=1, 2, 3$) are eigenvalues and corresponding eigenvectors of the coefficient matrix as $\eta\to-\infty$, $\zeta_i=\zeta_i(M)$ and $w_{ji}=w_{ji}(M)$ ($j=1, 2, 3$) are eigenvalues and corresponding eigenvectors of the coefficient matrix as $\eta\to+\infty$, $c_{ij}$'s and $d_{ij}$'s are constants. However, we are only interested in the stable solutions of Eq.(23)-(26) as $\eta\to\pm\infty$, we can solve $\lambda_i$ and $\zeta_i$ numerically to further determined the form of our solutions of Eq.(23)-Eq.(26). Figure 1 and Figure 2 summarize the numerical results of $\lambda_i$ and $\zeta_i$, respectively.

The solution is stable if $\lambda$ has positive value as $\eta\to-\infty$. In Fighure 1, two eigenvalues are complex numbers with negative real part for $1< M< 4.7$, they become real (still negative) as $M \geq 4.7$, only one is real and positive, namely $\lambda_1$, for all $M > 1$. Therefore, the stable solutions of Eq.(23)-(25) as $\eta\to-\infty$ have the form of $v_{i1}e^{\lambda_1\eta}$ (i=1, 2, 4), where $\begin{bmatrix} v_{11} \\ v_{21} \\ v_{31} \end{bmatrix}$ is the corresponding eigenvector of $\lambda_1$. In Figure 2, near M=1, two eigenvalues become unbounded near M=1.345, after that they become complex conjugate with negative real part, while only one eigenvalue ($\zeta_1$) is negative for all $M > 1$ (the solution is stable if $\zeta$ has negative value as $\eta\to+\infty$). Hence, the stable solutions as $\eta\to+\infty$ have the form of $w_{i1}e^{\zeta_1\eta}$ (i=1, 2, 4), where $\begin{bmatrix} w_{11} \\ w_{21} \\ w_{31} \end{bmatrix}$ is the corresponding eigenvector of $\zeta_1$. Therefore,

$$z_{i1}^- = z_{i1}^-(\eta, M) = v_{i1}e^{\lambda_1\eta}, \qquad i = 1, 2. 4, \qquad \text{(e)}$$

$$z_{i1}^+ = z_{i1}^+(\eta, M) = w_{i1}e^{\zeta_1\eta}, \qquad i = 1, 2. 3. \qquad \text{(f)}$$

Since $z_{i1}^-$ and $z_{i1}^+$ satisfy the homegeneous differential equations (23)-(25) as $\eta\to-\infty$ and $\eta\to+\infty$ , respectively, it implies that $v_{21}$ and $v_{31}$ can be written in terms of $v_{11}$ and $w_{21}$ and $w_{31}$ can be written in terms of $w_{11}$. Therefore, we have

$$z_i = z_i(\eta, M) = \begin{cases} Z_i^- & \text{as } \eta\to-\infty \\ & \qquad\qquad + O(\varepsilon^2), \\ Z_i^+ & \text{as } \eta\to+\infty, \end{cases} \qquad \text{(g)}$$

where $Z_i^- = z_i^- + z_{i1}^-(\eta,M)\varepsilon$, $\quad Z_i^+ = z_i^+ + z_{i1}^+(\eta,M)\varepsilon$, $\quad i = 1, 2, 3,$

$z_i^-$ and $z_i^+$ are defined in (b),

and $z_{11}^-(\eta, M) = v_{11}e^{\lambda_1\eta}$, $z_{21}^-(\eta, M) = \alpha v_{11}e^{\lambda_1\eta}$, $z_{31}^-(\eta, M) = \beta v_{11}e^{\lambda_1\eta}$,
$z_{11}^+(\eta, M) = w_{11}e^{\zeta_1\eta}$, $z_{21}^+(\eta, M) = \phi w_{11}e^{\zeta_1\eta}$, $z_{31}^+(\eta, M) = \varphi w_{11}e^{\lambda_1\eta}$,

and $\alpha$, $\beta$, $\phi$ and $\varphi$ are functions of M can be determined by substituting (g) into Eq.(23)-(25). We summarize the relationship of $\lambda_1$ and $\zeta_1$ with M in Figure 3, we note that $\lambda_1\to3.00$ and $\zeta_1\to-3.762$ as $M\to\infty$. These two eigenvalues are bounded as $M\to\infty$.

Figure 1. Upstream Eigenvalues

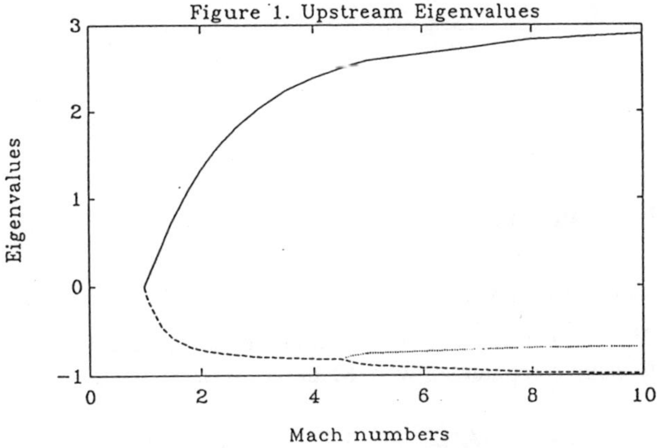


Figure 2. Downstream Eigenvalues

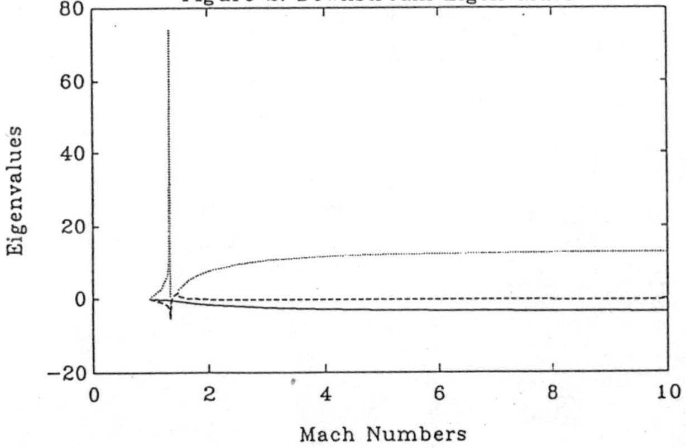

Figure 3. Valid Eigenvalues

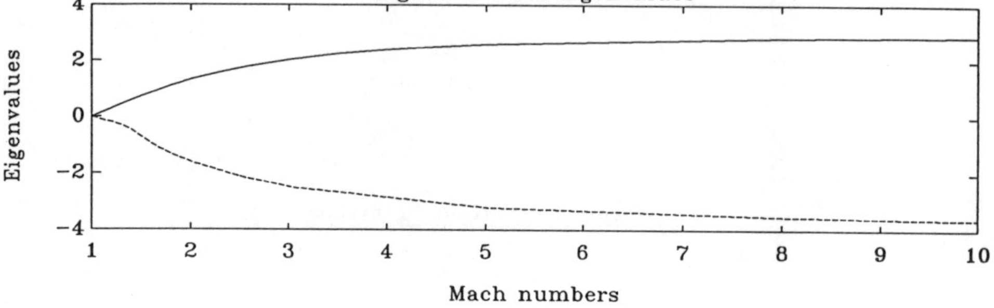


## 3. Shock Structure near M = 1.0

We now turn to study the shock structure as $M \to 1^+$. We first substitute $p_1 = \dfrac{3}{5M^2}$ into Eq.(23)-(25) and then rewrite the system in the matrix form (all the substitutions can be made simply using computer software Maple V). For $\eta \to -\infty$, we have

Eq.(27) $$A^-(Z^-)' = B^-Z^-, \qquad A^+(Z^+)' = B^+Z^+,$$

where $' = \dfrac{d}{d\eta}$, $\quad Z^- = \begin{bmatrix} z_{11} \\ z_{21} \\ z_{41} \end{bmatrix}$, $\quad Z^+ = \begin{bmatrix} z_{11} \\ z_{21} \\ z_{31} \end{bmatrix}$,

$$A^- = \begin{pmatrix} \dfrac{9}{5M^2} - 1 & 1 & 0 \\ \dfrac{12}{5M^2} - 2 & \dfrac{11}{3} - \dfrac{1}{M^2} & \dfrac{3q_1}{5M^2} - q_1 + q_2 \\ \dfrac{9}{25M^4} + \dfrac{6}{5M^2} - 1 & \dfrac{4}{3} & q_3 \end{pmatrix},$$

$$B^- = \begin{pmatrix} 1 - \dfrac{1}{M^2} & \dfrac{-5}{9} & \dfrac{-q_1}{3} \\ 2 - \dfrac{1}{M^2} & \dfrac{-20}{9} & \dfrac{-4q_1}{3} \\ 1 - \dfrac{1}{M^2} & \dfrac{-7}{9} & \dfrac{-q_1}{6} \end{pmatrix}, \quad A^+ = \begin{pmatrix} a_{11} & a_{12} & a_{13} \\ a_{21} & a_{22} & a_{23} \\ a_{31} & a_{32} & a_{33} \end{pmatrix},$$

$$B^{+}=\begin{pmatrix} \frac{4}{3}-\frac{(1+3/M^2)}{3(1/4+3/4M^2)} & \frac{-5}{9(\frac{1}{4}+\frac{3}{4M^2})} & \frac{-q_1}{3(\frac{1}{4}+\frac{3}{4M^2})} \\ \frac{8}{3}-\frac{2(1+3/M^2)}{(1/4+3/4M^2)} & \frac{-20}{9} & \frac{-4q_1}{3} \\ \frac{4}{3}-\frac{(1+3/M^2)}{3(1/4+3/4M^2)} & \frac{-7}{9} & \frac{-q_1}{6} \end{pmatrix},$$

wherc $a_{11}=-\frac{6}{5M^2}+2,\qquad a_{12}=1,\qquad a_{13}=0,$

$$a_{21}=1+\frac{3}{M^2}+6(\frac{3}{5M^2}+1)(\frac{1}{4}+\frac{3}{4M^2})-9(\frac{1}{4}+\frac{3}{4M^2})^2+\frac{(3-\frac{3}{5M^2})(1+\frac{3}{M^2})(\frac{3}{5M^2}+1)}{8(\frac{1}{4}+\frac{3}{4M^2})}$$

$$-\frac{9}{16}(3-\frac{3}{5M^2})(1+\frac{3}{M^2})-\frac{(3-\frac{3}{5M^2})^2(1+\frac{3}{M^2})^2}{128(\frac{1}{4}+\frac{3}{4M^2})^2},$$

$$a_{22}=(\frac{11}{12}-\frac{11}{4M^2})-\frac{5(3-\frac{3}{5M^2})(1+\frac{3}{M^2})}{48(\frac{1}{4}+\frac{3}{4M^2})},\ a_{23}=q_2+(\frac{1}{4}+\frac{3}{4M^2})q_1-\frac{q_1(3-\frac{3}{5M^2})(1+\frac{3}{M^2})}{16(\frac{1}{4}+\frac{3}{4M^2})},$$

$$a_{31}=4(1+\frac{3}{4M^2})(\frac{1}{4}+\frac{3}{4M^2})-5(\frac{1}{4}+\frac{3}{4M^2})^2+(\frac{3}{4}-\frac{3}{20M^2})(\frac{1}{4}+\frac{33}{20M^2}),$$

$$a_{32}=(\frac{1}{3}+\frac{1}{M^2}),\ a_{33}=q_3,$$

and $q_1$, $q_2$, and $q_3$ are same as in Section 2. We found that

$$(A^+, B^+)\rightarrow(A^-, B^-)\text{ as }M\rightarrow 1^+, \qquad (A)$$

$$(A^+, B^+)=(A^-, B^-)\text{ if }M=1.0, \qquad (B)$$

by $z_1^+=[(1+5p_1)/4]\rightarrow 1=z_1^-$, $T^b=[(3+p_1)(1+5p_1)/16]\rightarrow p_1$ as $M\rightarrow 1^+$. From the above numerical results, we have $\lambda_1\rightarrow 0$ and $\zeta_1\rightarrow 0$ as $M\rightarrow 1^+$ (cf. Figue 3). However, inverse $A^{-1}=(A^-)^{-1}=(A^+)^{-1}$ does not exist at M=1.0. In other words, it is singular at M=1.0. In order to study the case $M\rightarrow 1.0$, we substitute $M = 1+\sigma$ into $A^-$ and $B^-$, where $0<\sigma<<1$, and we do echoln form of the augment matrix $[A^- \mid B^-]$, then we have

$$A^{-}=\begin{pmatrix} \frac{4}{5}+2\sigma & 0 & O(\sigma^2) \\ 0 & \frac{1}{3}(\frac{-13-28\sigma}{-2+\sigma}) & O(\sigma^2) \\ 0 & 0 & O(\sigma^2) \end{pmatrix}, \qquad (C)$$

$$B^{-}=\begin{pmatrix} O(\sigma) & \frac{-5}{9} & O(\sigma^3) \\ O(\sigma) & \frac{-5}{9}\left(\frac{(-7+10\sigma)}{(-2+5\sigma)}\right) & O(\sigma^3) \\ O(\sigma) & \frac{-1}{3}\left(\frac{(47+69\sigma)}{(-13-54\sigma)}\right) & O(\sigma^3) \end{pmatrix}. \qquad \text{(D)}$$

From the results (C) and (D), it seems that we cannot remodel Eq.(27) to be a singular perturbation problem in such a form as $\sigma \to 0$ (cf.[7]). However, the solutions of eq.(27) are exist at M=1.0 , for example, we can choose

$$\begin{bmatrix} z_{11}(\eta,1) \\ z_{21}(\eta,1) \\ z^*(\eta,1) \end{bmatrix} = \begin{bmatrix} \alpha v_{11} e^{0\eta} \\ 0 \\ \beta v_{11} e^{0\eta} \end{bmatrix},$$

where $z^*(\eta,1) = z_{31}(\eta,1)$ or $z_{41}(\eta,1)$, $\alpha$ and $\beta$ are constants. Method of matching can be used to matching the solution from $\eta \to -\infty$ to the solution as $\eta \to +\infty$ at M =1.0, which will provide a continuous stable solution if other boundary conditions are given.

## 4. Conclusions

We study the asymptotic behavior of solutions of a first-order differential-algebraic system of equations (1)-(8) with the boundary conditions which is a model of highly nonequilibrium gas flow in the normal shock wave. Our results show that there exist the stable solutions for the first-order perturbation equations (23)-(26) for all the paramter M > 1 as $\eta \to \pm\infty$ inside the shock, there is a removable singular point at M = 1. While it means that for a monotonic gas, the solutions of fourteen-moment system exist for a normal shock wave in the upstream and the downstream for all Mach numer M > 1.

## References

1. H. Grad, *On the Kinetic of Rarefied Gases*, Comm. Pure Appl. Math., 2, no.4, (1949), 331-407.
2. H. Grad, *The Profile of a Steady Plane Shock Wave*, Comm. Pure Appl. Math., 5, no. 3, (1952), 257-300.
3. J. Kevorkian and J. D. Cole, *Perturbation Methods in Applied Mathematics*, Spring-Verlag, New York, 1981.
4. M. N. Kogan, *Rarefied Gas Dynamics*, trans. ed. L. Trilling, Plenum Press, 1969.
5. F. E. Lumpkin III and D. R. Chapman, *Accuracy of the Burnett Equations for Hypersonic Real Gas Flows*, AIAA 91-0771.
6. R. Marz and E. B. Weinmuller, *Solvability of Boundary Value Problems for Systems of Signular Differential-Algebraic Equations*, SIAM J. Math. Anal., 24 (1993), 200-215.
7. A. H. Nayfeh, *Perturbation Methods*, Wiley, New York, 1973.
8. R. E. O'Malley, Jr., *Singular Perturbation Methods for Ordinary Differential Equations*, Spring-Verlag, New York, 1991.
9. S. Shao and H.K. Cheng, *Maxwell Transfer Equations: Modifications of Grad's Thirteen-Moment System, I*, Partial differential equations, Joseph Wiener and Jack K. Hale eds., Pitman Research Notes in Math. 273 (1992), 216-220.

# Geometry of Resonance Poles for Three-Dimensional Schrödinger Operator with Nearly Coulomb Potential*

Marianna A. Shubov
Department of Mathematics, Texas Tech University
Lubbock, Texas 79409-1042, U.S.A.

## 1 Introduction

We consider resonances for the Hamiltonian operator

$$H = -(1/2)\Delta - Z/r + q(r) \tag{1}$$

acting on $L^2(\mathbb{R}^3)$. Here $Z > 0, r = |x| = (x_1^2 + x_2^2 + x_3^2)^{1/2}$, $x \in \mathbb{R}^3$ and a spherically symmetric function $q(r)$ has a compact support. The precise conditions on $q$ will be formulated later. (1) can be considered as an approximate Hamiltonian for an electron in the field of a positively charged atomic ion. $Z$ is the charge of the ion and $q$ represents the perturbation of a purely Coulomb field of the nucleus by an electron cloud.

Our main results include the existence of a low-energy sequence of resonances converging to zero and the high-energy asymptotics of resonances (see Section 2). A detailed exposition of these results will appear in our three forthcoming papers [18-20].

The behavior of resonances is well-known in the case of a spherically symmetric short-range potential, i.e. for the operator (1) with $Z = 0$ (see [1, 2]). Namely, the whole set of resonances splits into infinite sequence of infinite series corresponding to different values of an angular momentum $\ell$. Resonances in each series can be described

*Research partially supported by the National Science Foundation Grant DMS-9212037.

as zeroes of the corresponding Jost function. They have infinity as the only point of accumulation and are asymptotically close to a certain logarithmic curve on the second Riemann sheet of the spectral parameter. This curve depends on $\ell$. Moreover, resonances of the full 3-dimensional operator stay away from zero and have the only limiting point at infinity. An extensive literature is devoted to the generalization of these classical results for different classes of potentials (see references in [3-5]). However, the behavior of resonances in the case of Coulomb-like potentials was not clarified.

As it is well-known, at the present moment there is no universal definition of resonances. We use the definition, according to which resonances are poles of the analytic continuation of the quadratic form of the resolvent to the second "nonphysical" sheet of the Riemann surface of the spectral parameter through the branch cut along the continuous spectrum. More precisely, let $H_0 = -(1/2)\Delta - Z/r$ be the unperturbed Hamiltonian. Both $H$ and $H_0$ are correctly defined self-adjoint operators on $L^2(\mathbb{R}^3)$ and both have an absolutely continuous spectrum along the positive real semiaxis (see conditions on $q$ below).

**DEFINITION 1** *(see [3]) Suppose, there is a dense set of vectors $D \subset L^2(\mathbb{R}^3)$ such that for all $\psi \in D$ both $(\psi, (H-E)^{-1}\psi) = R_\psi(E)$ and $(\psi, (H_0-E)^{-1}\psi) = R_\psi^{(0)}(E)$ have analytic continuation onto the second sheet (across the real axis from the upper half-plane of the first sheet). If $R_\psi^{(0)}(E)$ is analytic at $E_0$ and $R_\psi(E)$ has a pole at $E_0$ for some $\psi$ we say that $E_0$ is a resonance pole.*

The Hamiltonian (1) is separable in spherical coordinates. In the non-Coulomb case $Z = 0$ the corresponding radial Schrödinger equation has the form

$$(H_\ell^{(0)} - E)\psi = \left(-\frac{d^2}{dr^2} + \frac{\ell(\ell+1)}{r^2} + q(r) - E\right)\psi = 0, \ \ell = 0, 1, 2, \ldots \tag{2}$$

In the general case $Z > 0$ and it has the form

$$(H_\ell - E)\psi = \left(H_\ell^{(0)} - \frac{2Z}{r} - E\right)\psi = 0, \ \ell = 0, 1, 2, \ldots. \tag{3}$$

The set of all resonances of the operator (1) splits into infinite sequence of series corresponding to the radial operators $H_\ell, \ell = 0, 1, 2, \ldots$ defined on $L^2(0, \infty)$.

# 2 Main Results

Our first two results are concerned with the behavior of resonances at low energies, i.e. in a neighborhood of zero on the second Riemann sheet of the spectral parameter $E$.

**THEOREM 1** *Assume that $q(r)$ is a measurable function, which satisfies the conditions:*

$$q(r) = 0 \text{ for } r > a; \int_0^a r|q(r)|dr < \infty. \tag{4}$$

*Then for every fixed value of $\ell$ there exists a disk centered at the origin on the second Riemann sheet of the spectral parameter $E$ which is free of resonances of the radial operator $H_\ell$. Resonances of $H_\ell$ can have the only point of accumulation at infinity.*

**THEOREM 2** *Assume that, in addition to (4), $q(r)$ satisfies the conditions:*

$$q(r) \geq 0 \text{ for } r \in [0, a]; \; 0 < \int_0^a rq(r)dr < \infty. \tag{5}$$

*Then the following statements hold.*

*a) The operator (1) has a series of resonances converging to zero.*

*b) Let $r(\ell)$ be the radius of the maximal resonance-free disk centered at the origin on the second Riemann sheet of the radial operator $H_\ell$. The following estimate takes place:*

$$r(\ell) \leq Z^2(m - C_0 e^{-m^2})^{-2}, \text{ where } m = [(\ell + 0.5)^{1/2}]. \tag{6}$$

*Here by $[t]$ we denote the greatest integer less or equal than $t$. The constant $C_0 > 0$ does not depend on $\ell$. It can be represented as a continuous function of the quantity $\left(\int_0^a rq(r)dr\right)^{-1}$.*

*(6) implies that for sufficiently large $\ell$*

$$r(\ell) \leq Z^2(\ell - 2\sqrt{\ell})^{-1}, \tag{7}$$

*which means, in particular, that $r(\ell) \leq C\ell^{-1}$, where $C$ does not depend on $\ell$.*

Our next result is the high-energy asymptotics of resonances. To formulate it we introduce a new spectral parameter $k$, which is related to $E$ by the formula: $k^2 = E$. Notice that the second Riemann sheet of $E$ corresponds to the lower half-plane of $k$.

**THEOREM 3** *Assume that $q(r)$ satisfies (4). Let $m \geq 0$ be an integer. Assume that, in addition, $q(r)$ has the following properties:*

*a) it is $(m-1)$ times continuously differentiable on $(0,\infty)$ (if $m \geq 1$);*

*b) it has continuous $m$-th and $(m+1)$-st derivatives on $(0,a]$ and its $(m+2)$-nd weak derivative on $[0,a]$ exists and is bounded a.e.;*

*c) the $m$-th derivative $q^{(m)}(r)$ has a nonzero jump at $r=a$ : $q^{(m)}(a) = \lim_{r\to a^-} q^{(m)}(r) \neq 0$ (in the case $m=0$ we assume $q^{(0)}(r) = q(r)$).*

*Let*

$$\alpha_{\ell n}^{(m)} = \frac{\pi n}{a} - \frac{n}{|n|}\frac{\pi}{2a}\left(\tilde{\ell} + \frac{m}{2}\right) - \frac{i}{2a}\ell n\left[\left(\frac{2\pi|n|}{a}\right)^{m+2}\frac{1}{|q^{(m)}(a)|}\right], \tag{8}$$

*where $\tilde{\ell} = \ell + 1$ for $q^{(m)}(a) > 0$ and $\tilde{\ell} = \ell$ for $q^{(m)}(a) < 0$, $n = \pm 1, \pm 2, \ldots$, and $\ell = 0, 1, 2, \ldots$.*

*Then for each $\ell = 0,1,2,\ldots$ the radial operator $H_\ell$ (see (3)) has a countable set of resonance poles. The set of these poles, except for maybe a finite number of them, can be split into two infinite sequences $\left\{E_{\ell n}^C = \left(k_{\ell n}^C\right)^2\right\}_{n=1}^{\infty}$ and $\left\{E_{\ell n}^C = \left(k_{\ell n}^C\right)^2\right\}_{n=-1}^{-\infty}$. For elements of these sequences the following estimate takes place*

$$\left|k_{\ell n}^C - \alpha_{\ell n}^{(m)}\right| \leq C_\ell \frac{\ell n|n|}{|n|}, \quad n = \pm 1, \pm 2, \ldots, \tag{9}$$

*where $C_\ell > 0$ is a constant, which depends only on $\ell$.*

*The operator $H_\ell^{(0)}$ (see (2)), also, has a countable set of resonance poles. The set of these poles, except for maybe a finite number of them, can be split into two sequences $\left\{E_{\ell n} = (k_{\ell n})^2\right\}_{n=1}^{\infty}$ and $\left\{E_{\ell n} = (k_{\ell n})^2\right\}_{n=-1}^{-\infty}$. For elements of these sequences a similar estimate takes place*

$$\left|k_{\ell n} - \alpha_{\ell n}^{(m)}\right| \leq \tilde{C}_\ell \frac{\ell n|n|}{|n|}, \quad n = \pm 1, \pm 2, \ldots, \tag{10}$$

*with another constant $\tilde{C}_\ell > 0$. Notice, that the numeration of resonances with respect to index $n$ is asymptotic and not absolute.*

**Remarks.** a) Theorem 1 shows that the behavior of resonances of the radial Schrödinger operator $H_\ell$ is the same in the Coulomb-like case and in the short-range case ($Z = 0$). Namely, in both cases resonances stay away from zero and accumulate only to infinity.

b) Theorem 2 shows that there is a dramatic difference in the behavior of low energy resonances for full 3-dimensional operators in the Coulomb-like and short-range cases. Namely, in the short-range case $r(\ell) \geq r_0 > 0$, while in the Coulomb-like case $\lim_{\ell\to\infty} r(\ell) = 0$. Moreover, in the short range case $\lim_{\ell\to\infty} r(\ell) = \infty$. The latter fact follows from the possibility to estimate the number of resonance poles in a disk of a given radius ([6, 7]). Theorem 2 shows that such estimates are impossible for Coulomb-like potentials.

c) Theorem 3 gives the high-energy asymptotics of resonances only in the case when either $q(r)$ or one of its derivatives has a jump on the boundary of the support. It is known that infinite series of resonances can exist for infinitely differentiable potentials. The asymptotics of resonances in this case is beyond the scope of our methods.

**Conjecture.** a) In the case of a nonspherically symmetric perturbation $q(x)$ satisfying: $q(x) = 0$ for $|x| \geq a; q(x) \geq 0$ for $|x| \leq a$; $0 < \int_{|x|\leq a} q(x)|x|^{-1}dx < \infty$; resonances of (1) also accumulate to zero.

b) If the perturbation $q(x)$ in (1) is not spherically symmetric but has a spherically symmetric support and a constant jump on the boundary of this support, then the serial structure of resonances appears at high energies and the asymptotics is given by (9). Such a result could be a generalization of the results obtained in [6-8] for a nonspherically symmetric short-range case.

# 3 Auxiliary results and outline of proofs.

All the above results are obtained based on a detailed asymptotic analysis at low and high energies of the solutions of Eq. (3). We describe below the main auxiliary theorems, which might be of interest in themselves. The starting point in the proofs of Theorem 1-3 is the fact that resonance poles of the radial operator $H_\ell$ can be obtained as roots of a suitably defined Jost function. We call this fact the *resonance selection principle.* It is well known for short-range potentials. However, in the case of Coulomb-like potentials it is nontrivial and requires a special proof.

To formulate the *resonance selection principle* we need some information about the solutions of Eq. (3). The exposition is divided in several steps.

1) Introduce in (2) and (3) a new variable $\rho = Zr$ and a new spectral parameter $s$, such that $1/s^2 = E/Z^2$. Notice, that the second sheet of $E$ corresponds to the upper half plane of $s$. We also rename $2q(r)/Z^2$ by $q(\rho)$ and $aZ$ by $a$. Eq. (2) has two linearly independent solutions $\Phi_{0\ell}(s,\rho)$ and $F_{0\ell}(s,\rho)$. They are regular and irregular Whittaker functions correspondingly. $\Phi_{0\ell}(s,\rho)$ has a zero of order $(\ell+1)$ at $\rho = 0$, $F_{0\ell}(s,\rho)$ has a singularity $\rho^{-\ell}$ at $\rho = 0$. Eq. (3) also has two linearly independent solutions. The regular solution $\Phi_\ell(s,\rho)$ has zero of order $(\ell+1)$ at $\rho = 0$. The singular solution $F_\ell(s,\rho)$ has a singularity $\rho^{-\ell}$ at $\rho = 0$ and is proportional to $F_{0\ell}(s,\rho)$ for $\rho > a$. $\Phi_\ell$ and $F_\ell$ satisfy certain Volterra integral equations. To obtain the properties of $\Phi_\ell$ and $F_\ell$ it is natural to study these equations by the method of successive approximations. However, there is an obstacle in this way. Namely, the kernel of the above integral equations is a combination of the Whittaker functions. It turns out that the information about the confluent hypergeometric function available from an extensive literature (see, for example, [11-13]) is insufficient for the application of the successive approximation method. To analyze the integral equations we prove certain new asymptotic formulae involving the confluent hypergeometric function. Using these formulae we obtain estimates for $\Phi_\ell(s,\rho)$ and $F_\ell(s,\rho)$. It turns out that it is necessary to consider separately three different regions of the parameter $s$. Let $A > 0$. We consider the low-energy region $\Omega_A = \{s \in \mathbb{C} : |s| \geq A\}$, which we divide into two sectors $\Omega_{A,\delta} = \{s \in \Omega_A : |\arg s + \pi/2| \geq \delta\}$ and $\tilde{\Omega}_{A,\delta} = \Omega_A \backslash \Omega_{A,\delta}$. We consider, also, the high-energy region $\hat{\Omega}_A = \{s \in \mathbb{C} : |s| \leq A, s \neq 0\}$. In all three regions we have different estimates. The case of $\tilde{\Omega}_{A,\delta}$ is important for the study of the low-energy asymptotics of the discrete spectrum. It was considered in [14, 15]. The above mentioned estimates imply, in particular, the following result.

**THEOREM 4** *For any $\rho \geq 0$ the solution $\Phi_\ell(s,\rho)$ is an analytic function of $s \in \mathbb{C}\backslash\{0\}$. For any $\rho > 0$ the solution $F_\ell(s,\rho)$ is an analytic function of $s \in \mathbb{C}\backslash\{0\}$ with poles on the negative imaginary semiaxis. These poles go to infinity and starting from certain number are all simple. For each $s \in \Omega_{A,\delta}$ the following estimates hold: $|F_\ell(s,\rho)| \leq C_\ell \rho^{-\ell}, |\Phi_\ell(s,\rho)| \leq \tilde{C}_\ell \rho^{\ell+1}$.*

We emphasize that the constants $C_\ell, \tilde{C}_\ell$ as well as the constants in the above mentioned more detailed estimates, are not uniform with respect to $\ell, A$ and $\delta$.

2) **THEOREM 5** *(*Resonance selection principle*). Resonances in the sense of Definition 1 corresponding to the operator (1) with $q(r)$ satisfying (4) are those and*

*only those points $E$ of the second Riemann sheet that can be represented in the form $E = Z^2/s^2$, where $s$ belongs to the upper half-plane and satisfy one of the equations*

$$\lim_{\rho \to 0} \rho^\ell F_\ell(s, \rho) = 0, \ \ell = 0, 1, 2, \ldots \tag{11}$$

We prove Theorem 5 in two steps. First of all we introduce a new definition of resonances.

**DEFINITION 2** *A resonance for the operator (1) with $q(r)$ satisfying (4) is a point of the spectral parameter $s$, such that for some $\ell$ there exists a growing solution of Eq. (3), which is bounded at $\rho = 0$ and proportional to $F_{0\ell}(s, \rho)$ for $\rho > a$. The corresponding resonance pole on the second sheet of $E$ is given by $E = Z^2/s^2$.*

We prove Theorem 5 based on Definition 2. This proof goes as follows. Using integral equations for $F_\ell$ and $\Phi_\ell$ and the above mentioned estimates for their solutions in the region $s \in \Omega_{A,\delta}$ or $s \in \hat{\Omega}_A$, we derive an auxiliary formula for the Wronskian

$$\{\Phi_\ell, F_\ell\} = (2\ell + 1) \lim_{\rho \to 0} \rho^\ell F_\ell(s, \rho). \tag{12}$$

(The derivation is a little different for $\Omega_{A,\delta}$ and $\hat{\Omega}_A$). Now if $s$ satisfies (11) for some $\ell$, then $\{\Phi_\ell, F_\ell\} = 0, \Phi_\ell$ and $F_\ell$ are linearly dependent and, therefore, $F_\ell(s, \rho)$ has no singularity at $\rho = 0$. Thus, it is the solution from Definition 2. If, conversely, the latter solution exists, then it is not difficult to show that $\Phi_\ell(s, \rho)$ and $F_\ell(s, \rho)$ are linearly dependent for the corresponding value of $s$ . Therefore, $\{\Phi_\ell, F_\ell\} = 0$ and $s$ satisfies (11) due to (12).

The second step is the proof of the equivalence of Definitions 1 and 2. The latter proof is based on the following facts. The resolvent $(H - E)^{-1}$ is an integral operator whose kernel is the Green's function $G(x, x'; s)$ of $H(E = Z^2/s^2)$. $G(x, x'; s)$ can be expressed through the Green's functions $G_\ell(\rho, \rho'; s)$ of the radial operators $H_\ell$ in the form of an infinite series ([16]). We know, that $G_\ell(\rho, \rho'; s) = \Phi_\ell(s, \rho) F_\ell(s, \rho')/\{\Phi_\ell, F_\ell\}$ for $\rho \leq \rho'$. So, zeros of $\{\Phi_\ell, F_\ell\}$ are poles of $G_\ell(\rho, \rho'; s)$ and, therefore, poles of $G(x, x'; s)$. Conversely, it is possible to show that poles of $G(x, x'; s)$ can be produced only by poles of some of $G_\ell(\rho, \rho'; s)$ and not by the divergence of the above mentioned series.

3) We derive Theorem 2 from the following statement about the roots of Eq. (11).

**THEOREM 6** *For each sufficiently small $\epsilon > 0$ there exists a positive integer $\ell(\epsilon)$ such that the following statement holds. If $\ell \geq \ell(\epsilon)$, then Eq. (11) has precisely one root $s_t^\ell$ in each disk $B_\epsilon(it) = \{z \in \mathbb{C} : |z - it| < \epsilon\}$, where $t$ is any positive integer from the interval: $(\ell + 0.5)^{1/4} < t < (\ell + 0.5)^{1/2}$. The behavior of $\ell(\epsilon)$ when $\epsilon \to 0$ can be characterized as follows. Let $\epsilon(\ell) = \sup\{\epsilon : \ell(\epsilon) = \ell\}$. Then $\epsilon(\ell) \leq C(\sqrt{e}/t)^t$, where $t = \ell + 0.5$ and $C > 0$ is a constant.*

The derivation Theorem 2 from Theorem 6 is straightforward. To prove Theorem 6 we consider the resonance equation (11). It turns out that the above mentioned estimates for $\Phi_\ell$ and $F_\ell$, obtained from the corresponding Volterra integral equations by the method of successive approximations, are insufficient for the proof of Theorem 6. The reason is that these estimates are not uniform with respect to $\ell$. We have to derive new estimates of the so-called Liouville-Green type (see [17]) for $\Phi_\ell$ and $F_\ell$. These estimates are uniform with respect to $\ell$. Using the latter estimates it is possible to reduce Eq. (11) to the following asymptotic form:

$$\frac{2ms^{2m}\Gamma(-m-0.5+is)}{\Gamma(m+0.5+is)} = i\left(\int_0^a \rho q(\rho)d\rho\right)\left(1+O\left(m^{-1/4}\right)\right), \tag{13}$$

where $m = \ell + 0.5$.

Now we apply the following lemma about the Euler gamma-function.

**LEMMA 1** *Let $E_m = \{s \in \mathbb{C} : |Res| \leq 1,\ m^{1/4} \leq Ims \leq m^{1/2}\}$. For all $s \in E_m$ the following estimate is valid: the left hand side of (13) is equal to $\sinh(\pi s)D(s,m)$, where $D(s,m)$ satisfies the inequality $C_1(e|s|m^{-1})^{2m} \leq |D(s,m)| \leq C_2(e|s|m^{-1})^{2m}$ with some constants $C_1 > 0, C_2 > 0$.*

Using the Lemma we rewrite (13) in the form

$$\sinh(\pi s) = -i\left(\int_0^a \rho q(\rho)d\rho\right)^{-1} D(s,m)\left(1+O\left(m^{-1/4}\right)\right), \tag{14}$$

where $s \in E_m$. Based on (14) and the above estimate for $D(s,m)$ it is easy to complete the proof using the Rouchet theorem and the fact that sinh $(\pi s)$ has simple roots at all points $it$ with integer $t$.

4) To prove Theorem 3 we introduce a new spectral parameter $k = Zs^{-1}$ (or $k^2 = E$) and consider Eq. (11) in the high-energy region $\{k \in \mathbb{C} : |k| \geq A\}$ for sufficiently large $A > 0$. We also consider an analogue of Eq. (11) for the operator (1)

with $Z = 0$. To study the resonance equation it is necessary to obtain estimates for $\Phi_\ell$ and $F_\ell$. These estimates are obtained from the above mentioned Volterra integral equations using the successive approximations method. The analysis in the high-energy region differs significantly from the corresponding analysis in the low-energy region.

# REFERENCES


[1] T. Regge, Analytic propertics of the scattering matrix, Nuovo Cimento 8, No. 5 (1958), 671.

[2] R.G. Newton, "Scattering Theory of Waves and Particles," Springer-Verlag, Berlin/New York/ Heidelberg, 1982.

[3] M. Reed and B. Simon, "Methods of Modern Mathematical Physics IV," Academic Press, New York/London, 1978.

[4] H.L. Cycon, R.G. Froese, W. Kirsh and B. Simon, "Schrödinger operators," Springer-Verlag, Berlin/Heidelberg/New York, 1987.

[5] I.M. Knowles, Y. Saito, "Differential Equations and Mathematical Physics". Proc. Internat. Conf., Birmingham, AL (1986).

[6] R.B. Melrose, Polynomial Bounds on the number of scattering poles, J. Funct. Anal. 53, (1983), 287-305.

[7] M. Zworski, Sharp polynomial bounds on the number of scattering poles of radial potentials, J. Funct. Anal. 82 (1989), 370-405.

[8] M.A. Shubova, Construction of resonance states for a three-dimensional Schrödinger operator, Problems of Mathematical Physics, Vol. 9, pp. 145-180, Leningrad University, 1979.

[9] M.A. Shubova, Resonance selection principle for a three-dimensional Schrödinger operator. Proc. Steklov Inst. Math. 159 (1984), 181-195.

[10] M.A. Shubova, Serial structure of resonances for the three-dimensional Schrödinger operator. Proc. Steklov Inst. Math. 159, (1984) 197-217.

[11] H. Buchholz, "The Confluent Hypergeometric Function," Springer-Verlag, Berlin/Heidelberg/New York, 1969.

[12] L.J. Slater, "Confluent Hypergeometric Functions," Cambrige University Press, 1960.

[13] A. Erdelyi and C.A. Swanson, "Asymptotic Forms of Whittacker's Confluent Hypergeometric Functions," Memoirs of the AMS, No. 25, 1957.

[14] M.A. Shubova, Construction of eigenfunctions of the discrete spectrum and principle for selecting eigenvalues for a radial Schrödinger operator with nearly Coulomb potential. Theor. and Math. Physics **76**, (1988) 379-392.

[15] M.A. Shubov, Asymptotics of the discrete spectrum for a radial Schrödinger operator with nearly Coulomb potential. Integral Eq. and Operator Theory, Vol. 14, (1991) 586-608.

[16] E.C. Titchmarsh, "Eigenfunction Expansions Associated with Second Order Ordinary Differential Equations, Part II", Oxford Univ., Oxford, (1972).

[17] F.W.J. Olver, "Asymptotics and Special Functions," Academic Press, New York/ London, 1974.

[18] M.A. Shubov, Resonance selection principle and low-energy resonances for a radial Schrödinger operator with nearly Coulomb potential. To appear in Journal of Math. Analysis and Appl.

[19] M.A. Shubov, Low-energy chain of resonances for three-dimensional Schrödinger operator with nearly Coulomb potential. To appear in Journal of Differential Equations.

[20] M.A. Shubov, High-energy asymptotics of resonances for three-dimensional Schrödinger operator with screened Coulomb potential. Submitted to Integral Equations and Operator Theory.

# The Gevrey Asymptotics and Exact Asymptotics

Yasutaka Sibuya
School of Mathematics, University of Minnesota
Minneapolis, Minnesota 55455, U.S.A.

**ABSTRACT**

We have been studying the Gevrey asymptotics since 1979. Recently, through these researches, multisummability of formal solutions of meromorphic differential equations was established. This means that the sum of formal solutions can be uniqely defined and the sums are actual solutions. Exact asymptotics is based on the summability of formal power series in this sense. In this paper, we shall explain basic facts concerning the Gevrey asymptotics and exact asymptotics. Also some simple applications will be given.

## 1. Preliminaries

First, we shall explain basic concepts in the Gevrey asymptotics and multisummability of formal power series. The original ideas of the theory of the Gevrey asymptotics were given by G.N. Watson [14] and F. Nevanlinna [8]. Recently J.-P.Ramis, B.Malgrange and Y.Sibuya have developed the theory systematically. (See, for example, Appendices of [10] .) The Gevrey asymptotics is based on the two definitions given below.

**Definition 1.1.** *Let $s$ be a nonnegative number. A formal power series $f = \sum_{m\geq 0} a_m z^m \in C[[z]]$ is said to be of the Gevrey order $s$ if there exist two nonnegative numbers $C$ and $A$ such that*

$$|a_m| \leq C(m!)^s A^m \quad for\ m = 0, 1, 2, \cdots , \tag{1.1}$$

*where $C[[z]]$ is the set of all formal power series in $z$ with coefficients in the field $C$ of complex numbers.*

We denote by $C[[z]]_s$ the set of all power series of the Gevrey order $s$.

**Definition 1.2.** *A functuion $\phi$ of $z$ is said to admit an asymptotic expansion of the Gevrey order $s$ $(s \geq 0)$ as $z \to 0$ on a sectorial domain*

$$\mathcal{D}(R_0, a, b) \;=\; \{\, z \;;\; 0 < |z| < R_0,\ \ a \;<\; \arg z \;<\; b \,\}$$

*if $\phi$ satisfies the following conditions:*

(i) *$\phi$ is holomorphic on $\mathcal{D}(R_0, a, b)$,*

(ii) *there exists a formal power series $p = \sum_{m \geq 0} a_m z^m \in C[[z]]_s$ such that an inequality*

$$\left| f(z) - \sum_{m=0}^{N-1} a_m z^m \right| \leq K_{R,\alpha,\beta} \, (N!)^s \, (B_{R,\alpha,\beta})^N \, |z|^N$$

*holds on every proper subsector $\mathcal{D}(R, \alpha, \beta)$ of $\mathcal{D}(R_0, a, b)$ for every positive integer $N$, where $K_{R,\alpha,\beta} > 0$ and $B_{R,\alpha,\beta} > 0$.*

We denote by $\mathcal{A}_s(R_0, a, b)$ the set of all functions admitting asymptotic expansions of the Gevrey order $s$ as $z \to 0$ on the sectorial domain $\mathcal{D}(R_0, a, b)$. Also set $J(f) = p$ . The map

$$J : \mathcal{A}_s(R_0, a, b) \;\to\; C[[z]]_s \tag{1.2}$$

is a homomorphism of commutative differential algebras over $C$. The kernel of homomorphism (1.2) is denoted by $\mathcal{A}_{s,0}(R_0, a, b)$ which is characterized by the theorem below.

**Theorem 1.3.** *A function $\phi$ is contained in $\mathcal{A}_{s,0}(R_0, a, b)$ if and only if*

(i) *$\phi$ is holomorphic on $\mathcal{D}(R_0, a, b)$,*

(ii) *an inequality $|f(z)| \leq \mu_{R,\alpha,\beta} \; exp \left\{ -\lambda_{R,\alpha,\beta} \frac{1}{|z|^{\frac{1}{s}}} \right\}$ holds on every proper subsector $\mathcal{D}(R, \alpha, \beta)$ of $\mathcal{D}(R_0, a, b)$, where $\mu_{R,\alpha,\beta} > 0$ and $\lambda_{R,\alpha,\beta} > 0$.*

The following theorem is very useful in the Gevrey asymptotics.

**Theorem 1.4.** *Assume that a covering $\{ \mathcal{S}_1, \cdots, \mathcal{S}_N \}$ at $z = 0$ is good and that $N$ functions $\phi_1(z), \cdots, \phi_N(z)$ satisfy the conditions:*

(1) *$\phi_\ell(z)$ is holomorphic and bounded on $\mathcal{S}_\ell$ for each $\ell$,*

(2) *we have $|\phi_\ell(z) - \phi_{\ell+1}(z)| \leq \gamma \; exp \left[ -\lambda \frac{1}{|z|^k} \right]$ on $\mathcal{S}_\ell \cap \mathcal{S}_{\ell+1}$, where $\gamma \geq 0$, $\lambda > 0$ and $k > 0$ are suitable numbers.*

*Then $\phi_\ell \in \mathcal{A}_{1/k}(\mathcal{S}_\ell)$ for every $\ell$ where $\mathcal{A}_s(\mathcal{S})$ denotes the set of all functions admitting asymptotic expansions of the Gevrey order $s$ as $z \to 0$ on the sectorial domain $\mathcal{S}$.*

**Remark 1.5**: We call $\{ \mathcal{S}_1, \cdots, \mathcal{S}_N \}$ a good covering at $z = 0$ if

(i) for each $\ell$, $\mathcal{S}_\ell = \{z;\ 0 < |z| < R,\ a_\ell < arg\ z < b_\ell\}$,

(ii) $a_\ell < a_{\ell+1}$, $\quad b_\ell - a_\ell < \pi$,

(iii) $\mathcal{S}_\ell \cap \mathcal{S}_{\ell+1} \neq \emptyset$ $(\ell = 1, \cdots, N)$ and $\mathcal{S}_\ell \cap \mathcal{S}_h = \emptyset$ otherwise if $\ell \neq h$,
(iv) $\bigcup_{1 \leq \ell \leq N} \mathcal{S}_\ell = \{ z;\ 0 < |z| < R \}$,
where $R$ is a positive number.

Theorem 1.3 implies that homomorphism (1.2) is injective if and only if $s < \frac{b-a}{\pi}$. It is also known that $J$ is surjective if and only if $s \geq \frac{b-a}{\pi}$. In the case when $s < \frac{b-a}{\pi}$, power series belonging to the image of homomorphism (1.2) are called k-summable power series ($k = \frac{1}{s}$) in the direction $\arg z = d$ which bisects the sectorial domain $\mathcal{D}(R_0, a, b)$. We denote by $C\{z\}_{k;d}$ the set of all power series in $z$ which are $k$-summable in the direction $d$. For an $f \in C\{z\}_{k;d}$, there exist two positive numbers $R_0$ and $\epsilon$ and a unique function $\phi \in \mathcal{A}_{\frac{1}{k}}(R_0, d - \frac{\pi}{2k} - \epsilon, d + \frac{\pi}{2k} + \epsilon)$ such that $J(\phi) = f$. We call $\phi$ the $k$-sum of $f$ in the direction $d$.

Let $\mathcal{K} = \{k_1, k_2, \cdots, k_N\}$ be a set of a finite number of positive numbers such that

$$0 < k_N < k_{N-1} < \cdots < k_1 < +\infty. \tag{1.3}$$

We assume that $\mathcal{K}$ is fixed but arbitrarily given. Also let $\arg z = d$ be a fixed but arbitrarily given direction. Set

$$C\{z\}_{\mathcal{K};d} = \left\{ f \in C[[z]];\ f = \sum_{\ell=1}^{N} f_\ell\ ,\ \ f_\ell \in C\{z\}_{k_\ell;d} \right\}. \tag{1.4}$$

Then we can prove the following theorem.

**Theorem 1.6.** *If $k_j > \frac{1}{2}$ for every $j$, then $C\{z\}_{\mathcal{K};d}$ is a commutative differential subalgebra of $C[[z]]$.*

**Remark 1.7**: In case when the condition of Theorem 1.6 is not satisfied, we use a new independent variable $t = z^{1/p}$ for a suitable positive integer $p$ so that the condition be satisfied.

For $f = \sum_{1 \leq \ell \leq N} f_\ell$ $\ (f_\ell \in C\{z\}_{k_\ell;d},\ \ell = 1, 2, \cdots, N)$, we define the $\mathcal{K}$-sum of $f$ in the direction $d$ by $\phi = \sum_{1 \leq \ell \leq N} \phi_\ell$, where $\phi_\ell$ is the $k_\ell$-sum of $f_\ell$ in the direction $d$. It can be proved that this sum is independent of the representation of $f$ as a sum $f = \sum_{1 \leq \ell \leq N} f_\ell$ (cf. J.Ecalle [5] and J.Maltinet and J.-P.Ramis [7]). Thus we arrive at the following definition.

**Definition 1.8.** *A formal power series $f$ is said to be $\mathcal{K}$-multisummable in the direction $d$ if $f \in C\{z\}_{\mathcal{K};d}$.*

## 2. Exact asymptotics and its motivation

If a power series $p = \sum_{m \geq 0} a_m z^m$ is convergent, the sum of $p$ can be well defined without any ambiguity. If $p$ is a divergent power series, then, for any given sectorial domain $\mathcal{D}(R_0, a, b) = \{z; 0 < |z| < R_0, a < \arg z < b\}$, there exists a function $f$ such that $f$ is holomorphic in $\mathcal{D}(R_0, a, b)$ and that $f$ admits $p$ as its asymptotic expansion in Poincaré's sense as $z \to 0$ in $\mathcal{D}(R_0, a, b)$. It is one of main difficulties in the Poincaré asymptotics that $f$ is not uniquely determined by $p$. As we explained in §1, in the Gevrey asymptotics, map (1.2) ( i.e. $J : \mathcal{A}_s(R_0, a, b) \to C[[z]]_s$) is surjective if and only if $s \geq \frac{b-a}{\pi}$, while $J$ is injective if and only if $s < \frac{b-a}{\pi}$. Hence, if the opening $b - a$ of the sectorial domain $\mathcal{D}(R_0, a, b)$ is large enough, $f$ is determined uniquely by $p$, and hence we can regard $f$ a <u>sum</u> of $p$. In this way, we define a $k$-summable series in a direction $d$ and its sum in the direction $d$ (cf. §1). More generally speaking, as we explained in §1, if $p$ is multisummable in a direction $d$, we can define its unique sum in the direction $d$. In particular, we already know that formal power series solutions of a (linear and nonlinear) meromorphic differential equation are multisummable in a direction $d$ and their sums are actual solutions if $d$ is not singular (cf. B.L.J.Braaksma [3,4], W.Balser, B.L.J.Braaksma, J.-P.Ramis and Y.Sibuya [1] and Y.Sibuya [13]). This is the starting point of <u>*Exact Asymptotics*</u>.

In classical analysis, it is frequently important to verify that a given quantity $f(z)$ is holomorphic in the neighborhood of a point $z = a$. The function $f(z)$ is holomorphic at $z = a$ if and only if $f(z)$ is the sum of a convergent power series $p = \sum_{m \geq 0} a_m (z - a)^m$. In exact asymptotics, it is also frequently important to verity that a given quantity $f(z)$ is the sum of a multisummable power series $p = \sum_{m \geq 0} a_m z^m$ in a given direction $d$. To derive results of this nature, the following characterization of multisummability is very useful (cf. W.Balser and A.Tovbis [2] and B.Malgrange and J.-P.Ramis [6]).

**Theorem 2.1.** *A formal power series $p = \sum_{m \geq 0} a_m z^m$ is $\{k_1, k_2, \cdots, k_\nu\}$-summable in a direction $d$, where $\frac{1}{2} < k_\nu < k_{\nu-1} < \cdots < k_1$, if and only if there exist $\nu$ sectorial domains $\mathcal{S}_1 \subset \mathcal{S}_2 \subset \cdots \subset \mathcal{S}_\nu$ such that*

*(1) $\mathcal{S}_\ell = \{z; \ |\arg z - d| < \frac{\pi}{2k_\ell} + \epsilon_\ell, \ 0 < |z| < r\}$, where the $\epsilon_\ell$ and $r$ are some positive numbers,*

*(2) there exists a good covering $\{\mathcal{U}_1, \cdots, \mathcal{U}_N\}$ of $\mathcal{D}(r) = \{z; \ 0 < |z| < r\}$ such that*

*(a) $\mathcal{S}_1 = \mathcal{U}_{h_0}$ for some $h_0$,*

*(b) openings of sectors $\mathcal{U}_h$ are strictly smaller than $\frac{\pi}{k_1}$ if $h \neq h_0$,*

(c) *for each* $\ell$, *there exists a subset* $\mathcal{J}_\ell$ *of the set* $\{1, \cdots, N\}$ *such that* $\mathcal{S}_\ell = \bigcup_{h \in \mathcal{J}_\ell} \mathcal{U}_h$,

(d) *for each* $h$, *there exists* $f_h \in \mathcal{A}_{\frac{1}{k_\nu}}(\mathcal{U}_h)$ *such that* $J(f_h) = p$ *and that* $f_h - f_{h+1} \in \mathcal{A}_{s_h, 0}(\mathcal{U}_h \cap \mathcal{U}_{h+1})$, *where* $\frac{1}{s_h} \geq k_{\ell-1}$ *if* $\mathcal{U}_h \cup \mathcal{U}_{h+1} \subset \mathcal{S}_\ell$.

*Furthermore* $f_{h_0}$ *is the sum of* $p$ *in the direction* $d$.

**Note**: Since $f_h$ satisfies (d), we have $\frac{1}{s_h} \geq k_\nu$.

If $f$ is the sum of a $\{k_1, k_2, \cdots, k_\nu\}$-summable series $p$, then we can find sectorial domains $\mathcal{S}_1 \subset \mathcal{S}_2 \subset \cdots \subset \mathcal{S}_\nu$ from a representation $f = \sum_{1 \leq \ell \leq \nu} g_{k_\ell}$, where, for each $\ell$, $g_\ell$ is the sum of a $k_\ell$-summable series. To prove the sufficiency of the conditions of Theorem 2.1 for $\{k_1, k_2, \cdots, k_\nu\}$-summability, we use the following formula:

$$f_{h_0} = -\sum_{h=1}^{N} \frac{1}{2\pi i} \int_{\gamma_h} \frac{f_{h+1}(\zeta) - f_h(\zeta)}{\zeta - z} \, d\zeta \; + \; \phi(z) \qquad \text{for} \qquad z \in \mathcal{U}_{h_0} = \mathcal{S}_1 \, , \tag{2.1}$$

where the path $\gamma_h$ is a line segment starting from $z = 0$ to a point on the boundary of $\mathcal{U}_h \cap \mathcal{U}_{h+1}$ and $\phi(z)$ is the sum of a convergent power series (cf. Y.Sibuya [11,12]). Conditions of Theorem 2.1 implies that each integral

$$\int_{\gamma_h} \frac{f_{h+1}(\zeta) - f_h(\zeta)}{\zeta - z} \, d\zeta \tag{2.2}$$

is the sum of a $k_j$-summable series in the direction $d$ for a suitable $k_j$. In exact asymptotics, we frequently utilize integrals of form (2.2) (cf. Y.Sibuya [11]).

For example, by utilizing Theorem 2.1, we can prove the following results.

**Theorem 2.2.** *Suppose that*

(i) $h(\vec{y})$ *is a convergent power series in the entries of a vector* $\vec{y}$,

(ii) $\vec{f}(z)$ *is the multisum of* $J(\vec{f})$ *in a direction* $d$,

(iii) $\vec{f}(0) = \vec{0}$.

*Then* $h(\vec{f}(z))$ *is also the multisum of* $h(J(\vec{f}))$ *in the direction* $d$. *In particular, if* $f$ *is the multisum of* $J(f)$ *in a direction* $d$ *and if* $f(0) \neq 0$, *then* $\frac{1}{f}$ *is also the multisum of* $\frac{1}{J(f)}$ *in the direction* $d$.

(Cf. Proposition (2.3) of B.Malgrange and J.-P.Ramis [6].)

## 3. Another application of Theorem 2.1

As an application of Theorem 2.1, we shall prove the following theorem.

**Theorem 3.1.** *Suppose that a function $f(z)$ satisfies the following conditions:*
*(1) $f(z)$ is holomorphic and bounded in a sectorial domain $\mathcal{D}(R_0, a, b)$,*
*(2) $f(z)$ is the $\{k_1, \cdots, k_\nu\}$–sum of a multisummable series $p = \sum_{m\geq 0} a_m z^m$ in two directions $d_1$ and $d_2$ such that $a < d_1 < d_2 < b$, where $0 < k_\nu < \cdots < k_1$.*

*We also assume that*

$$d_2 - d_1 + \frac{\pi}{k_\nu} < 2\pi . \tag{3.1}$$

*Then $f(z)$ is the sum of $p$ in every direction $d$ such that $d_1 \leq d \leq d_2$.*

PROOF:

Set

$$S_\ell = \mathcal{D}\left(R_0, d_1 - \frac{\pi}{2k_\ell} - \epsilon, d_2 + \frac{\pi}{2k_\ell} + \epsilon\right) \qquad (\ell = 1, \cdots, \nu),$$

where $\epsilon$ is a sufficiently small positive number. Choose a good covering $\{\mathcal{U}_{-N}, \cdots, \mathcal{U}_{-1}, \mathcal{U}_0, \mathcal{U}_1, \cdots, \mathcal{U}_N\}$ of $\mathcal{D}(R_0) = \{z;\ 0 < |z| < R_0\}$ so that
(I) $\mathcal{U}_0 = S_1$,
(II)

$$\begin{cases} \mathcal{U}_h \subset \mathcal{D}\left(R_0, d_2 + \dfrac{\pi}{2k_1} - \epsilon, \dfrac{d_1 + d_2}{2} + \pi + \epsilon\right) & \text{for} \quad h = 1, \cdots, N, \\ \mathcal{U}_{-h} \subset \mathcal{D}\left(R_0, \dfrac{d_1 + d_2}{2} - \pi - \epsilon, d_1 - \dfrac{\pi}{2k_1} + \epsilon\right) & \text{for} \quad h = 1, \cdots, N, \end{cases}$$

(III) the openings of the $\mathcal{U}_{\pm h}$ are smaller than $\frac{\pi}{k_1}$ for $h = 1, \cdots, N$,
(IV) the sectorial domain $\mathcal{U}_{-N} \cap \mathcal{U}_N$ contains the direction $\dfrac{d_1 + d_2}{2} + \pi$ (mod $2\pi$),
(V) there exist functions $f_{-N}(z), \cdots, f_{-1}(z), f_0(z), f_1(z), \cdots, f_N(z)$ such that
(V-a) $f_0(z) = f(z)$ on $\mathcal{U}_0$,
(V-b) $f_{\pm h} \in \mathcal{A}_{\frac{1}{k_\nu}}(\mathcal{U}_{\pm h})$, and $J(f_{\pm h}) = p \qquad (h = 1, \cdots, N)$,
(V-c) $f_h - f_{h+1} \in \mathcal{A}_{s_h, 0}(\mathcal{U}_h \cap \mathcal{U}_{h+1})$, where $\frac{1}{s_h} \geq k_{\ell-1}$ if $\mathcal{U}_h \cup \mathcal{U}_{h+1} \subset S_\ell$ for $h = -N, \cdots, N-1$,
(V-c') $f_N - f_{-N} \in \mathcal{A}_{\frac{1}{k_\nu}, 0}(\mathcal{U}_{-N} \cap \mathcal{U}_N)$.

Existance of such a covering $\{\mathcal{U}_{-N}, \cdots, \mathcal{U}_{-1}, \mathcal{U}_0, \mathcal{U}_1, \cdots, \mathcal{U}_N\}$ of $\mathcal{D}(R_0)$ is guaranteed by Theorem 2.1. Now we can write $f(z)$ in the form:

$$\begin{aligned} f(z) = &- \sum_{h=-N}^{-1} \frac{1}{2\pi i} \int_{\gamma_h} \frac{f_{h+1}(\zeta) - f_h(\zeta)}{\zeta - z} d\zeta \\ &- \sum_{h=0}^{N-1} \frac{1}{2\pi i} \int_{\gamma_h} \frac{f_{h+1}(\zeta) - f_h(\zeta)}{\zeta - z} d\zeta \\ &- \frac{1}{2\pi i} \int_{\gamma_N} \frac{f_{-N}(\zeta) - f_N(\zeta)}{\zeta - z} d\zeta + \phi(z) \qquad \text{for} \qquad z \in \mathcal{U}_0 = \mathcal{S}_1 , \end{aligned} \tag{3.2}$$

where, for each $h = -N, \cdots, N-1$, the path $\gamma_h$ is a line segment starting from $z = 0$ to a point on the boundary of $\mathcal{U}_h \cap \mathcal{U}_{h+1}$, while the path $\gamma_N$ is a line segment starting from $z = 0$ to a point on the boundary of $\mathcal{U}_{-N} \cap \mathcal{U}_N$, and $\phi(z)$ is the sum of a convergent power series. We can show easily that each integral on the right-hand side of (3.2) is the sum of a $k_j$-summable power series in all directions $d$ such that $d_1 \leq d \leq d_2$ for a suitable $k_j$. This completes the proof of Theorem 3.1.

**Remark 3.2:** The author does not know whether Condition (3.1) can be removed. It must be noticed that if $f(z)$ is the sum of $p = J(f)$ in all directions such that $d_1 \leq d \leq d_2$ and $d_2 - d_1 \geq 2\pi$, then $p$ must be a convergent power series. In fact, under this situation, we have

$$d_1 \leq \frac{d_1 + d_2}{2} - \pi < \frac{d_1 + d_2}{2} + \pi \leq d_2 .$$

Hence, the uniqueness of the sum of $p$ implies that $f(z)$ is holomorphic and bounded in a neighborhood of $z = 0$.

## 4. Linear systems with multisummable coefficients

In this section, utilizing Theorem 2.1, we shall prove the following theorems:

**Theorem 4.1.** *Suppose that*

*(i) $\vec{f}(z)$ is the $\{k_1, k_2, \cdots, k_\nu\}$-sum of $J(\vec{f})$ in a direction $d$, where $\frac{1}{2} < k_\nu < k_{\nu-1} < \cdots < k_1$,*

*(ii) $A(z)$ is an $n \times n$ matrix whose entries are convergent power series in $z$,*

*(iii) $\sigma$ is a nonnegative integer.*

*Then, there exists a vector $\vec{g}(z)$ such that*

*(1)* $\vec{g}(z)$ *is the* $\{k_1, k_2, \cdots, k_\nu\}$*-sum of* $J(\vec{g})$ *in the direction* $d$*,*

*(2) the entries of the vector* $z^\sigma \dfrac{d\vec{g}}{dz} - A(z)\vec{g}(z) - \vec{f}(z)$ *are convergent power series in* $z$*.*

**Theorem 4.2.** *Suppose that*

*(i)* $A(z)$ *is an* $n \times n$ *matrix whose entries are* $\{k_1, k_2, \cdots, k_\nu\}$*-sums in a direction* $d$*, where* $\frac{1}{2} < k_\nu < k_{\nu-1} < \cdots < k_1$*,*

*(ii)* $\sigma$ *is a nonnegative integer.*

*Then, there exists an* $n \times n$ *matrix* $P(z)$ *such that*

*(1) the entries of* $P(z)$ *and* $P(z)^{-1}$ *are* $\{k_1, k_2, \cdots, k_\nu\}$*-sums in the direction* $d$*,*

*(2) the entries of the matrix* $P(z)^{-1}A(z)P(z) - z^\sigma P(z)^{-1}\dfrac{dP(z)}{dz}$ *are convergent power series in* $z$*.*

**Theorem 4.3.** *Suppose that*

*(i)* $\vec{f}(z)$ *is the* $\{k_1, k_2, \cdots, k_\nu\}$*-sum of* $J(\vec{f})$ *in a direction* $d$*, where* $\frac{1}{2} < k_\nu < k_{\nu-1} < \cdots < k_1$*,*

*(ii)* $A(z)$ *is an* $n \times n$ *matrix whose entries are* $\{k_1, k_2, \cdots, k_\nu\}$*-sums in a direction* $d$*,*

*(iii)* $\sigma$ *is a nonnegative integer.*

*Then, there exists an* $n \times n$ *matrix* $P(z)$ *and a vector* $\vec{g}(z)$ *such that*

*(1) the entries of* $P(z)$ *and* $P(z)^{-1}$ *are* $\{k_1, k_2, \cdots, k_\nu\}$*-sums in the direction* $d$*,*

*(2)* $\vec{g}(z)$ *is the* $\{k_1, k_2, \cdots, k_\nu\}$*-sum of* $J(\vec{g})$ *in the direction* $d$*,*

*(3) The linear transformation:*

$$\vec{y} = P(z)\,(\,\vec{w} + \vec{g}(z)\,) \tag{4.1}$$

*changes the differential equation:*

$$z^\sigma \frac{d\vec{y}}{dz} = A(z)\vec{y} + \vec{f}(z) \tag{4.2}$$

*to*

$$z^\sigma \frac{d\vec{w}}{dz} = E(z)\vec{w} + \vec{h}(z)\,, \tag{4.3}$$

*in such a way that the entries of the* $n \times n$ *matrix* $E(z)$ *and the vector* $\vec{h}(z)$ *are convergent power series in* $z$*.*

Theorem 4.3 follows immediately from Theorems 4.1 and 4.2.

PROOF OF THEOREM 4.1:

Setting

$$\vec{p} = J(\vec{f}) \qquad \text{and} \qquad \vec{f}_{h_0}(z) = \vec{f}(z)\,,$$

choose $\nu$ sectorial domains $\mathcal{S}_1 \subset \mathcal{S}_2 \subset \cdots \subset \mathcal{S}_\nu$, a good covering $\{\mathcal{U}_1, \cdots, \mathcal{U}_N\}$ of $\mathcal{D}(r) = \{z;\ 0 < |z| < r\}$ and $N$ vectors $\vec{f}_h \in \left(\mathcal{A}_{\frac{1}{k_\nu}}(\mathcal{U}_h)\right)^n$ $(h = 1, \cdots, N)$ in such a way that Conditions (1) and (2) of Theorem 2.1 are satisfied.

Consider, for each $h$, the differential equation:

$$z^\sigma \frac{d\vec{v}_h}{dz} = A(z)\vec{v}_h + \vec{f}_h(z) - \vec{f}_{h+1}(z) \qquad \text{in} \quad \mathcal{U}_h \cap \mathcal{U}_{h+1}\,. \tag{4.4-h}$$

We can fix $\mathcal{S}_1, \cdots, \mathcal{S}_\nu$ and $\mathcal{U}_1, \cdots, \mathcal{U}_N$ in such a way that the openings of $\mathcal{U}_h \cap \mathcal{U}_{h+1}$ are small enough so that, for each $h$, Equation (4.4-h) admits a solution

$$\vec{v}_h(z) \in (\mathcal{A}_{s_h,0}(\mathcal{U}_h \cap \mathcal{U}_{h+1}))^n$$

(cf. J.-P.Ramis and Y.Sibuya [9]). Then there eixst $\vec{\phi}_h(z) \in \mathcal{A}_{\frac{1}{k_\nu}}(\mathcal{U}_h)$ $(h = 1, \cdots, N)$ such that

$$\vec{v}_h(z) = \vec{\phi}_h(z) - \vec{\phi}_{h+1}(z) \qquad \text{in} \qquad \mathcal{U}_h \cap \mathcal{U}_{h+1}$$

(cf. Y.Sibuya [12]). This implies that

$$z^\sigma \frac{d\vec{\phi}_h}{dz} - A(z)\vec{\phi}_h - \vec{f}_h(z) = z^\sigma \frac{d\vec{\phi}_{h+1}}{dz} - A(z)\vec{\phi}_{h+1} - \vec{f}_{h+1}(z) \quad \text{in} \quad \mathcal{U}_h \cap \mathcal{U}_{h+1}\,.$$

Hence setting $h = h_0$ and $\vec{g}(z) = \vec{\phi}_{h_0}(z)$, we conclude that the entries of the vector $z^\sigma \dfrac{d\vec{g}}{dz} - A(z)\vec{g} - \vec{f}(z)$ are convergent power series in $z$.

Applying Theorem 2.1 to $\mathcal{S}_1, \cdots, \mathcal{S}_\nu$, $\mathcal{U}_1, \cdots, \mathcal{U}_N$ and $\vec{\phi}_1(z), \cdots \vec{\phi}_N(z)$, we also conclude that $\vec{g}(z)$ is the $\{k_1, \cdots, k_\nu\}$–sum of $J(\vec{g})$. This completes the proof of Theorem 4.1.

PROOF OF THEOREM 4.2:

Setting

$$p = J(A) \qquad \text{and} \qquad f_{h_0}(z) = A(z),$$

choose $\nu$ sectorial domains $\mathcal{S}_1 \subset \mathcal{S}_2 \subset \cdots \subset \mathcal{S}_\nu$, a good covering $\{\mathcal{U}_1, \cdots, \mathcal{U}_N\}$ of $\mathcal{D}(r) = \{z;\ 0 < |z| < r\}$ and $n \times n$ matrices $f_1, \cdots, f_N(z)$ whose entries are in $\mathcal{A}_{\frac{1}{k_\nu}}(\mathcal{U}_h)$ in such a way that Conditions (1) and (2) of Theorem 2.1 are satisfied.

Consider, for each $h$, the differential equation:

$$z^\sigma \frac{dQ_h}{dz} = f_h(z)Q_h - Q_h f_{h+1}(z) + f_h(z) - f_{h+1}(z) \qquad \text{in} \quad \mathcal{U}_h \cap \mathcal{U}_{h+1} \tag{4.5-h}$$

for an $n \times n$–matrix $Q_h(z)$. Again we can assume that the openings of $\mathcal{U}_h \cap \mathcal{U}_{h+1}$ are small enough so that, for each $h$, Equation (4.5-h) admits a solution $Q_h(z)$ whose entries belong to $\mathcal{A}_{s_h,0}(\mathcal{U}_h \cap \mathcal{U}_{h+1})$ (cf. J.-P.Ramis and Y.Sibuya [9]). Then there eixst $n \times n$–matrices $P_h(z)$ $(h = 1, \cdots, N)$ such that the entries of $P_h(z)$ and $P_h(z)^{-1}$ belong to $\mathcal{A}_{\frac{1}{k_\nu}}(\mathcal{U}_h)$ and that

$$I + Q_h(z) = P_h(z)P_{h+1}(z)^{-1} \quad \text{in} \quad \mathcal{U}_h \cap \mathcal{U}_{h+1}\,, \tag{4.6-h}$$

where $I$ is the $n \times n$–identity matrix (cf. Chapter VI and Appendices of Y.Sibuya [10]). This implies that

$$\begin{aligned} &P_h(z)^{-1}f_h(z)P_h(z) - z^\sigma P_h(z)^{-1}\frac{dP_h(z)}{dz} \\ &= P_{h+1}(z)^{-1}f_{h+1}(z)P_{h+1}(z) - z^\sigma P_{h+1}(z)^{-1}\frac{dP_{h+1}(z)}{dz} \quad \text{in} \quad \mathcal{U}_h \cap \mathcal{U}_{h+1}\,. \end{aligned}$$

Hence setting $h = h_0$ and $P(z) = P_{h_0}(z)$, we conclude that the entries of the matrix $P(z)^{-1}A(z)P(z) - z^\sigma P(z)^{-1}\dfrac{dP(z)}{dz}$ are convergent power series in $z$.

Now observe that, since $P_h(z) - P_{h+1}(z) = Q_h(z)P_{h+1}(z)$, the entries of $P_h(z) - P_{h+1}(z)$ belong to $\mathcal{A}_{s_h,0}(\mathcal{U}_h \cap \mathcal{U}_{h+1})$. Therefore applying Theorem 2.1 to $S_1, \cdots, S_\nu$, $\mathcal{U}_1, \cdots, \mathcal{U}_N$ and $P_1(z), \cdots P_N(z)$, we also conclude that $P(z)$ is a $\{k_1, \cdots, k_\nu\}$–sum . Then utilizing Theorem 2.2, we can also show that $P(z)^{-1}$ is a $\{k_1, \cdots, k_\nu\}$–sum. This completes the proof of Theorem 4.2.

**Acknowledgements:** The main part of this paper was finished at University of Groningen, Netherlands while the author was a guest researcher supported partially by the Netherlands Organisation for Scientific Research and the Department of Mathematics, University of Groningen. The author wishes to express his appreciation for the hospitality of these institutes. In particular, he thanks Professor B.L.J.Braaksma for many useful discussions of multisummability and the Stokes phenomena. The author is also partially supported by a grant from National Science Foundation.

## REFERENCES

[1] Balser,W., Braaksma,B.L.J., Ramis,J.-P. and Sibuya,Y. (1991), Multi-summability of formal power series solutions of linear ordinary differential equations, Asymp. Anal., 5, 27-45.

[2] Balser,W. and Tovbis,A. (1992), Multisummability of iterated integrals, preprint.

[3] Braaksma,B.L.J. (1991), Multisummability and Stokes multipliers of linear meromorphic differential equations, J. Differential Equations 92, 45-75.

[4] Braaksma,B.L.J. (1992), Multisummability of formal power series solutions of nonlinear meromorphic differential equations, Ann. Inst. Fourier, Grenoble, 42, 517–540.

[5] Ecalle,J. (1990), The acceleration operators and their applications to (a) differential equations, (b) quasi-analytic functions, (c) the constructive proof of Dulac's conjecture, an invited address ICM at Tokyo.

[6] Malgrange,B. and Ramis,J.-P. (1992), Fonctions multisommables, Ann. Inst. Fourier, Grenoble, 42, 353-368.

[7] Maltinet,J. and Ramis,J.-P. (1991), Elementary acceleration and multisummability, Ann. Inst. Henri Poincaré, Physique Théorique, 54, 331-401.

[8] Nevanlinna,F. (1919), Zur Theorie der asymptotischen Potenzreihen, Ann.Acad. Sci. Fenn. Ser.A I Math. Dissertationes 12.

[9] Ramis,J. -P. and Sibuya,Y. (1989). Hukuhara domains and fundamental existence and uniqueness theorems for asymptotic solutions of Gevrey type, Asymp. Anal. 2, 39–94.

[10] Sibuya,Y. (1990), Linear Differential Equations in the Complex Domain: Problems of Analytic Continuation, Translations of Mathematical Monographs, vol. 82, American Mathematical Society.

[11] Sibuya,Y. (1991), Gevrey asymptotics and Stokes multipliers, Differential Equations and Computer Algebra, Academic Press, 131-147

[12] Sibuya,Y. (1991), Gevrey asymptotics, Ordinary and delay differential equations, edited by J.Wiener and J.K.Hale, Pitman Research Notes in Math. Series 272, 207–217.

[13] Sibuya,Y. (1992), Normal forms and Stokes multipliers of nonlinear meromorphic differential equations, Computer Algebra and Differential Equations, Cambridge Univ. Press, to appear.

[14] Watson,G.N. (1911), A theory of asymptotic series, Philos. Trans. Royal Soc. London, Ser. A, 211, 279–313.

# UNIVERSAL SHAPES FOR ROTATING INCOMPRESSIBLE FLUID DROPS

DONALD R. SMITH

*Department of Mathematics, University of California at San Diego*
*La Jolla, CA 92093, USA*

and

JAMES E. ROSS

*Department of Mathematical Sciences, San Diego State University*
*San Diego, CA 92182, USA*

Dedicated to W. A. HARRIS, Jr., M. IWANO, and Y. SIBUYA

ABSTRACT

Axisymmetric shapes of Beer [1855], Rosenthal [1962], D. Ross [1968] and others provide universal solutions for a problem related to Plateau's rotating drops, for the material class of Noll's simple incompressible fluid subject to a frequently used linear constitutive relation between pressure jump and mean curvature at the free boundary. A bifurcation from spheroidal to toroidal shapes occurs within this family, leading initially to a decrease in angular momentum. The spheroidal and toroidal axisymmetric shapes connect tangentially in the angular speed–angular momentum plane, and the corresponding bifurcation slope is calculated. Techniques borrowed from singular perturbation theory are useful in related problems for nonaxisymmetric shapes as studied in Ross and Smith [1993].

## 1. Preliminary Results

The steady rigid rotation of a homogeneous incompressible fluid drop which is surrounded by a rigidly rotating incompressible fluid is considered; see Ross and Smith [1993] for a discussion of relevant experiments of Plateau [1843] and Wang, Trinh, Croonquist and Elleman [1986] and others. An interfacial surface separates the drop from the surrounding fluid, and interfacial surface tension balances the pressure gradients due to the rotational acceleration. The rotating drop occupies a domain $\mathcal{D}_1 = \mathcal{D}_1(t)$ at time $t$ and the surrounding fluid occupies an outer domain $\mathcal{D}_2 = \mathcal{D}_2(t)$ enclosing $\mathcal{D}_1$. The fluids are incompressible homogeneous simple fluids of respective mass densities $\rho_1$ and $\rho_2$, and they undergo separate steady rigid rotations about a common axis with respective angular rotation speeds $\omega_1$ and $\omega_2$. (A **simple fluid** is a simple material (in the sense of Noll) that has maximal symmetry, with symmetry group consisting of the entire unimodular group; cf. Smith [1993].)

A steady rigid rotation about an axis takes initial points $X = (X_1, X_2, X_3)^T$ into points $x = (x_1, x_2, x_3)^T$ at time $t$ with

$$x \equiv x(X,t) = F(t)X, \qquad F(t) = \begin{pmatrix} \cos\omega t & \sin\omega t & 0 \\ -\sin\omega t & \cos\omega t & 0 \\ 0 & 0 & 1 \end{pmatrix}$$

where $\omega$ is the angular speed and Cartesian coordinates are used with the $x_3$-axis taken to coincide with the axis of rotation, and the origin is placed on the axis of rotation. For brevity the notations $x$ and $X$ are used for points and also for the corresponding vectors that translate the origin to the given points. A routine calculation based on the balance of linear momentum shows that any such steady rigid rotation with constant angular speed $\omega$ is dynamically possible for any homogeneous incompressible simple fluid of constant mass density $\rho$ in the absence of body force, with resulting Cauchy stress $T(x) = -p(x)I$ with pressure $p(x) = \frac{1}{2}\rho\omega^2 r^2 + p_0$ for a suitable constant $p_0$ and where $r$ is distance from the axis of rotation (cf. *Exercise 8.4.3* of Smith [1993]). For the rotating drop and the surrounding rotating fluid this result yields for the respective Cauchy stresses $T_1(x)$ and $T_2(x)$,

$$T_i(x) = -p_i(x)I \quad \text{with} \quad p_i(x) = \frac{1}{2}\rho_i\omega_i^2 r^2 + p_0^{(i)}, \quad \text{for} \quad i = 1, 2, \tag{1.1}$$

for suitable constants $p_0^{(1)}$ and $p_0^{(2)}$, where $r$ denotes distance from the common axis of rotation.

The two rotating fluids are separated by an interfacial surface membrane modelled by the smooth boundary surface $\partial\mathcal{D}_1$ along with the following linear constitutive relation between the membrane pressure jump and mean curvature

$$p_1(x) - p_2(x) = 2\sigma H(x) \tag{1.2}$$

for points $x$ on the boundary surface $\partial\mathcal{D}_1$, where $H(x)$ is the mean curvature of the surface at $x$, and $\sigma$ is a fixed positive constant which is the coefficient of interfacial surface tension. From Eqs. (1.1), (1.2) follows the free-boundary equation

$$2\sigma H(x) = p_0 + \frac{1}{2}\left(\rho_1\omega_1^2 - \rho_2\omega_2^2\right) r^2 \tag{1.3}$$

for points on the surface, with constant $p_0 := p_0^{(1)} - p_0^{(2)}$. Eq. (1.3) can also be obtained as the Euler/Lagrange equation for a suitable variational problem; cf. Ross and Smith [1993]. For brevity we often identify the drop and its domain, referring to the "drop $\mathcal{D}_1$".

A theorem of Wente [1982] shows that the drop has a plane of symmetry perpendicular to the rotational $x_3$-axis, while a result of Smith and Ross [1993] shows that the center of gravity of the drop is on the axis of rotation. We take the origin of coordinates to be placed on the rotational axis at the center of mass of the drop, and then Wente's theorem implies that the drop is symmetric across the $x_1x_2$-plane $x_3 = 0$.

Any suitable drop region $\mathcal{D} = \mathcal{D}_1$ with boundary satisfying the free-boundary Eq. (1.3) corresponds to a universal solution for the stated mathematical problem in the class of simple incompressible fluids. From a practical standpoint it seems necessary that the two fluids either be rotating with a common rotational speed $\omega_1 = \omega_2$ (with possibly different mass densities) or else at least one of the fluids should be inviscid—otherwise, unwanted dynamical viscous effects at the free-boundary will be expected to interfere physically with the possibility of steady rigid rotation.

The volume of the incompressible drop is taken to be fixed

$$\text{volume}(\mathcal{D}_1) = V_1 = \frac{4}{3}\pi r_1^3 \tag{1.4}$$

for a given positive constant $V_1$, where $r_1 > 0$ is the radius of a reference sphere with the same volume. It is also convenient to introduce a normalized angular speed $\hat{\omega}$ as

$$\hat{\omega} := \begin{cases} +\left[\left(\rho_1\omega_1^2 - \rho_2\omega_2^2\right)/8\sigma\right]^{1/2} r_1^{3/2} & \text{if } \rho_1\omega_1^2 \geq \rho_2\omega_2^2 \\ -\left[\left(\rho_2\omega_2^2 - \rho_1\omega_1^2\right)/8\sigma\right]^{1/2} r_1^{3/2} & \text{if } \rho_1\omega_1^2 < \rho_2\omega_2^2 \end{cases} \tag{1.5}$$

where $r_1$ is the radius of a reference sphere as in Eq. (1.4).

The total kinetic energy $T = (1/2)\int \rho|v|^2 dV$ for the rotating system is $T = T_1 + T_2 \equiv \frac{1}{2}\rho_1\omega_1^2\int_{\mathcal{D}_1} r^2 dV + \frac{1}{2}\rho_2\omega_2^2\int_{\mathcal{D}_2} r^2 dV = T_0 + T^*$ with

$$T_0 := \frac{1}{2}\rho_2\omega_2^2\int_{\mathcal{D}_1\cup\mathcal{D}_2} r^2 dV, \quad T^* := \frac{1}{2}\left(\rho_1\omega_1^2 - \rho_2\omega_2^2\right)\int_{\mathcal{D}_1} r^2 dV. \tag{1.6}$$

$T_0$ gives the kinetic energy that the system would have if the drop had the same density and rotational speed as the surrounding fluid (with $T_0$ independent of the interfacial free-boundary), while $T^*$ represents an incremental kinetic energy associated with the free-boundary. Following D. Ross [1968] we use the following nondimensionalized version $\hat{T}$ of $T^*$ as a measure of the kinetic energy for the system,

$$\hat{T} := \frac{|T^*|}{4\pi\sigma r_1^2}, \tag{1.7}$$

where $4\pi\sigma r_1^2$ is the potential energy of a reference sphere at rest with the radius $r_1$ as in Eq. (1.4), so that $\hat{T}$ gives a measure of the kinetic energy as a fraction of the potential energy that the drop would have at rest. For a corresponding measure of the angular momentum we use

$$\hat{L} := \frac{2\hat{T}}{\hat{\omega}} = \frac{2\hat{\omega}}{\pi r_1^5} \int_{\mathcal{D}_1} r^2 dV \tag{1.8}$$

where the normalized angular speed $\hat{\omega}$ is given by Eq. (1.5).

## 2. Axisymmetric Shapes

Consider now the special case of a smooth *axisymmetric* drop that occupies a time-independent domain $\mathcal{D} = \mathcal{D}_1$ which is rotationally symmetric about the axis of rotation. We introduce cylindrical coordinates $(r, \theta, z)$ with origin placed at the center of mass of the drop and with the cylindrical $z$-axis taken to coincide with the axis of rotation. Following Beer [1855, 1869] we seek axisymmetric drop shapes that are also symmetric across the equatorial plane $x_3 \equiv z = 0$. By Wente's result it suffices to describe the "bottom" half of the symmetric drop in the negative half-space $z \leq 0$. The surface of the bottom half-drop is generated for $0 \leq \theta \leq 2\pi$ by the trace of the surface in an $rz$-plane which can be represented by the graph

$$z = Z(r) \quad \text{for} \quad r_- \leq r \leq r_+ \quad (\text{with } r^2 = x_1^2 + x_2^2) \tag{2.1}$$

for suitable numbers $r_-$ and $r_+$ with $0 \leq r_- < r_+$ and for a nonpositive-valued function $Z \leq 0$ satisfying $Z(r_+) = 0$.

The mean curvature of the axisymmetric surface generated by Eq. (2.1) can be given as (cf. Beer [1869] or Gulliver [1984])

$$H = \frac{1}{2}\left(\frac{dv}{dr} + \frac{v}{r}\right) \quad \text{with} \quad v(r) = \frac{Z'(r)}{\sqrt{1 + (Z'(r))^2}}. \tag{2.2}$$

The expression for $H$ from Eq. (2.2) can be inserted into the left side of Eq. (1.3), and the resulting equation can be integrated to give

$$rv(r) = ar^4 + br^2 + c \tag{2.3}$$

with

$$a = \frac{\rho_1\omega_1^2 - \rho_2\omega_2^2}{8\sigma}, \quad b = \frac{p_0}{2\sigma}, \quad c = \text{integration constant.} \tag{2.4}$$

Inserting the expression for $v$ from Eq. (2.2) into Eq. (2.3) yields directly

**Theorem 2.1.** *(Beer [1869], Gulliver [1984]). The surface obtained by rotating Eq. (2.1) about the $z$-axis satisfies the free-boundary Eq. (1.3) if and only if $Z$ satisfies*

$$rZ'(r) = (ar^4 + br^2 + c)\sqrt{1 + (Z'(r))^2} \tag{2.5}$$

*with constants $a, b, c$ as in Eq. (2.4).*

Regarding the constant $c$ of integration, the case $c = 0$ leads to the spheroidal family studied by Beer [1855, 1869] for $a \geq 0$ and by Rosenthal [1962] for $a < 0$, while the case $c \neq 0$ leads to the toroidal family studied by D. Ross [1968] and Gulliver [1984] (see also Rayleigh [1914]).

Eq. (2.5) can be solved for $Z'$ and integrated to yield (with $Z(r_+) = 0$)

$$Z(r) \equiv Z(r; r_+, \mu, \lambda) = -r_+ \int_{r/r_+}^{1} \frac{\psi(t, \mu, \lambda)}{\sqrt{t^2 - \psi(t, \mu, \lambda)^2}} dt \quad \text{for} \quad 0 \leq r_- \leq r \leq r_+ \tag{2.6}$$

with

$$\psi(t, \mu, \lambda) \equiv \psi(t) := \mu \left(t^2 - 1\right)\left(t^2 - \lambda^2\right) + \frac{t^2 - \lambda}{1 - \lambda} \tag{2.7}$$

and with

$$0 \leq \lambda = \frac{r_-}{r_+} < 1, \tag{2.8}$$

where $a, b, c$ of Eq. (2.5) have been written in terms of new parameters $r_+, \mu, \lambda$ as

$$a = \frac{\mu}{r_+^3}, \quad b = \frac{1 - \mu}{r_+} + \frac{\lambda}{r_+}\left[\frac{1}{1 - \lambda} - \mu\lambda\right], \quad \text{and} \quad c = -\lambda r_+ \left[\frac{1}{1 - \lambda} - \mu\lambda\right]. \tag{2.9}$$

The integral in Eq. (2.6) can be expressed in terms of canonical hyperelliptic integrals, but Eq. (2.6) is convenient for our purposes. The parameter $a$ can be eliminated between the relevant equations of Eq. (2.4) and Eq. (2.9) to give

$$\mu = \begin{cases} +\left(r_+/r_1\right)^3 \hat{\omega}^2 & \text{if } \rho_1\omega_1^2 \geq \rho_2\omega_2^2 \\ -\left(r_+/r_1\right)^3 \hat{\omega}^2 & \text{if } \rho_1\omega_1^2 < \rho_2\omega_2^2. \end{cases} \tag{2.10}$$

Negative values of $\mu < 0$ correspond to negative values of the angular speed $\hat{\omega} < 0$ as might occur in the case $\omega_1 = \omega_2$ if the outer fluid is denser than the drop fluid with $\rho_2 > \rho_1$.

The quantity $\lambda = r_-/r_+$ serves as the primary parameter in the toroidal case with $0 < \lambda < 1$ and $c \neq 0$. In this case $Z(r_-) = Z(r_+) = 0$ with $Z'(r_\pm) = \infty$ for $0 < r_- < r_+$, and the drop does not intersect the $z$-axis of rotation. The spheroidal case occurs when $\lambda = 0$, $r_- = 0$ and $c = 0$, and in this case the drop intersects the $z$-axis with $Z'(0) = 0$ and $Z(r_+) = 0$ for $r_+ > 0$. The quantity $\mu$ serves as the primary parameter in the spheroidal case.

A routine argument shows that the integral giving $Z(r; r_+, \mu, \lambda)$ in Eqn. (2.6) exists for $\lambda \leq r/r_+ \leq 1$ for every fixed $0 \leq \lambda < 1$ and for every $\mu$ satisfying $-\frac{1}{2}(1-\lambda)^{-2} < \mu < 4/[(1-\lambda^2)(1+\lambda)]$. In the spheroidal case ($\lambda = 0, r_- = 0$) the function $Z(0; r_+, \mu, 0)$ is differentiable with respect to $\mu$ with negative $\mu$-derivative for $-1/2 < \mu < 4$, and $Z(0; r_+, \mu, 0) \to -\infty$ as $\mu \to 4$. Hence the equation $Z(0; r_+, \mu, 0)/r_+ = 0$ has precisely one root $\mu^*$ on the interval $-1/2 < \mu < 4$, and the approximate value of this root can be easily obtained with Newton's method as

$$\mu^* \doteq 2.32911. \tag{2.11}$$

In this case $Z(0) < 0$ for $-1/2 < \mu < \mu^*$.

Similarly, in the toroidal case, the equation $Z(\lambda r_+; r_+, \mu, \lambda)/r_+ = 0$ uniquely determines $\mu = \mu^{**}(\lambda)$ in terms of $\lambda$,

$$\mu = \mu^{**}(\lambda) \quad \text{for} \quad 0 \le \lambda < 1. \tag{2.12}$$

The value of the function $\mu^{**}(\lambda)$ coincides at $\lambda = 0$ with the earlier value of (2.11),

$$\mu^{**}(0) = \mu^* \doteq 2.32911 \tag{2.13}$$

in the case $\lambda = 0$.

**3. The Bifurcation**

Plateau [1843] devised and performed a version of a rotating drop experiment in which a viscous liquid drop is driven by an enclosed rotating spindle while being immiscibly levitated in another surrounding liquid with approximately the same density but with a much smaller viscosity. The surrounding liquid supports the rotating drop so as to cancel most of the effects of the earth's gravity. In his experiments Plateau observed axisymmetric shapes symmetric about the rotational axis of the drop, beginning with a sphere when the rotation speed is zero. For small positive rotational speeds Plateau's axisymmetric shapes were oblate spheroids that bulge near the equator and flatten near the poles. The shapes became flatter with increasing rotational speed until the polar thickness contracted and the shapes became biconcave with depressions at the poles, similar to the shapes obtained from Eq. (2.6) in the spheroidal case $\lambda = 0$, $r_- = 0$ for $0 < \mu < \mu^*$, $\hat{\omega} \ge 0$.

Plateau also observed *bubbles* similar to those obtained from Eq. (2.6) in the spheroidal case for $-1/2 < \mu < 0$, $\hat{\omega} < 0$ and he observed other shapes as well, including axisymmetric toroidal profiles that did not meet the axis of rotation, similar to the shapes obtained from Eq. (2.6) in the toroidal case $0 < \lambda < 1$. The latter toroidal shapes were obtained by Plateau by driving the rotating spindle harder with a larger angular speed and momentum. The nature of the presumed bifurcation from the axisymmetric spheroidal to toroidal shapes has been the subject of some uncertainty with different authors asserting that a bifurcation occurs either with an increase in angular momentum, an increase in angular speed, or an increase

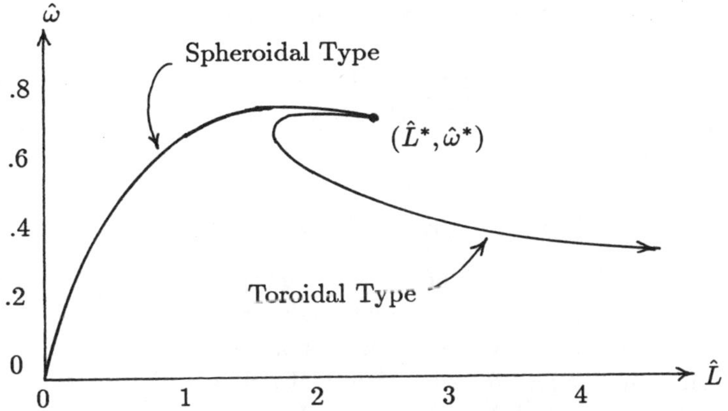


Figure 1

**Angular Speed vs Angular Momentum**

in the parameter $\mu$ of Eq. (2.10). Rayleigh [1914] states that the toroidal shapes should arise at *"higher rotations"*, while Swiatecki [1974, p. 62] asserts that if a (liquid) *"globe with surface tension is rotated, it flattens at first into an oblate pseudospheroid which, with increasing angular momentum, eventually goes over into a torus."* Chandrasekhar [1965] remarks on *"the emergence of a sequence of toroidal figures"* for parameter values $\mu > 2.3291$. These statements are generally open to ambiguous interpretation and indeed Tagg, Cammack, Croonquist and Wang [1980] note simply that it *"is less certain where the toroidal shapes arise"*. We find that the bifurcation from spheroidal to toroidal shapes, which occurs at $\hat{\omega} = \hat{\omega}^* = 1/\sqrt{2}$ and $\hat{L} = \hat{L}^* \doteq 2.8506$, leads initially to a *decrease* in angular momentum, a decrease in the parameter $\mu$, and an increase in angular speed. After the bifurcation as one proceeds further along the toroidal family, the angular momentum and the parameter $\mu$ eventually increase without bound while the angular speed eventually decreases toward zero.

The toroidal and spheroidal families connect at the bifurcation point with a common slope in the $(\hat{L}, \hat{\omega})$-plane as indicated in Figure 1. The resulting bifurcation slope can be computed in terms of the functions

$$h(\mu) := \int_0^1 \frac{\phi(t,\mu)}{\sqrt{1-\phi(t,\mu)^2}}dt, \quad J(\mu) := 2\int_0^1 t\sqrt{1-\phi(t,\mu)^2}dt \qquad (3.1)$$

with $\phi(t,\mu) := (1-\mu+\mu t^2)t = \psi(t,\mu,0)/t$. Then we have the following theorem which is proved in the *Appendix*.

**Theorem 3.1**. *The spheroidal and toroidal axisymmetric shapes connect tangen-*

*tially in the $(\hat{L},\hat{\omega})$-plane at the bifurcation point $(\hat{L}^*,\hat{\omega}^*)$ with common slope*

$$\left.\frac{d\hat{\omega}}{d\hat{L}}\right|_{(\hat{L}^*,\hat{\omega}^*)} = \frac{27(\mu^*/4)^{4/3}(\mu^*-1)h'(\mu^*)}{6(2\mu^*+1)+\mu^*(\mu^*-1)(-h'(\mu^*))\left[\mu^*-1+21\mu^* J(\mu^*)\right]} \doteq -0.070757 \tag{3.2}$$

*where $\mu^*$ is the root of $Z(0;r_+,\mu,0)/r_+$ of Eq. (2.11).*

The axisymmetric family which begins with the sphere at $(\hat{L},\hat{\omega})=(0,0)$ bifurcates at $(\hat{L},\hat{\omega})=(\hat{L}^*,\hat{\omega}^*)$ from spheroidal to toroidal shapes as indicated in Figure 1 where angular speed is plotted versus angular momentum for the toroidal family superimposed on a similar plot for the spheroidal family with $\hat{L}>0,\hat{\omega}>0$. The bifurcation results in an initial decrease in angular momentum and a corresponding increase in angular speed. The problem of the *driven drop* (specified angular speed) has multiple distinct axisymmetric solutions for angular speeds $\hat{\omega}$ in a certan range, and likewise the *free drop* problem (specified angular momentum) has multiple axisymmetric solutions for angular momentum $\hat{L}$ in a suitable range. For large angular momentum the free drop problem has only one axisymmetric solution and it is of toroidal type. Nothing is implied here regarding the stability or instability of these various solutions.

The case in which the surrounding fluid is *compressible* is discussed in a forthcoming work.

Acknowledgments

We thank Roland Bulirsch for making available to us a copy of the reference Beer [1869], Robert Gulliver for useful remarks, and we particularly thank William A. Harris, Jr. for encouragement and early discussions of techniques.

## Appendix

**Proof of Theorem 3.1.** The angular speeds and momentums are obtained from Eqs. (1.5), (1.8), (2.6), (2.7), (2.10), (2.13) in the toroidal and spheroidal cases as

$$\text{toroidal case:} \quad \begin{cases} \hat{\omega} = f(\mu,\lambda)\big|_{\mu=\mu^{**}(\lambda)} \\ \hat{L} = g(\mu,\lambda)\big|_{\mu=\mu^{**}(\lambda)} \quad \text{for} \quad 0 \le \lambda < 1, \end{cases} \tag{A.1}$$

and

$$\text{spheroidal case:} \quad \begin{cases} \hat{\omega} = f(\mu,0) \\ \hat{L} = g(\mu,0) \quad \text{for} \quad 0 \le \mu \le \mu^*, \end{cases} \tag{A.2}$$

where

$$f(\mu,\lambda) := \left[\frac{3}{2}\mu I_2(\mu,\lambda)\right]^{\frac{1}{2}}, \quad g(\mu,\lambda) := 2\left(\frac{2}{3}\right)^{\frac{7}{6}} \mu^{\frac{1}{2}} I_2(\mu,\lambda)^{-\frac{7}{6}} I_4(\mu,\lambda) \tag{A.3}$$

with $I_k(\mu,\lambda) := \int_\lambda^1 (t^k - \lambda^k)\psi(t,\mu,\lambda)/\sqrt{t^2 - \psi(t,\mu,\lambda)^2}dt$ for $k = 2, 4$ and for suitable values of $0 < \mu \leq \mu^*, 0 \leq \lambda < 1$, with $\mu^{**}(\lambda)$ the function of Eq. (2.12). The slope in the $(\hat{L},\hat{\omega})$-plane along the toroidal family is obtained from Eq. (A.1) as $d\hat{\omega}/d\hat{L} = [d\hat{\omega}/d\lambda]/[d\hat{L}/d\lambda]$ which gives for the toroidal bifurcation slope

$$\left.\frac{d\hat{\omega}}{d\hat{L}}\right|_{\text{tor.}} = \lim_{\lambda\to 0+}\left[\left.\frac{f_\lambda(\mu,\lambda) + f_\mu(\mu,\lambda)(d\mu^{**}/d\lambda)}{g_\lambda(\mu,\lambda) + g_\mu(\mu,\lambda)(d\mu^{**}/d\lambda)}\right|_{\mu=\mu^{**}(\lambda)}\right], \tag{A.4}$$

and similarly for the spheroidal family from Eq. (A.2)

$$\left.\frac{d\hat{\omega}}{d\hat{L}}\right|_{\text{sph.}} = \lim_{\mu\to\mu^*}\frac{f_\mu(\mu,0)}{g_\mu(\mu,0)} = \frac{f_\mu(\mu^*,0)}{g_\mu(\mu^*,0)} \tag{A.5}$$

where the subscripts on $f, g$ denote partial derivatives.

Implicit differentiation of the equation defining $\mu^{**}(\lambda)$ shows that the function $\mu^{**}$ of Eq. (2.12) is of class $C^1$ for $0 < \lambda < 1$, with first derivative satisfying

$$\lim_{\lambda\to 0+}\frac{d\mu^{**}(\lambda)}{d\lambda} = -\infty. \tag{A.6}$$

The first equality of Eq. (3.2) follows directly from (A.4), (A.5) and (A.6), while the numerical value −0.070757... follows by numerical evaluations of the relevant integrals for $h'(\mu^*)$ and $J(\mu^*)$. ■

## REFERENCES

Beer, A. 1855 Pogg. Ann., **96**: 210.

Beer, A. 1869 *Einleitung in der mathematische Theorie der Elasticität und Capillarität*, part 2, A. Gissen Verlag, Leipzig.

Chandrasekhar, S. 1965 The stability of a rotating liquid drop, *Proc. R. Soc. London* A **286**: 1-26.

Gulliver, R. 1984 Tori of prescribed mean curvature and the rotating drop, *Soc. Math. de France Astérique*, **118**: 167–179.

Plateau, J. A. F. 1843 Mémoire sur les phénomènes que présente une masse liquide libre et soustraite à l'action de la pesanteur, *Mémoires de l'Acad. Bruxelles* **16**: 1-35. English translation 'Experimental and theoretical researches on the figures of equilibrium of a liquid mass withdrawn from the action of gravity', *The Annual Report of the Board of Regents of the Smithsonian Institute*, 1863 (pp. 207-225), Government Printing Office, Washington D.C.

Rayleigh, Lord 1914 The equilibrium of revolving liquid under capillary force, *Phil. Mag.*, Series 6, **28**: 161-170.

Rosenthal, D. K. 1962 The shape and stability of a bubble at the axis of a rotating liquid, *J. Fluid Mech.*, **12**: 358-366.

Ross, D. K. 1968 The shape and energy of a revolving liquid mass held together by surface tension, *Aust. J. Phys.*, **21**: 823-835.

Ross, J. E. and Smith, D. R. 1993 On an ODE model for narrow-necked rotating liquid drops, *Advances in Applied Math.*, to appear.

Smith, D. R. 1993 *An Introduction to Continuum Mechanics*, Kluwer Academic Publishers, The Netherlands.

Smith. D. R. and Ross, J. E. 1993 Universal shapes and bifurcation for rotating incompressible fluid drops, *Methods and Applications of Analysis*, to appear.

Swiatecki, W. J. 1974 The rotating, charged or gravitating liquid drop and problems in nuclear physics and astronomy, pp. 52-78 of *Proceedings of the International Colloquium on Drops and Bubbles* (ed. D. J. Collins, M. S. Plesset & M. M. Saffren), Jet Propulsion Laboratory, Pasadena, California.

Tagg, R., Cammack, L., Croonquist, A. and Wang, T. G. 1980 *Rotating Liquid Drops: Plateau's Experiment Revisited*, National Aeronautics and Space Administration, JPL Publication 80-66 (Jet Propulsiom Laboratory, California Institute of Technology, Pasadena, CA).

Wang, T. G., Trinh, E. H., Croonquist A. P., and Elleman, D. D. 1986 Shapes of rotating free drops: Spacelab experimental results, *Phy. Rev. Letters* **56**, 452–455.

Wente, H. C. 1982 The symmetry of rotating fluid bodies, *Manuscripta Mathematica* **39**, 287–296.

# A GENERALIZATION OF THE AIRY INTEGRAL FOR $f'' - z^n f = 0$

GARY G. GUNDERSEN
*Department of Mathematics, University of New Orleans, New Orleans, Louisiana 70148 USA*

and

ENID M. STEINBART
*Department of Mathematics, University of New Orleans, New Orleans, Louisiana 70148 USA*

It is well known that the Airy integral is a solution of the Airy differential equation $f'' - zf = 0$ and that the Airy integral is a contour integral function with special properties. We show that there exist analogous special contour integral solutions of the more general equation $f'' - z^n f = 0$ where $n$ is any positive integer. Details and related results are provided in [GS].

[GS] G. Gundersen and E. Steinbart, *A Generalization of the Airy Integral for $f'' - z^n f = 0$*, Trans. Amer. Math. Soc. **337** (1993), 737–755.

# Stokes Multipliers for a Generalized Weber Equation

T. J. TABARA

*Department of Mathematics, Golden Gate University, San Francisco, CA 94105-2968*

## 0. Introduction

In the paper of M. Faierman ([1]), he considers the differential equation of the following type

$$y'' - (\alpha^{-2}z^{\alpha} - (\nu + \alpha^{-1}))y = 0, \tag{E}$$

which arises in the study of non self-adjoint elliptic boundary value problems involving weight function, where $\nu$ is a complex parameter and $\alpha$ is a real parameter with $\alpha \geq 2$. As we know, (E) is of the Weber type when $\alpha = 2$,

$$y'' - \left( \frac{z^2}{4} - (\nu + \frac{1}{2})\right)y = 0. \tag{0.1}$$

Hence he calls (E) the generalized Weber equation. Let $D_{\nu}(z)$ be the particular solution for (0.1) that is an entire function of $(z,\nu)$. His objective is to find a solution for (E), defined as $D_{\nu,\alpha}(z)$, a generalized parabolic cylinder function, and to study some of its basic properties similar to those of $D_{\nu}(z)$ that have important applications to physical problems.

This note is one of the two papers studying (E) along with his definition, where we shall investigate $D_{\nu,\alpha}(z)$, when $\alpha$ is a rational number, from a different point of view using results of the so - called Sibuya-Lin-Tabara's equation and give similar results to his in terms of asymptotic expansions. The section 1 will explain necessary information on the Sibuya-Lin-Tabara's equation in order to discuss the equation (E). In the section 2, we shall derive results proved by Faierman differently. In the section 3, we shall study properties of Stokes multipliers. The second paper will discuss somewhere else the case with an irrational number $\alpha$ and asymptotic behavior of Stokes multipliers as $\nu \to \infty$.

## 1. The Equation of Sibuya-Lin-Tabara type

In a series of papers ([2], [3], etc), Y. Sibuya, C. H. Lin and T. Tabara considered the existence of the entire solutions for the differential equation

$$(\delta^2 - p\delta)y - (x^m + a_1x^{m-1} + \cdots + a_{m-1}x)y = 0, \tag{1.1}$$

where $\delta = x\frac{d}{dx}$, $m$ and $p$ are fixed integers with $m \geq 2$ and $1 \leq p \leq m-1$ and $a_1, \ldots, a_{m-1}$ are complex parameters under the assumption that $x = 0$ is an apparent singular point.

In order to satisfy this condition, $p$ must be chosen with $1 \le p \le m-1$, and $a_{m-p}$ must be expressed in terms of other parameters for each pair of $(m, p)$ such as

$$a_{m-1} = 0 \ (p = 1),$$
$$a_{m-2} = a_{m-1}^2 \ (p = 2),$$
$$a_{m-3} = a_{m-1}\, a_{m-2} - \tfrac{1}{4} a_{m-1}^3 \ (p = 3), \ etc.$$

In general,

$$a_{m-p} = -\sum_{j=1}^{p-1} a_{m-(p-j)} l_j \,, \tag{SC}$$

where $j(j-p)l_j = a_{m-j} + \sum_{k=1}^{j-1} a_{m-(j-k)} l_k$, $1 \le j \le p-1$. This condition will give us two linearly independent solutions that are entire in both $x$ and the parameters. Thus from now on, we shall set $a = (a_1, \dots, a_{m-(p+1)}, a_{m-(p-1)}, \dots, a_{m-1}) \in C^{m-2}$ due to (SC). We constructed a solution of (1.1) possessing a certain asymptotic expansion about the irregular singular point $x = \infty$ under (SC). Here when $p = 1$, (1.1) is of the Sibuya's type ([3]). The following existence theorem of the particular solution for this equation can be derived by utilizing the fundamental lemma of Hsieh-Sibuya ([3]).

**THEOREM 1.1.** *The differential equation* (1.1) *under* (SC) *has a solution*

$$y = \mathcal{Y}_m(x,a) \tag{1.2}$$

*such that*

(i) $\mathcal{Y}_m(x,a)$ *is an entire function of* $(x,a)$

(ii) $\mathcal{Y}_m(x,a)$ *admits an asymptotic representation*

$$\mathcal{Y}_m(x,a) = x^{r_m(a)} [\,1 + O(x^{-\frac{1}{2}})\,] \exp(-E_m(x,a)) \tag{1.3}$$

*uniformly on each compact set in the a-space as* $x \to \infty$ *in any closed subsector of the open sector*

$$|\arg x| < \frac{3\pi}{m}, \tag{1.4}$$

*where* $r_m(a)$ *is a polynomial in* $a$. □

If we define $b_h(a)$ by

$$\left[1 + \sum_{h=1}^{m-1} a_h x^{-h} + \frac{1}{4}(p^2 - 1)x^{-m}\right]^{\frac{1}{2}} = 1 + \sum_{h=1}^{\infty} b_h(a)\, x^{-h} \tag{1.5}$$

with $|x|$ sufficiently large, $b_h(a)$ is well-defined such as

$$b_1(a) = \frac{1}{2}a_1,$$
$$b_2(a) = \frac{1}{8}(4a_2 - a_1^2),$$
$$b_3(a) = \frac{1}{16}(8a_3 - 4a_1a_2 + a_1^3),$$
$$b_4(a) = \frac{1}{128}(-5a_1^4 + 24a_1^2a_2 - 16a_2^2 - 32a_1a_3 + 64a_4),$$
$$\cdots\cdots \qquad \cdots\cdots$$

Then $r_m(a)$ and $E_m(x,a)$ are determined by

$$r_m(a) = -\frac{1}{4}(m - 2p) - b(a), \text{ where } b(a) = \begin{cases} 0 & \text{if } m \text{ is odd} \\ b_{\frac{1}{2}m}(a) & \text{if } m \text{ is even} \end{cases} \tag{1.6}$$

and

$$E_m(x,a) = \frac{2}{m}x^{\frac{1}{2}m} + \sum_{1 \le h < m/2} \frac{2}{m - 2h} b_h(a) x^{\frac{1}{2}(m-2h)}. \tag{1.7}$$

The uniqueness of this particular solution for (1.1) is also shown as follows ([4]):

**THEOREM 1.2.** *The differential equation* (1.1) *under* (SC) *has one and only one solution* $y = \mathcal{Y}_m(x,a)$ *that satisfies all of the conditions given in Theorem* 1.1. □

## 2. The Existence and Uniqueness of a generalized parabolic cylinder solution of (E)

If we set $\alpha = \frac{q}{p}$ with $q \ge 2p$ and $z = \left(\frac{\alpha}{p}\right)^{\frac{2}{2+\alpha}} x^p$ such that $p$ and $q$ are positive integers, not necessarily relative prime, and then (E) becomes

$$(\delta^2 - p\delta)y - \left(x^{2p+q} - \left(\frac{\alpha}{p}\right)^{\frac{4}{2+\alpha}} p^2(\nu + \alpha^{-1})x^{2p}\right)y = 0. \tag{2.1}$$

This is of the form (1.1), and we write

$$(\delta^2 - p\delta)y - (x^{2p+q} + a_q x^{2p})y = 0, \tag{2.2}$$

where $a_q = a_q(\nu) = -\left(\frac{\alpha}{p}\right)^{\frac{4}{2+\alpha}} p^2(\nu + \frac{1}{\alpha})$. Thus (SC) condition and $1 \le p < 2p + q$ are always satisfied. Hence (E) can be analyzed in terms of (1.1) completely when $\alpha$ is a rational number with $\alpha \geq 2$. When Theorems 1.1 and 1.2 are applied to the equation (2.2), we shall have the following results for the existence and uniqueness theorems:

**THEOREM 2.1.** *The differential equation* (2.2) *has a solution*

$$y = \mathcal{Y}(x, a_q) = \mathcal{Y}(x,\nu) \tag{2.3}$$

*such that*

(i) $\mathcal{Y}(x,\nu)$ *is an entire function of* $(x,\nu)$

(ii) $\mathcal{Y}(x,\nu)$ *admits an asymptotic representation*

$$\mathcal{Y}(x,\nu) = x^{r(\nu)}\,[\,1 + O(x^{-\frac{1}{2}})\,]\exp(-E(x,\nu)) \tag{2.4}$$

*uniformly for* $\nu$ *lying in each compact set in* $C$ *as* $x \to \infty$ *in any closed subsector of the open sector*

$$|\arg x| < \frac{3\pi}{2p+q}, \quad i.e. \quad |\arg z| < \frac{3p\pi}{2p+q} = \frac{3\pi}{2+\alpha}. \tag{2.5}$$

□

From (1.5) we set $b_h(\nu)$ by

$$\left[1 + a_q x^{-q} + \frac{1}{4}(p^2-1)x^{-(2p+q)}\right]^{\frac{1}{2}} = 1 + \sum_{h=1}^{\infty} b_h(\nu)\,x^{-h}$$

$$= 1 + \frac{1}{2}a_q x^{-q} - \frac{1}{8}a_q{}^2 x^{-2q} + \frac{1}{8}(p^2-1)x^{-(2p+q)}$$

$$+ \frac{3}{48}a_q{}^3 x^{-3q} - \frac{1}{16}(p^2-1)a_q x^{-2(p+q)} - \cdots. \tag{2.6}$$

We see that when $\alpha \geq 2$, $q \geq \frac{1}{2}(2p+q)$, so there is no term of $x^{-\frac{1}{2}(2p+q)}$ appeared in (2.6). If $1 \leq h < q$, then $b_h(\nu) \equiv 0$. Hence $r(\nu)$ and $E(x,\nu)$ are given, independent of $\nu$, by

$$r(\nu) = -\frac{1}{4}q \tag{2.7}$$

and

$$E(x,\nu) = \frac{2}{2p+q}\,x^{\frac{1}{2}(2p+q)}. \tag{2.8}$$

If we combine this observation with Theorem 1.2, we can show exactly what Theorem 2.1 of Faierman's claim is in a different form:

**Theorem 2.2.** *There exists one and only one solution of* (2.2) *such that*

$$x^{\frac{1}{4}q}\exp\Big[\frac{2}{2p+q}x^{\frac{1}{2}(2p+q)}\Big]\mathcal{Y}(x,\nu) \to 1, \tag{2.9}$$

*uniformly for* $\nu$ *lying in each compact set in* $C$ *as* $x \to \infty$ *in any closed subsector of the open sector*

$$|\arg x| < \frac{3\pi}{2p+q}, \quad i.e. \quad |\arg z| < \frac{3p\pi}{2p+q} = \frac{3\pi}{2+\alpha}. \tag{2.10}$$

*Moreover, by the uniqueness theorem* 1.2, *we can write* $\mathcal{Y}(x,\nu)$ *for* $D_{\nu,\alpha}(x)$ *and the asymptotic expression will be expressed by* (2.4). □

**Lemma 2.3.** *The derivative* $\mathcal{Y}'(x,\nu)$ *of* $\mathcal{Y}(x,\nu)$ *with respect to* $x$ *admits an asymptotic representation*

$$x^{\frac{1}{4}q-\frac{1}{2}(2p+q-2)}\exp\Big[\frac{2}{2p+q}x^{\frac{1}{2}(2p+q)}\Big]\mathcal{Y}'(x,\nu) \to -1, \tag{2.11}$$

*uniformly for* $\nu$ *lying in each compact set in* $C$ *as* $x \to \infty$ *in any closed subsector of the open sector given in* (2.10). □

Next we set for each integer $k$,

$$f_k(x,\nu) = \mathcal{Y}(\omega^{-k}x, G^{-k}(\nu)), \tag{2.12}$$

where $\omega = \exp\left(\frac{2\pi}{2p+q}i\right)$ and $G^{-k}(\nu) = G^{-k}(a_q(\nu)) = \omega^{-kq}a_q(\nu)$. Then we obtain the following solutions for (E):

**Lemma 2.4.** *For each integer* $k$, $f_k(x,\nu)$ *is a solution of the differential equation* (E) *and*

$$f_k(x,\nu) = \omega^{\frac{k}{4}q}x^{-\frac{1}{4}q}[1 + O(x^{-\frac{1}{2}})]\exp\left[(-1)^{k+1}\frac{2}{2p+q}x^{\frac{1}{2}(2p+q)}\right] \tag{2.13}$$

*uniformly for* $\nu$ *lying in each compact set in* $C$ *as* $x \to \infty$ *in any closed subsector of the open sector*

$$\left|\arg x - \frac{2k\pi}{2p+q}\right| < \frac{3\pi}{2p+q}. \tag{2.14}$$

□

Moreover set $S_k = \{x \in C : \left|\arg x - \frac{2k\pi}{2p+q}\right| < \frac{\pi}{2p+q}\}$ for $k \in \mathcal{M} = \{0, 1, \cdots, (2p+q)-1\}$, then due to the coefficient of $E(x,\nu)$ the solution $f_k(x,\nu)$ tends to 0 as $x \to \infty$

along any direction in the sector $S_k$ for which we call $f_k(x,\nu)$ a *subdominant* (*recessive*) *solution* for (2.2) and it tends to the infinity as $x \to \infty$ along any direction in the sectors $S_{k-1}$ and $S_{k+1}$, called a *dominant solution* for (2.2). The sector given in (2.14) is $S_{k-1} \cup \overline{S}_k \cup S_{k+1}$. In order to study properties of Stokes multipliers, we need the local solutions of (E) at $x = 0$ which can be derived by the Frobenius method:

**LEMMA 2.5.** *There exist two linearly independent solutions of* (E) :

$$\begin{cases} \varphi_1(x,\nu) = 1 - \frac{1}{2}\left(\frac{\alpha}{p}\right)^{\frac{4}{2+\alpha}} p^2(\nu + \frac{1}{\alpha})x^{2p} + \sum_{N=2}^{\infty} \varphi_{1,2pN}(\nu)\, x^{2pN} \\ \varphi_2(x,\nu) = \qquad x^p + \sum_{N=1}^{\infty} \varphi_{2,(2N+1)p}(\nu) x^{(2N+1)p}, \end{cases} \tag{2.15}$$

*which are entire in* $(x,\nu)$. □

## 3. Properties of Stokes Multipliers

As you see that $f_{k+1}(x,\nu)$ and $f_{k+2}(x,\nu)$ are linearly independent, $f_k(x,\nu)$ can be expressed as

$$f_k(x,\nu) = C_k(\nu)\, f_{k+1}(x,\nu) + \tilde{C}_k(\nu)\, f_{k+2}(x,\nu) \quad \text{for all } k \in \mathcal{M}, \tag{3.1}$$

where the coefficients $C_k(\nu)$ and $\tilde{C}_k(\nu)$ are called the *Stokes multipliers* for $f_k(x,\nu)$ with respect to $f_{k+1}(x,a)$ and $f_{k+2}(x,\nu)$. The relation (3.1) is called a *lateral connection formula* for $f_k(x,a)$. If we set for all $j,\, k \in \mathcal{M}$

$$\Phi_{j\,k}(x,\nu) = \begin{bmatrix} f_j(x,\nu) & f_k(x,\nu) \\ \delta f_j(x,\nu) & \delta f_k(x,\nu) \end{bmatrix} \text{ and } W_{j\,k}(x,\nu) = det\,[\Phi_{j\,k}(x,\nu)\,], \tag{3.2}$$

then the matrix $\Phi_{k\,k+1}(x,\nu)$ is a fundamental matrix of the equation (2.2) in the sector $\mathcal{D}_k = \{ x \in C : S_k \cup S_{k+1} \}$. The sectors $\mathcal{D}_k$ cover $C^* = \{ x : 0 < |x| < +\infty \}$.

**LEMMA 3.1.** *For every integer* $k \in \mathcal{M}$,

$$W_{k\,k+1}(x,\nu) = 2\omega^{-kp}\omega^{\frac{1}{4}q} x^p$$

$$= \frac{px^p}{p!} \begin{vmatrix} f_k(0,\nu) & f_{k+1}(0,\nu) \\ f_k^{(p)}(0,\nu) & f_{k+1}^{(p)}(0,\nu) \end{vmatrix} . \tag{3.3}$$

□

From the above, we may set $W_{j\,k}(x,\nu) = W_{j\,k}(\nu)x^p$ .

**LEMMA 3.2** *The Stokes multipliers $C_k(\nu)$ and $\tilde{C}_k(\nu)$ are expressed in terms of $W_{j\,k}(x,\nu)$:*

$$W_{k\,k+2}(x,\nu) = C_k(\nu)\, W_{k+1\,k+2}(x,\nu),$$

$$W_{k\,k+1}(x,\nu) = -\,\tilde{C}_k(\nu)\, W_{k+1\,k+2}(x,\nu),$$

*and*

$$C_k(\nu) = C_0(G^{-k}(\nu)) \text{ and } \tilde{C}_k(\nu) = \tilde{C}_0(G^{-k}(\nu)) , \tag{3.4}$$

$$\begin{cases} C_k(\nu) = \frac{1}{2}\omega^{-\frac{1}{4}q}\omega^p W_{k\,k+2}(\nu) \\ \tilde{C}_k(\nu) = -\,\omega^p. \end{cases} \tag{3.5}$$

□

**LEMMA 3.4.** *If we set for all $k \in \mathcal{M}$*

$$S_k(\nu) = \begin{bmatrix} C_k(\nu) & 1 \\ \tilde{C}_k(\nu) & 0 \end{bmatrix}, \tag{3.6}$$

*then*

$$S_{(2p+q)-1}(\nu)S_{(2p+q)-2}(\nu) \cdots S_1(\nu)S_0(\nu) = \begin{bmatrix} 1 & 0 \\ 0 & 1 \end{bmatrix}. \tag{3.7}$$

□

**LEMMA 3.5.** *For each fixed $\nu \in \mathcal{C}$, there exists an integer $k_0 \in \mathcal{M}$ such that*

$$C_{k_0}(\nu) \neq 0. \tag{3.8}$$

□

As an immediate corollary to the above, we have :

**LEMMA 3.6.** *For all $\nu \in C$ the set $\mathcal{M}$ contains three integers $j$, $k$ and $l$ such that $f_j(x,\nu)$, $f_k(x,\nu)$ and $f_l(x,\nu)$ are pairwise linearly independent where $j$, $k$ and $l$ depend on $\nu$.* □

The solution $\mathcal{Y}(x,\nu)$ can also be expressed as a linear combination of $\varphi_1(x,\nu)$ and $\varphi_2(x,\nu)$ given by Lemma 2.5,

$$\mathcal{Y}(x,\nu) = \alpha(\nu)\varphi_1(x,\nu) + \beta(\nu)\varphi_2(x,\nu), \tag{3.9}$$

where $\alpha(\nu)$ and $\beta(\nu)$ are constants depending on the parameter $\nu$. This is called the *central connection formula* for $\mathcal{Y}(x,\nu)$. Like the lateral connection formula, (3.9) will give valuable information for the study of Stokes Multipliers.

By (2.15), we get for all $k \in \mathcal{M}$, as $x \to 0$,

$$\begin{aligned} f_k(x,\nu) &= \alpha(G^{-k}(\nu))[\,1 + \tfrac{1}{2}a_q(\nu)x^{2p}\,] + \omega^{-kp}\beta(G^{-k}(\nu))x^p + O(x^{3p}), \\ f_k{}'(x,\nu) &= \alpha(G^{-k}(\nu))[\,pa_q(\nu)\,x^{2p-1}\,] + p\omega^{-kp}\beta(G^{-k}(\nu))x^{p-1} + O(x^{3p-1}). \end{aligned} \tag{3.10}$$

We note that we have used the following identities

$$\varphi_{12p}(G^{-k}(\nu))(\omega^{-k}x)^{2p} = \varphi_{12p}(\nu)x^{2p}. \tag{3.11}$$

Since $f_k(0,\nu) = \alpha(G^{-k}(\nu))$ and $f_k^{(p)}(0,\nu) = p!\ \omega^{-kp}\beta(G^{-k}(\nu))$, by (3.9) and (3.10) the central connection formula is also expressed by

$$f_k(x,\nu) = f_k(0,\nu)\varphi_1(x,\nu) + \frac{f_k^{(p)}(0,\nu)}{p!}\varphi_2(x,\nu) \quad \text{for all } k \in \mathcal{M}.$$

Moreover, $W_{0\,1}(x,\nu)$ is given by

$$W_{0\,1}(x,\nu) = \begin{vmatrix} 1 + \frac{1}{2}a_q(\nu)x^{2p} & x^{2p} \\ \frac{1}{2}a_q(\nu)x^{2p} & px^{2p} \end{vmatrix} \begin{vmatrix} \alpha(\nu) & \alpha(G^{-1}(\nu)) \\ \beta(\nu) & \omega^{-p}\beta(G^{-1}(\nu)) \end{vmatrix} + O(x^{2p}), \text{ as } x \to 0.$$

Evaluating this at $x = 0$, we have

$$W_{0\,1}(x,\nu) = \begin{vmatrix} \alpha(\nu) & \alpha(G^{-1}(\nu)) \\ \beta(\nu) & \omega^{-p}\beta(G^{-1}(\nu)) \end{vmatrix} px^p.$$

By the similar method, we can obtain $W_{j\,k}(x,\nu)$ as

$$W_{j\,k}(x,\nu) = \begin{vmatrix} \alpha(G^{-j}(\nu)) & \alpha(G^{-k}(\nu)) \\ \omega^{-jp}\beta(G^{-j}(\nu)) & \omega^{-kp}\beta(G^{-k}(\nu)) \end{vmatrix} px^p. \tag{3.12}$$

From (3.3), we get

$$\begin{vmatrix} \alpha(\nu) & \alpha(G^{-1}(\nu)) \\ \beta(\nu) & \omega^{-p}\beta(G^{-1}(\nu)) \end{vmatrix} = \frac{2}{p}\omega^{\frac{1}{4}q}, \tag{3.13}$$

and

$$W_{k\,k+1}(x,\nu) = \frac{1}{p!}\begin{vmatrix} \varphi_1(x,\nu) & \varphi_2(x,\nu) \\ \delta\varphi_1(x,\nu) & \delta\varphi_2(x,\nu) \end{vmatrix}\begin{vmatrix} f_k(0,\nu) & f_{k+1}(0,\nu) \\ f_k^{(p)}(0,\nu) & f_{k+1}^{(p)}(0,\nu) \end{vmatrix}. \tag{3.14}$$

Now due to (3.4), we shall set $C_0(\nu) = C(\nu)$, and $\tilde{C}_0(\nu) = \tilde{C}(\nu)$, then we will study properties of the Stokes multipliers $C(\nu)$ and $\tilde{C}(\nu)$ with respect to the parameter $a_q(\nu)$ at the origin in the $a_q$ - space:

**LEMMA 3.7.**

$$C_k(\nu) = \frac{1}{2}p\omega^{-kp}\omega^{\frac{1}{4}q}\omega^{p}\begin{vmatrix} \alpha(G^{-k}(\nu)) & \alpha(G^{-(2+k)}(\nu)) \\ \beta(G^{-k}(\nu)) & \omega^{-2p}\beta(G^{-(2+k)}(\nu)) \end{vmatrix}. \tag{3.15}$$

□

**LEMMA 3.8.** *The entire functions $C(\nu)$ and $\tilde{C}(\nu)$ satisfy*

$$\begin{cases} C(-1/\alpha) = C_k(-1/\alpha) = 1 + \omega^p \\ \tilde{C}(-1/\alpha) = \tilde{C}_k(-1/\alpha) = -\omega^p \end{cases} . \tag{3.16}$$

□

We observe the following simple fact for the derivatives of $C_k(\nu)$ at $a_q(\nu) = 0$, *i.e.*, $\nu = -1/\alpha$ for all $k \in \mathcal{M}$ ([5]):

$$\left.\frac{\partial C_k(\nu)}{\partial a_q}\right|_{\nu=-1/\alpha} = \left.\frac{\partial C(G^{-k}(\nu))}{\partial a_q}\right|_{\nu=-1/\alpha} = \omega^{-qk}\left.\frac{\partial C(\nu)}{\partial a_q}\right|_{\nu=-1/\alpha}. \tag{3.17}$$

**THEOREM 3.9.** *For each of the derivatives of $C_k(\nu)$ at $a_q(\nu) = 0$,*

$$\left.\frac{\partial C(\nu)}{\partial a_q}\right|_{\nu=-1/\alpha} = \frac{1}{2} p\omega^{-\frac{1}{4}q}\omega^{p}(\omega^{-(p+q)} - 1)(\omega^{-q} - 1)$$

$$\times \frac{\omega^{-q} - \omega^{-p}}{\omega^{-q} + 1} \int_0^{+\infty} \tau^{p-1}\varphi_0(\tau)^2 d\tau , \tag{3.18}$$

*and*

$$\left.\frac{\partial C(\nu)}{\partial a_q}\right|_{\nu=-1/\alpha} \neq 0 , \tag{3.19}$$

*where* $\omega = \exp\left(\frac{2\pi}{2p+q} i\right)$, *and* $\varphi_0(x)$ *is a solution of*

$$(\delta^2 - p\delta)y - x^{2p+q}y = 0 \text{ with } \varphi_0(x) \to 0 \text{ as } x \to +\infty . \tag{3.20}$$

□

The author gratefully acknowledges that this work is partially supported by the Faculty Development Fund of the Golden Gate University.

## References

[1] Faierman, M., Generalized parabolic cylinder functions, J. of Asymptotic Analysis 5 (1992), pp 517-531.

[2] Lin, C. H., Sibuya,Y., and Tabara, T. J., Zeros of solutions of a Second Order Linear Ordinary Differential Equations with a Polynomial Coefficient, Funkcial. Ekvac. (1993), to appear.

[3] Sibuya, Y., *Global Theory of a Second Order Linear Ordinary Differential Equations with a Polynomial Coefficient,* North-Holland Mathematics Studies 18, North-Holland/American Elsevier, 1975.

[4] Tabara, T. J., An asymptotic analysis and Stokes multipliers for a second order linear differential equation with polynomial coefficients, Ph.D. Dissertation, University of Minnesota, 1989.

[5] Tabara, T. J., A locally prescribed Stokes phenomenon, Funkcial. Ekvac. 35 (1992), pp. 429-450.

[6] Tabara, T. J., Local uniqueness of a boundary value problem in the complex plane, J. of Asymptotic Analysis (1993), to appear.

# LIMITING ABSORPTION PRINCIPLE FOR SCHRÖDINGER OPERATORS WITH OSCILLATING POTENTIALS

PETER REJTO
*School of Mathematics*
*University of Minnesota*
*Minneapolis, Minnesota 55455*
and
MARIO TABOADA
*Department of Mathematics*
*University of Southern California*
*Los Angeles, CA 90089*

*Dedicated to Bill Harris and Yasutaka Sibuya*

**Abstract**

We consider Schrödinger operators with potentials which are short-range perturbations of von Neumann-Wigner type potentials. The main result is the validity of a limiting absorption principle for this class of operators on intervals not containing possible embedded eigenvalues.

## 1. Introduction.

Consider the radial potential

$$p_0(\rho) = c\frac{\sin b\rho}{\rho\beta}, \quad \rho = |x|, x \in \mathbb{R}^n \tag{1.1}$$

where $c, b, \beta$ are constants, and let $H_0$ be the operator defined as the closure of $-\Delta + p_0$, defined on $C_0^\infty(\mathbb{R}^3)$, with respect to the $L^2(\mathbb{R}^3)$ norm. In order to define perturbations of $H_0$, define the weight

$$m_s(x) = (1+|x|^2)^{\frac{s}{2}}, \quad x \in \mathbb{R}^3, \quad s \in \mathbb{R}$$

and the weighted $L^2$ and Sobolev spaces associated with $m_s$:

$$\begin{aligned} L^{2,s}(\mathbb{R}^3) &= \{f : \|f\|_s^2 = \int_{\mathbb{R}^3} |f(x)|^2 m_s(x)dx < \infty\} \\ H^{2,-s}(\mathbb{R}^3) &= \{f : \|f\|_{2,-s}^2 = \textstyle\sum_{|\alpha|\le 2} \|\partial^\alpha f\|_{-s}^2 < \infty\}. \end{aligned}$$

The potential $V(x)$ is *short range* if the map $V : H^{2,-s}(\mathbb{R}^3) \to L^{2,s}(\mathbb{R}^3)$ is *compact.* It is well-known that if $V$ is short range, then the operator $H = H_0 + V$ is essentially self-adjoint.

The following result is our main theorem:

**Theorem 1.1** Let $V$ be a short-range potential, and let $H = H_0 + V, \quad s > \frac{1}{2}$. Assume also that $\frac{4}{5} < \beta \leq 1$, and denote the resolvent of $H$ by $R(z) = (H - z)^{-1}$.

Then there exists a discrete set $\{\lambda_i\} \in \mathbb{R} \setminus \{0\}$ such that the limits:

$$R^{\pm}(\lambda) = \lim_{\epsilon \to 0+} R(\lambda \pm i\varepsilon), \quad \lambda \neq \lambda_i, 0,$$

exist in the norm topology of bounded operators from $L^2(\mathbb{R}^n)$ into $H^{2,-s}(\mathbb{R}^n)$.

Previous results regarding the validity of a limiting absorption principle for the operator $H$ include Devinatz and Rejto [3], [4], who prove a result for restricted intervals, Devinatz, Moeckel and Rejto [2] who analyze the case $\beta = 1$, and Saito [17]. The potential (1.1), with $\alpha = \beta = 1$, was introduced by von Neumann and Wigner [19] in 1929. In this classic paper, the authors construct a perturbation $V(x)$ such that the perturbed operator has an eigenvalue embedded in the continuous spectrum. The only restriction we impose in our main result is that the intervals considered do not contain such a possible embedded eigenvalue.

## 2. Reduction of the Problem

In this section we indicate several steps which allow us to reduce the problem to the asymptotic analysis of an ordinary differential operator with a large parameter.

### *2.1 Reduction to the Free Operator*

By a result of Agmon [1], extended by Devinatz and Rejto [4], it suffices to prove the validity of Thm. 1.1 for the "free" operator $H_0$.

### *2.2 Reduction to Wronskian Estimates*

The operator $H_0$ admits a complete family of reducing subspaces, on which it acts like an ordinary differential operator. Specifically, for $j \in \mathbf{Z}$, let

$$p_{0,j}(\rho) = \frac{j(j+1)}{p^2} + p_0(\rho), \quad \rho \in \mathbb{R}^+.$$

Then, on the reducing subspaces, $H_0$ is unitarily equivalent to $2j + 1$ copies of the closure of the ordinary differential operator given by

$$L(p_{0,j})f(\rho) = -f''(\rho) + p_{0,j}(\rho)f(\rho)$$

$$f \in C^\infty(\mathbb{R}^+) \cap L^2(\mathbb{R}^+), \quad f(0) = 0.$$

To study the resolvent of this family of operators, consider the auxiliary equation

$$(*) \qquad f''(\rho) + (\mu - p_{0,j}(\rho))f(\rho) = 0, \quad Im(\mu) \neq 0.$$

We next introduce the parameter $\nu = j + \frac{1}{2}$, and consider two special solutions $f_\ell$ and $f_r$ of $(*)$, satisfying

$$(2.1) \qquad f_\ell \sim (4\pi)^{-\frac{1}{2}} \exp(\frac{i\pi}{4}) \cdot (\mu - \nu^2 \rho^{-2})^{\frac{1}{4}} \exp[-i \int_{\nu\mu^{\frac{1}{2}}}^{\rho} (\mu^2 - \nu^2\sigma^{-2})^{\frac{1}{2}} d\sigma],$$

as $\rho \to 0$. Here we choose the branch of the square root satisfying $Re(z^{\frac{1}{2}}) > 0$ if $z \in \mathbf{C} \setminus (-\infty, 0]$.

$$(2.2) \qquad f_r(\rho) \sim \mu^{-\frac{1}{2}} \exp(i\mu^{\frac{1}{2}}\rho), f_r'(\rho) \sim i \exp(i\mu^{\frac{1}{2}}\rho)$$

as $\rho \to \infty$.

The existence of $f_\ell(\rho)$ can be shown by mapping the point $\rho = 0$ into $\rho = \infty$ by means of a substitution, together with the known behavior at $\infty$ of solutions of the Airy equation (cf. [5] for details).

The existence of $f_r$ follows by asymptotic integration (cf. [8]).

Consider the Wronskian of $f_\ell, f_r$:

$$W(f_\ell, f_r) = f_\ell f_r' - f_r f_\ell'.$$

Then the resolvent kernel of $L(p_{0,j})$ is given by:

$$(\mu - L(p_{0,j}))^{-1}(\xi, \eta) = \frac{1}{W(f_\ell, f_r)} \begin{cases} f_\ell(\eta) f_r(\xi), \eta < \xi, \\ f_r(\eta) f_\ell(\xi), \eta > \xi. \end{cases}$$

By the results of Devinatz and Rejto [4], it follows that the following result implies the main Thm. 1.1:

**Theorem 2.1** Let $I$ be a compact subinterval of $\mathbb{R}^+$ such that $\frac{b^2}{4} \notin I$, and let $f_\ell, f_r$ be the basic solutions defined above, so that the asymptotic formulas hold as $\mu \to \lambda \in I$ from above.

Then:

$$\inf_{\lambda \in I} \inf_{j \in \mathbf{Z}^+} |W(f_\ell(\lambda, j), f_r(\lambda, j)| \neq 0.$$

To illustrate one of the main difficulties, notice that one has:

For each $j \in \mathbf{Z}^+, \lambda \in \mathbb{R}^+$,

$$\lim_{\rho \to 0} (\lambda - p_{0,j}(\rho)) = -\infty,$$

$$\lim_{\rho \to \infty} (\lambda - p_{0,j}(\rho)) = \lambda.$$

Therefore, there exists at least one point where the function $\lambda - p_{0,j}$ changes sign. In other words, there exists a *turning point.* The turning point is approximately given by

$$\tau = \tau(\lambda, j) = \nu \lambda^{-\frac{1}{2}} > 0.$$

As is well-known, at the turning point the solutions change qualitatively, which creates difficulties in the proof of the estimates for the Wronskian required in Thm. 2.1.

### *2.3 Reduction to Estimates on $f_\ell$*

By using the fact that $f_r$ is somewhat simple, together with elementary estimates on the Wronskian, one can prove that Thm. 2.1 is implied by the following:

**Theorem 2.2** Let $I$ be a compact subinterval of $\mathbb{R}^+$ such that $\frac{b^2}{4} \notin I$, and let $f_\ell$ by defined as before.

Then, for each $j^0 \in \mathbf{Z}^+$, there exists a sequence $\rho_k \to \infty$ such that

$$\inf_{\lambda \in I} \inf_{j > j^0} \liminf_{k \to \infty} | \vec{f}_\ell (\rho_k)| \neq 0,$$

where $\vec{f}_\ell = (f_\ell, f_\ell')$.

**Remark:** This last step reduces the problem to that of obtaining estimates on a single function. As pointed above, the main difficulty is due to the turning point, which goes to $+\infty$ with $j$.

## 3. Approximate Solutions

In this section we give an overview of the method used to prove the estimates required by Thm. 2.2 on $f_\ell$ and its derivative.

First, we define left, middle, and right intervals $J_\ell, J_m, J_r$ by:

$$\begin{aligned} J_\ell &= [0, \lambda^{-\frac{1}{2}}(\nu - \nu^\kappa)], \\ J_m &= [\lambda^{-\frac{1}{2}}(\nu - \nu^\kappa), \lambda^{-\frac{1}{2}}(\nu + \nu^\kappa)], \\ J_r &= [\lambda^{-\frac{1}{2}}(\nu + \nu^\kappa), \infty) \end{aligned}$$

where $0 < \kappa < 1$ is an appropriately chosen number.

**Remark:** Note that the middle interval $J_m$ is centered at the approximate turning point $\tau = \lambda^{-\frac{1}{2}}\nu$. The estimates on $J_m$ are thus qualitatively different (harder) than those on the left and right intervals. Observe also that the length of the middle interval goes to $\infty$, but *more slowly than* $\nu$. This is a key feature in the success of this subdivision.

The steps involved in obtaining the estimates on $f_\ell$ on the various intervals are the following:

(i) Define an *approximate solution.*
(ii) Show that the desired estimate holds for the approximate solution.
(iii) Estimate the difference between the approximate solution and the exact solution

Regarding approximate potentials, we apply the following general methodology: let $J$ be an interval and let $\psi$ be a function such that $(\lambda - \psi)^{-1} \in C^2(J)$. Define the two functions

$$\theta_\pm = \theta_\pm(\psi,\lambda) = \pm(\lambda-\psi)^{\frac{1}{2}} + \frac{1}{4i}(\lambda-\psi)^{-1}\psi'. \tag{3.1}$$

Clearly, for any smooth function $\theta_-$, the function

$$y(\rho) = \exp(i\int_{1/2\cdot\nu\lambda^{-1/2}}^{\rho} \theta_-(\sigma)d\sigma) \tag{3.2}$$

satisfies the equation

$$y'' - (i\theta'_- - \theta_-^2)y = 0.$$

One can also check that the function

$$z(\rho) = \exp(i\int_{1/2\cdot\nu\lambda^{-1/2}}^{\rho} \theta_+(\sigma)d\sigma)$$

also satisfies the same differential equation. Adding and subtracting the potential $i\theta'_- - \theta_-^2$ to the basic differential equation

$$f'' - (\lambda - p_{0,j})f = 0, \tag{3.3}$$

we obtain

$$f'' - (i\theta'_- - \theta_-^2)f = \{-i\theta'_- + \theta_-^2 - (\lambda - p_{0,j})\}f \tag{3.4}$$

Notice that $y$ and $z$ were chosen as solutions of the homogeneous equation corresponding to the operator on the left. This suggests an integral equation for $f$. To this end, let the *error potential* $e$ be defined by

$$e = e(\lambda, j) = -i\theta'_- + \theta_-^2 - (\lambda - p_{0,j})$$

and let the kernel be defined by

$$Q(\xi,\eta) = \begin{cases} (y(\eta)z(\xi) - z(\eta)y(\xi))W^{-1}(y,z)e(\eta), \eta < \xi \\ 0, \quad \eta > \xi. \end{cases}$$

Then it follows that if $f$ satisfies the integral equation

$$f = y + Qf, \tag{3.5}$$

where $Q$ denotes the Volterra operator with the kernel $Q(\xi,\eta)$ defined above, then it also satisfies the basic differential equation (3.4).

In order to estimate the difference between $f$ and $y$, the basic tool is a result on resolvent estimates for Volterra operators in weighted spaces; in order to state this result, for which we refer to [16], we need some definitions.

Let $J$ be a given subinterval of $(0,\infty)$, and let $w$ be a given positive continuous function on $J$. Define the weighted norm

$$\|f\|_w = \sup_{\rho\in J} |f(\rho)|w(\rho)^{-1},$$

and let $B(w)$ be the space of continuous functions $f$ on $J$ for which $\|f\|_w < \infty$. Then $B(w)$ is a Banach space; let $B(B(w))$ be the algebra of bounded operators on $B(w)$. Next, let $V$ be a Volterra operator with kernel $V(\xi,\eta)$ acting on $B(w)$. In other words,

$$Vf(\xi) = \int_{\substack{\eta\in J\\ \eta\le\xi}} V(\xi,\eta)f(\eta)d\eta, \quad \xi\in J, \quad f\in B(w).$$

Finally, define the Lover-Erdelyi-Olver bound ("LEO" for short) by

$$\|V\|_{(LEO,w)} = \int_J \sup_{\xi>\eta} w(\xi)^{-1}w(\eta)|V(\xi,\eta)|d\eta.$$

**Lemma 3.1** Let $w$ be a given weight function, let $V$ be a Volterra operator acting on $B(w)$, and let the LEO bound by defined as above. Suppose that

$$\|V\|_{(LEO,w)} < \infty.$$

Then

$$V\in B(B(w)) \text{ and } (I-V)^{-1}\in B(B(w)).$$

Furthermore,

$$\|V\|_w \le \|V\|_{(LEO,w)} \text{ and } \|(I-V)^{-1}\| \le \exp\|V\|_{(LEO,w)}.$$

We refer to [16] for a proof of this lemma, which allows us to obtain estimates for the difference between the approximate and the exact solution of (3.3), where the weight is given by the error potential in the corresponding interval $J_\ell$, $J_m$, or $J_r$.

## 4. Construction of Approximate Solutions on $J_\ell$

In this short section we describe, without motivation, our approximate solutions on the interval $J_\ell$. For this purpose, first define the function,

$$q_0(j)(\rho) = \nu(j)^2\rho^{-2} + p_0(\rho), \quad j=0,1,2,\cdots. \tag{4.1}$$

Second, with the help of this function define

$$c(\lambda,j) = c = b^2[4(\lambda - q_0(j) + p_0) - b^2]^{-1}. \tag{4.2}$$

Third, with the help of the definitions (4.1) and (4.2), define

$$\psi_\ell = q_0(j) + c(\lambda,j)p_0. \tag{4.3}$$

Fourth, we define the approximate solution $y_\ell$ by replacing the function $\psi$ by this function $\psi_\ell$ in the definition (3.1). Then using the definition (3.2) and the elementary formula,

(4.4)
$$\exp\left(\int_{1/2\cdot\nu\lambda^{1/2}}^{\rho}\frac{1}{4}\cdot(\lambda-\psi_\ell(\sigma))^{-1}\psi_\ell'(\sigma)d\sigma\right)=(\lambda-\psi_\ell(1/2\cdot\nu\lambda^{-1/2}))^{1/4}\cdot(\lambda-\psi_\ell(\rho))^{-1/4},$$

we find,

(4.5)
$$y_\ell(\rho)=(\lambda-\psi_\ell(1/2\cdot\nu\lambda^{-1/2}))^{1/4}\cdot(\lambda-\psi_\ell(\rho))^{-1/4}\cdot\exp[-i\int_{1/2\cdot\nu\lambda^{-1/2}}^{\rho}(\lambda-\nu^2\sigma^{-2})^{1/2}d\sigma].$$

Notice that the lower limit of integration in this formula was chosen independently of the asymptotic formula (2.1). Hence there is no reason to expect that this approximate solution $y_\ell(\rho)$ satisfies it. Therefore, we define the constants,

(4.6)
$$\alpha_\ell(j)=(4\pi)^{-1/2}\exp(i\pi/4)(\lambda-\psi_\ell(1/2\cdot\nu\lambda^{-1/2}))^{-1/4}\cdot\exp[-i\int_{\nu\lambda^{-1/2}}^{1/2\cdot\nu\lambda^{-1/2}}(\lambda-\nu^2\sigma^{-2}))^{1/2}d\sigma].$$

Then we can show that $\alpha_\ell(j)y_\ell(\rho)$ does satisfy the asymptotic formula (2.1).

We notice that this construction is different from the one in [3, formula (5.5)], inasmuch as it includes the oscillating potential $p_0$.

The procedure on each of the remaining intervals $J_m, J_r$ is similar in principle but the technical difficulties are different. Thus, for example, the construction of the approximate solution $y_m$ in the middle interval $J_m$ requires the use of a transformation mapping the approximate turning point of our basic equation into the turning point $z=0$ of the Airy equation. This allows us to exploit classical estimates for the Airy function, in the manner of [2], [3].

## 5. Concluding Remarks

In this paper, we have presented the main ideas involved in the proof of the main Thm. 1.1. The estimates involving approximate solutions are technical and limit the range of the parameter $\beta$. A full presentation of these results will appear elsewhere.

*ACKNOWLEDGMENT: It is a pleasure to thank A. Devinatz, W. Jäger, and Y. Sibuya for valuable conversations, and Lesley Newton and Chaunté Williams for their expert typing.*

## References


[1] Agmon, S., *Spectral properties of Schrödinger operators and scattering theory*, Ann. Scuola Norm. Sup. Pisa Sci. Mat. (1975), 151-218.

[2] Devinatz, A., Moeckel, R., and Rejto, P., *A limiting absorption principle for Schrödinger operators with Von Neumann-Wigner type potentials*, Integral Eq. and Operator Theory 14 (1991), 13-68.

[3] Devinatz, A. and Rejto, P., *A limiting absorption principle for Schrödinger operators with oscillating potentials I*, J. Diff. Eqs. 49 (1983), 29-84.

[4] ________, *A limiting absorption principle for Schrödinger operators with oscillating potentials II*, J. Diff. Eqs. 49 (1983), 85-104.

[5] Erdelyi, A., *Asymptotic Expansions*, Dover, New York, 1956.

[6] ________, *Asymptotic solutions of differential equations with transition points or singularities*, J. Math. Phys. 1 (1960), 16-26.

[7] ________, *The integral equations of asymptotic theory*, in "Asymptotic Solutions of Differential Equations," Wilcox, C.H. ed., John Wiley 1964, pp. 211-229.

[8] Harris, W.A. and Lutz, D.A., *Asymptotic integration of adiabatic oscillators*, J. Math. Anal. Appl. 51 (1975), 76-93.

[9] ________, *A unified theory of asymptotic integration*, J. Math. Anal. Appl. 57 (1977), 571-586.

[10] Jäger, W. and Rejto, P., *On the absolute continuity of the spectrum of Schrödinger operators with long range potentials*, Oberwolfach Tagungbericht, 17, 7-23.7, 1977.

[11] ________, *Limiting absorption principle for some Schrödinger operators with exploding potentials I*, J. Math. Anal. Appl. 91 (1983), 192-228.

[12] ________, *Limiting absorption principle for some Schrödinger operators with exploding potentials II*, J. Math. Anal. Appl. 95 (1983), 169-194.

[13] Love, C.E., *Singular integral equations of the Volterra type*, Trans. Amer. Math. Soc. 15 (1914), 467-476.

[14] Mochizuki, M. and Uchiyama, J., *Radiation conditions and spectral theory for 2-body Schrödinger operators with "oscillating" long range potentials I*, J. Math. Kyoto Univ. 18 (2) (1978), 377-408.

[15] Olver, F.T.W., *Asymptotics and Special Functions*, Academic Press, 1974.

[16] Rejto, P. and Taboada, M., *Weighted resolvent estimates for Volterra operators on unbounded intervals*, J. Math. Anal. Appl. 160, 1 (1991), 223-235.

[17] Saito, Y., *On the asymptotic behavior of the solutions of the Schrödinger equation*, Osaka J. Math. 14 (1977), 11-35.

[18] ________, *Schrödinger operators with a nonspherical radiation condition*, Pacific J. Math. 126 (1987), 331-359.

[19] von Neumann, J. and Wigner, E., *Über die merkwürdige diskrete Eigenwerte*, Phys. Z. 30 (1929), 465-467.

# EXPONENTIAL ASYMPTOTICS AND THE STOKES PHENOMENON FOR NONLINEAR ORDINARY DIFFERENTIAL EQUATIONS

**A. TOVBIS** *

*Department of Theoretical Mathematics*
*The Weizmann Institute of Science*
*Rehovot 76100, Israel*

ABSTRACT

The subject of the paper is exponential asymptotics to nonlinear differential equations or in the other terminology "asymptotics beyond all orders". We present the explicit expression for the leading term of exponential correction to the power series asymptotic. We also discuss the Stokes phenomenon and show that the exponential "jump" for solutions with the same power series asymptotics does not depend on the concrete choice of the solution.

## 0. Introduction.

It is known (see [PRG], [GJ]) that the study of travelling wave solutions to the singularly perturbed fifth order Kortveg-de Vries equation

$$\epsilon^2 u_{xxxxx} + u_{xxx} + 6uu_x + u_t = 0 \tag{1}$$

is reducible to the study of solutions of the ordinary differential equation

$$v''''(x) + v''(x) + v^2(x) = 0 \ , \tag{2}$$

which posssess the asymptotics

$$v(x) = O(x^{-2}) \ , \quad x \to \infty \ , \tag{3}$$

* Supported by a grant of the Ministry of Science and Technology of Israel.

in any closed proper subsector of the upper half-plane $\hat{S}$ of the complex $x$-plane.

Actually, the problem is to study the asymptotic behavior of $v(x)$ near the real semiaxes. It turns out that the corresponding asymptotics contain respectively exponential terms $c_1 e^{ix}$ and $c_2 e^{ix}$, where $c_{1,2} \in \mathbb{C}$. These terms are exponentially small in any proper subsector of $\hat{S}$, but become oscillatory and therefore "detectable" near the real semiaxes.

It is known that a solitary wave solution to (1) consists of the central core of the wave accompanied by co-propagating trailing oscillations. The constant $c_2 - c_1$, which we call the *Stokes constant of equation (2) in* $\hat{S}$, relates the amplitude and the phase shift of trailing oscillations and, therefore, is an important characteristic of the solitary wave solutions to (1) ([GJ]).

The main results of the present paper, applied to the equation (2), are as follows:

a) The mappings $c_{1,2} : \Psi_{\hat{S}} \mapsto \mathbb{C}$ are global coordinates on the manifold $\Psi_{\hat{S}}$ of the solutions to (2), satisfying (3);

b) The Stokes constant $c = c_2 - c_1$ does not depend on $v \in \Psi_{\hat{S}}$;

c) We present an explicit formula (see (20) and also (11) for the general case) for the Stokes constant $c$.

These results are obtained for a system of nonlinear ordinary equations of the general type

$$x^{1-r} z'(x) = f(x, z) , \quad x \in \bar{\mathbb{C}} , \quad z \in \mathbb{C}^n , \quad n, r \in \mathbb{N} , \tag{4}$$

which possesses a formal power series solution

$$\hat{z}(x) = \sum_{k=n_0}^{\infty} z_k x^{-k/p} , \qquad n_0, p \in \mathbb{N} . \tag{5}$$

Here the vector-valued function $f(x, z)$ is holomorphic at $(0, \infty) \in \bar{\mathbb{C}} \times \mathbb{C}^n$.

Once the formal solution (5) to (4) is fixed, one can define the *characteristic exponentials of (4)*, related to (5) (see **2**). For example, the second Painleve equation

$$v''(x) = 2v^3 + xv + \alpha , \qquad \alpha \in \mathbb{C} , \tag{6}$$

has two characteristic exponentials $x^{-1/4} e^{\pm 2/3 x^{3/2}}$ related to the formal power series solution with the leading term $-\frac{\alpha}{x}$ (see (18)). The other two characteristic exponentials $x^{-1/4} e^{\pm \frac{2\sqrt{2}}{3} i x^{3/2}}$ are related to the other formal power series solutions with the leading terms $\pm \frac{i}{\sqrt{2}} x^{\frac{1}{2}}$ (see [T1], §4.2).

In this paper we consider the equation (4) of the generic type, i.e. satisfying the following assumptions:

a) The equation (4) is a *one-level equation* of some order $\nu$, i.e. the characteristic exponentials of (4) are of the order $\nu$;

b) There exists only one characteristic exponential, which is decreasing in the considered sector $\hat{S}$ of opening $\pi/\nu$.

In fact, this work is a generalization of [GM] to the case of nonlinear equations.

## 1. Prenormalized equation.

In order to simplify the exposition (and without any loss of generality) we assume that the formal power series (5), as well as any other formal power series in what follows, contain only integer powers of $x$. For the same reason we also assume the number $n_0 = 1$.

Let deg of a series in negative powers of $x$ denote the exponent of the highest power of $x$ occurring in the series; we set $\deg a(x) = -\infty$ if $a(x) \equiv 0$. Let us denote $\eta = 2n(r+1) + \delta - 1$, where $\delta \in \mathbb{Z}^+$ is to be defined below. On the next step we introduce the transformations: $z \mapsto z - \sum_1^\eta z_k x^{-k}$, $z \mapsto T(x)z$ and $z \mapsto S(x)z$. Here $z_k$ are defined in (5), $T(x)$ is a polynomial in $x^{-1}$ of degree not more than $(n-1)(r+1)+1$ so that $\det T(\infty) \neq 0$ and finally

$$S(x) = \mathrm{diag}(x^{-n(r+1)-\delta}, \cdots, x^{-(2n-1)(r+1)-\delta}) .$$

Then, by means of these transformations, the equation (4) is reduced to (see [T3])

$$y' = P(x)y + \tilde{f}(x, y) , \tag{7}$$

where:

a) $\tilde{f}(x, y)$ is holomorphic at $(\infty, 0) \in \bar{\mathbb{C}} \times \mathbb{C}^n$;

b) $\deg \tilde{f}(x, 0) \leq -2$;

c) $\deg \frac{\partial \tilde{f}}{\partial y}\Big|_{y=0} \leq -2$;

d) the coefficient of any higher order monomial (in entries of the vector $y$) of $\tilde{f}(x, y)$ has degree not more than $-2-\delta$:

e) $xP(x)$ is a diagonal matrix which is polynomial in $x$ of order not more than $r$. The free term of $xP(x)$ is equal to $P_r = B_r + [n(r+1)+\delta]I + \mathrm{diag}(0, r+1, \cdots, (n-1)(r+1))$, where $I$ is the identity matrix and $B_r = \mathrm{diag}(b_1, \cdots, b_n)$ is the leading coefficient of the polynomial (in $x^{-1}$) diag $B(x)$. We choose the minimal $\delta \in \mathbb{Z}^+$, such that the entries of the free term of $xP(x)$ have nonnegative real parts.

The equation (7) is called a *prenormalized form* of the equation (4). The matrix $xP(x) - P_r$ is uniquely determined by the equation (4).

**Lemma 1.** ([T3]) *Suppose that the first entry of the diagonal matrix $xP(x) - P_r$ does not coincide with any other diagonal entry of this matrix. Then the first entry $b_1$ of the matrix $B_r$ and the first column $h$ of the matrix $T(\infty)$, if we additionally assume $\|h\| = 1$ in some norm $\|\cdot\|$, are uniquely determined by the equation (4).*

## 2. Volterra-type integral equations.

The equation (7) can be considered as a perturbation of the linear equation $y' = P(x)y$ and hence can be written as

$$y(x) = e^{Q(x)}(C + \int_{\Gamma(x)} e^{-Q(t)} \tilde{f}(t,y)dt) \tag{8}$$

where $Q'(x) = P(x)$, $C \in \mathbb{C}^n$ and $\Gamma(x)$ is a collection of $n$ individual contours of intergration for each entry of the vector integrand. These contours $\Gamma^j(x)$ of $\Gamma(x)$ have the common endpoint $x$.

The entries $q_j(x)$ of the diagonal matrix $Q(x)$ consist of polynomial and of logarithmic parts. Let $J$ denote the set of indicies $j$, such that $q_j(x)$ has a non-trivial polynomial part, and let $\lambda_j x^{\nu_j}$, where $\lambda_j \in \mathbb{C}$ and $\nu_j \in \{1, \cdots, r\}$, denote the leading term of $q_j$.

**Definition 1.** *The exponentials $e^{q_j(x)}$, where $j \in J$, are called characteristic exponentials of the equation (4).*

**Definition 2.** *A ray $\tau$, issuing from the origin, is called a Stokes ray of the equation (4) if $Re(\lambda_j x_j^{\nu}) = 0$ when $x \in \tau$.*

According to the assumption a) all $\nu_j = \nu$, where $j \in J$. Let us consider a sector $\hat{S}$, bounded by two successive Stokes rays $\tau_1$ and $\tau_2$, which corresponds to the exponential $e^{q_1(x)}$. Suppose $1 \in J$ and, according to the assumption b), suppose $e^{q_1(x)}$ is the only characteristic exponent, decreasing in $\hat{S}$. Moreover, for the sake of simplicity we assume that $q_1(x) = \lambda_1 x^{\nu} + d \ln x$, where $d = b_1 + n(r+1) + \delta$ is the first entry of the diagonal matrix $P_r$. It is a non-restrictive assumption since we can always find an appropriate change of the independent variable $x$, which brings $q_1(x)$ to the required form.

**Definition 3.** *For arbitrary $\nu > 0$ and $\alpha > 0$ the formal $\nu$-Borel transform $\mathcal{B}_\nu$ of $x^{-\alpha}$ is defined as*

$$\mathcal{B}_\nu x^{-\alpha} = \frac{p^{\alpha-\nu}}{\Gamma(\frac{\alpha}{\nu})},$$

*where Γ denotes the Gamma-function. This definition is extendable by linearity on arbitrary formal power series in negative powers of x.*

## 3. Main Theorem.

Let $\Psi_S$ denote the set of the solutions $z(x)$ of the equation (4), which are holomorphic in a given open sector $S$ on the Riemann surface of $\ln x$ for sufficiently large $|x|$ and which have the asymptotic expansion $\hat{z}(x)$ in any proper closed subsector of $S$. We denote this by

$$z(x) \sim \hat{z}(x)\,, \qquad x \to \infty\,, \qquad x \in S\,. \tag{9}$$

Generally, it is known that $\Psi_S$ is nonempty if the opening of $S$ is less than $\pi/r$ ([RS]) and that either (9) is convergent (has a nonzero radius of convergence) or $\Psi_S = \emptyset$ when the opening of $S$ is sufficiently large.

A solution $z \in \Psi_{\hat{S}}$ may have infinitely many singularities on the boundary rays $\tau_1$ and $\tau_2$ of $\hat{S}$. (Consider , for instance, $z' = z^2 + 1$ as the equation (4), where $\hat{z} = i$. Then solutions $\tan^{\pm 1}(x + a) \in \Psi_{\hat{S}}$, where $a \in \mathbb{R}$ and $\hat{S}$ is the upper half-plane.) Therefore we will consider asymptotics of solutions $z \in \Psi_{\hat{S}}$ not along the boundary rays $\tau_1$ and $\tau_2$ of $\hat{S}$, but along the curves $\ell_1$ and $\ell_2$, which are asymptotically closed relatively to $\tau_1$ and $\tau_2$ and which are defined by $|e^{q_1(x)}| = M|x|^{-\mu}$ . Here $M > 0$ and $\mu \in (0, 1)$ are some numbers.

The following Main Theorem gives, in some sense, a complete description of $\Psi_{\hat{S}}$ under the assumptions a) and b).

**Theorem.** *Let $z \in \Psi_{\hat{S}}$. Then:*

*a) $z(x)$ possesses the asymptotics*

$$z(x) = \sum_{k \le -\Re b_1,\ k \in \mathbb{N}} z_k x^{-k} + (c_j h + o(1)) x^{b_1} e^{\lambda_1 x^\nu}\,, \qquad x \to \infty\,, \qquad x \in \ell_j\,, \tag{10}$$

*where $j = 1, 2$, $c_j \in \mathbb{C}$ and $h$ is defined in Lemma 1;*

*b) $\Psi_{\hat{S}}$ is a complex analytic manifold, isomorphic to $\mathbb{C}$. Both functions $c_j(z)$ : $\Psi_{\hat{S}} \mapsto \mathbb{C}$, defined by (10), are global coordinates on $\Psi_{\hat{S}}$;*

*c) The difference $c = c_2(z) - c_1(z)$ does not depend on $z \in \Psi_{\hat{S}}$ and is given by*

$$ch = -2\pi i \lim_{p \to \sqrt[\nu]{-\lambda_1}} (p^\nu + \lambda_1) \mathcal{B}_\nu [x^{-b_1} \hat{z}(x)](p)\,, \tag{11}$$

*where the limit is taken along any curve which is nontangent to the ray* $\arg(p - \sqrt[\nu]{-\lambda_1}) = \arg(\sqrt[\nu]{-\lambda_1})$. *The choice of the root of $-\lambda_1$ is determined by the sector $\hat{S}$.*

Formula (11) relates the value $c$ of the "exponential jump" in the asymptotics of a solution $z \in \Psi_{\hat{S}}$ on the different sides of $\hat{S}$. This value does not depend on $z$ and can be determined via the formal power series solution $\hat{z}$ of (4). To evaluate $c$, one has to find the "residue" of the $\nu$-Borel sum $\mathring{Z}(p) = \mathcal{B}_\nu[x^{-b_1}\hat{z}](p)$ at the singular point $p = \sqrt[\nu]{-\lambda_j}$. Here $j$ is the index of the characteristical exponent, decreasing in $\hat{S}$. In general, it is possible to show that $\mathcal{B}_\nu \hat{z}$ is analytic on some Riemann surface, which has logarithmic branching at the points $m\sqrt[\nu]{-\lambda_j}e^{\frac{2\pi i k}{\nu}}$, where $j \in J$, $k = 0, \cdots, \nu - 1$, $m = 1, 2, \cdots$. By analogy with the linear case, $c$ is called a Stokes constant. In principle, it is possible to reconstruct any solution $z \in \Psi_{\hat{S}}$ from $\mathcal{B}_\nu \hat{z}$ by means of some "generalized" direct $\nu$-Laplace tarnsform.

## 4. Sketch of the proof.

**1).** The prenormalized equation (7) possesses the formal solution

$$\hat{y}(x) = S^{-1}(x)T^{-1}(x)(\hat{z}(x) - \sum_{k=1}^{\eta} z_k x^{-k}) \, .$$

By $\Phi_{\hat{S}}$ we denote the set of solutions $y(x)$ of (7), such that

$$y(x) \sim \hat{y}(x) \, , \qquad x \to \infty \, , \qquad x \in \hat{S} \, . \tag{12}$$

Let us pick a point $x_0 \in \hat{S}$ and define the collection of contours $\hat{\Gamma} = \{\hat{\Gamma}^j\}_{j=1}^n$ as: the contour $\hat{\Gamma}^j(x)$, $j \neq 1$, is a ray, issuing from $x$, such that $e^{-q_j(t)}$ is decreasing along $\hat{\Gamma}^j(x)$; the contour $\hat{\Gamma}^1(x)$ is the segment $[x_0, x]$. Then it is known (see [Wa], §14.3) that $\Phi_{\hat{S}}$ coincides with the set of solutions of (8), where $x_0$ varies in $\hat{S}$, $\Gamma = \hat{\Gamma}$ and $C$ is a vector with an arbitrary first entry $C_{(1)}$ (here and henceforth $f_{(j)}$ denotes the $j$-th entry of the vector $f$) and all other entries equal to zero. Moreover, $\Phi_{\hat{S}}$ is a complex analytic manifold, and for a fixed $x_0$ the value $C_{(1)}$ is a local coordinate on $\Phi_{\hat{S}}$. However it is not a global coordinate, since for a fixed $x_0$ the solution of (8) may not exist if $\|C\|$ is sufficiently large.

**2).** Let $\mathring{S}$ be an arbitrary proper subsector of $S$. In order to construct a global coordinate on $\Phi_{\hat{S}}$ let us change the contour $\hat{\Gamma}^1(x)$ to the contour $\Gamma_j^1(x)$, $j = 1, 2$, which consists of two parts: the segment $[x, x_j]$, where $x_j \in \ell_j$, and the part of the curve $\ell_j$ from $x_j$ to infinity. For a given $x \in \mathring{S}$ the point $x_j \in \ell_j$ is defined so that $e^{-q_1(t)}$ is decreasing along the segment $[x, x_j]$.

Let us introduce collections of contours $\Gamma_{1,2}$, where the contours $\Gamma_{1,2}^1$ are defined above, and other contours $\Gamma_{1,2}^k$, $k > 1$, coincide with $\hat{\Gamma}^k$. Let also denote by

$L$, $L \subset \mathbb{C}^n$, the subspace of vectors with all but the first zero entries. In [T3], §1, we showed that for any solution $y(x)$ of (8) with some $x_0 \in \hat{S}$, $C \in L$ and $\Gamma = \hat{\Gamma}$, there exist some vectors $C^{1,2} \in L$, so that $y(x)$ satisfies (8) with respectively $\Gamma = \Gamma_{1,2}$ and $C = C^{1,2}$. Conversely, if $y(x)$ satisfies (8) with $\Gamma = \Gamma_j$ and an arbitrary $C = C^j$ (where $j = 1$ or $j = 2$), and if $y(x)$ is defined at a point $x_0 \in \hat{S}$, then there exists some $C_{x_0} \in L$, so that $y(x)$ also satisfies (8) with $\Gamma = \hat{\Gamma}$ and $C = C_{x_0}$. Therefore, we showed that the mappings $F_j : \Phi_{\hat{S}} \mapsto \mathbb{C}$, $j = 1,2$, defined by $F_j y = C^j_{(1)}$, are one to one mappings. Thus the first two assertions of the Theorem for the prenormalized equation (7) are proved (the analyticity of the global coordinates $F_{1,2}$ is shown below).

**3).** Let us now prove that the transition function $\varphi_{\hat{S}} = F_2 \circ F_1^{-1} : \mathbb{C} \mapsto \mathbb{C}$ is a linear function. (Actually the fact that $\varphi_{\hat{S}} \equiv id$ for all $e^{q_j}$, $j \in J$ and all corresponding sectors $\hat{S}$ is equivalent to the convergence of the formal power series solution (5) of (4).)

Let $y(x)$ be the solution of the equation (8), where $\Gamma = \Gamma_j$ and $C = C^j$. Then $\frac{\partial y}{\partial C^j_{(1)}}$ satisfies the linear integral equation, obtained from (8), which can be considered in the same way as (8). So, there exists some $x_0 \in \hat{S}$, such that both $y$ and $\frac{\partial y}{\partial C^j_{(1)}}$ are analytic in $x_0$. This implies the analyticity of $F_{1,2}$. So, the transition function $\varphi_{\hat{S}}(\zeta) = a\zeta + c$, since linear functions are the only entire and schlicht functions on $\mathbb{C}$.

**4).** Let $\lambda_j x^\nu$ be the leading terms of $q_j(x)$, $j \in J$, and let $\sqrt[\nu]{-\lambda_j} = |\lambda_j|^{1/\nu} e^{\frac{i}{\nu}(\arg\lambda_j - \pi)}$. By $U$ we denote the complex $p$-plane with the cuts from points $\sqrt[\nu]{-\lambda_j} e^{\frac{2\pi i k}{\nu}}$, $j \in J$, $k = 0, \cdots, \nu - 1$, in the directions $\arg p = \frac{1}{\nu}(\arg\lambda_j + (2k-1)\pi)$ respectively.

In what follows, we use the standard direct and inverse $\nu$-Laplace transforms defined by the formulas

$$(\mathcal{L}_\nu Y(p))(x) = \nu \int_0^\infty e^{-x^\nu p^\nu} p^{\nu-1} Y(p) dp , \tag{13}$$

$$(\mathcal{L}_\nu^{-1} y(x))(p) = \frac{\nu}{2\pi i} \int_{\gamma_0} e^{p^\nu x^\nu} x^{\nu-1} y(x) dx \tag{14}$$

respectively. Here the contour $\gamma_0 = \Pi_{1/\nu}(\tilde{\gamma})$, where $\tilde{\gamma}$ is a vertical line in the right halfplane and $\Pi_u(x) = x^u$. Note that $\mathcal{L}^{\pm 1} = \mathcal{L}_1^{\pm 1}$ are the usual direct and inverse Laplace transforms.

Let $\tilde{S}$ be an arbitrary sector with opening more than $\pi/\nu$ and $y \in \Psi_{\tilde{S}}$. Then by $\mathcal{L}_\nu^{-1} y$ we mean the integral (14) along the contour $\gamma_0$, turned around in such a way that $\gamma_0 \subset \tilde{S}$.

**Theorem 1.** ([T3]). *Let (7) be a one-level equation of the order $\nu$ and $y \in \Phi_{\tilde{S}}$. Then $Y(p) = [\mathcal{L}_\nu^{-1} y](p)$ is holomorphic in $U$.*

**5).** Let us define

$$c(p) = \int_\gamma e^{p^\nu t^\nu} t^{-d} \tilde{f}_{(1)}(t, y) dt \ , \tag{15}$$

where the contour $\gamma$ consists of the part of $\ell_1$ from infinity to some fixed point $\xi_1 \in \ell_1$, a certain smooth path from $\xi_1$ to some $\xi_2 \in \ell_2$ and then the part of $\ell_2$ from $\xi_2$ to infinity. Then $F_2(y) - F_1(y) = c(\sqrt[\nu]{-\lambda_1})$ follows from the definition of $F_{1,2}$.

Suppose $F_1(y) = 0$. That means that $y \in \Phi_{\tilde{S}}$, where $\tilde{S}$ is some extension of the sector $\hat{S}$ in the negative direction. So, $y$ satisfies the integral equation (8), where $C = 0$ and $\Gamma$ coincides with $\hat{\Gamma}$ except for the first contour, which now is a ray directed to the sector $\tilde{S} \setminus \hat{S}$. Applying the transformation $\mathcal{L}_\nu^{-1}$ in the sector $\tilde{S}$ we get, after some calculations,

$$-\nu(p^\nu + \lambda_1)\mathring{Y}_{(1)}(p) = \mathcal{L}_\nu^{-1}[t^{-d-\nu+1} \tilde{f}_{(1)}(t, y)](p) \ , \tag{16}$$

where $\mathring{Y}_{(1)}(p) = [\mathcal{L}_\nu^{-1} x^{-d} y_{(1)}(x)](p)$. In [T3], §2, it was proved that $c(p)$ is continuous at the point $p = \sqrt[\nu]{-\lambda_1}$ and

$$c(p) = \frac{2\pi i}{\nu} \mathcal{L}_\nu^{-1}[t^{-d-\nu+1} \tilde{f}_{(1)}(t, y)](p) \ , \quad p \in U \ .$$

(In fact the problem here is that the contour $\gamma$ in (15) is not the same as the contour $\gamma_0$ in the definition of the $\mathcal{L}_\nu^{-1}$ transform.) Since $F_1(y) = 0$ implies $F_2(y) = c$, hence

$$c = -2\pi i \lim_{p \to \sqrt[\nu]{-\lambda_1}} (p^\nu + \lambda_1)\mathring{Y}_{(1)}(p) \ .$$

According to Theorem 1, $\mathring{Y}_{(1)}(p) = \mathcal{B}_\nu[x^{-d}\hat{y}_{(1)}(x)]$. Therefore the formula (11), where $h = Col(1, 0, \cdots, 0)$, is proven for the equation (7).

Let us now choose the other $y \in \Phi_{\hat{S}}$ so that $F_2(y) = 0$. Then by similar arguments we get $F_1(y) = -c$. Therefore the constant $a$ in the transition function $\varphi_{\hat{S}}$ is equal to one. So, $c = F_2(y) - F_1(y)$ does not depend on $y$, and for the prenormalized equation (7) the proof is complete.

**6).** If $z \in \Psi_{\hat{S}}$ then

$$z(x) = \sum_{k=1}^{\eta} z_k x^{-k} + T(x)S(x)y(x) \ , \tag{17}$$

where $y \in \Phi_{\hat{S}}$. Therefore, the first two assertions of the Theorem, where $c_j(z) = F_j y$, $j = 1, 2$, as well as the fact that $c = c_2(z) - c_1(z)$ does not depend on $z \in \Psi_{\hat{S}}$, follows from the corresponding assertions to the prenormalized equation (7).

For a scalar function $u(x)$, holomorphic in some sector $\tilde{S}$ of opening more than $\pi/\nu$, we introduce the operator

$$\mathcal{R}u = -2\pi i \lim_{p \to \sqrt[\nu]{-\lambda_1}} (p^\nu + \lambda_1)[\mathcal{L}_\nu^{-1} u](p) ,$$

where the limit is taken as in (11). It is clear that $\mathcal{R}y = 0$ if $\mathcal{L}_\nu^{-1} y$ exists and is continuous at $p = \sqrt[\nu]{-\lambda_1}$.

The following Lemmas from [T3] are utilized to complete the proof.

**Lemma 2.** *If $\mathcal{R}y$ exists, then $\mathcal{R}\left(\frac{y}{x^\alpha}\right) = 0$ for any $\alpha > 0$.*

**Lemma 3.** *If $j \neq 1$, then $\mathcal{R}y_{(j)} = 0$.*

According to these Lemmas, the formula (11) follows from

$$\mathcal{R}[x^{-b_1} z(x)] = \mathcal{R}[T^{(1)}(x) x^{-b_1 - n(r+1) - \delta} y_{(1)}(x)] = h\mathcal{R}[x^{-d} y_{(1)}(x)] = ch .$$

□

## 5. Stokes constants for Painleve II.

The formal series

$$\hat{v}(x) = -\frac{\alpha}{x} + \frac{2\alpha(\alpha^2 - 1)}{x^4} + \cdots = \frac{\alpha}{x} \sum_{k=0}^{\infty} v_k x^{-3k} , \tag{18}$$

where the coefficients $v_k$ are polynomials in $\alpha$, satisfies the second Painleve equation (6). The characteristic exponential $x^{-1/4} e^{-\frac{2}{3} x^{3/2}}$ of (6) is decreasing in the sector $\hat{S} = \{x : |\arg x| < \frac{\pi}{3}\}$. Then, according to the Theorem, the Stokes constant $c$ of the second Painleve equation in $\hat{S}$ is given by

$$c = -2\pi i \lim_{p \to (\frac{2}{3})^{\frac{2}{3}}} (p^{\frac{3}{2}} - \frac{2}{3}) \mathcal{B}_{\frac{3}{2}}^{-1} [x^{\frac{1}{4}} \hat{v}(x)](p) = -2\pi i \alpha \lim_{p \to (\frac{2}{3})^{\frac{2}{3}}} p^{-\frac{3}{4}} W(p) ,$$

where the power series $W(p) = \sum_{k=0}^{\infty} w_k p^{\frac{3}{2}k}$ has radius of convergence $\frac{2}{3}^{\frac{2}{3}}$. Here $w_{2k} = -\frac{2}{3} \frac{v_k}{\Gamma(2k + \frac{1}{2})}$, $w_{2k+1} = \frac{v_k}{\Gamma(2k + \frac{1}{2})}$.

In the particular case $\alpha = 0$ the Stokes constant $c = 0$. This demonstrates that if a solution $v(x) \in \Psi_{\hat{S}}$ (i.e. if $v(x) \sim 0$ in $\hat{S}$), then it has the same exponential corrections on both sides of $\hat{S}$.

## 6. The equation (1).

The equation (0.1) is a one-level equation of order 1, which have characteristic exponentials $e^{\pm ix}$. This equation possesses a formal power series solution

$$\hat{v}(x) = -\frac{6}{x^2} + \frac{90}{x^4} - \cdots = \sum_{k=1}^{\infty} v_k x^{-2k} .$$

Let us define the solution $v(x)$ by

$$v(x) \sim \hat{v}(x) , \qquad x \longrightarrow \infty, \qquad |\arg x| < \pi . \tag{19}$$

It can be shown directly that the exponential corrections to the asymptotics (19) near the rays $\arg x = \pm\pi$ are $c_\pm e^{\pm ix}$ (see also [GJ]). According to the Theorem

$$c_\pm = -2\pi i \lim_{p \to \pm i} (p \pm i) V(p) , \tag{20}$$

where $V(p) = \sum_{k=1}^{\infty} \frac{v_k}{(2k-1)!} p^{2k-1}$. It is easy to check that $c_+ = c_-$ and $c_+$ is a purely imaginary (but nonzero, see [T2]) number. The evaluation of $c_\pm$ is discussed in [GJ].

### References

[GJ] R. Grimshaw and N. Joshi, Weakly non-local solitary waves in a singularly perturbed Kortweg-de Vries equation, SIAM J. Appl. Math., submitted.

[GM] V. Gurarii and V. Matsaev, Stokes multipliers for systems of linear ordinary differential equations of first order, Soviet Math. Dokl. **31** (1), 52-56 (1985).

[PRG] Y. Pomeau, A. Ramani and B. Grammaticos, Structural stability of the Kortweg-de Vries solutions under a singular perturbation, Physica D **31**, 127-134 (1988).

[RS] J.P. Ramis and Y. Sibuya, Hukuhara domains and fundamental existence and uniqueness theorems for asymptotic solutions of Gevrey type, Asymptotic Analysis **2**, 39-94 (1989).

[T1] A. Tovbis, Nonlinear ordinary differential equations resolvable with respect to an irregular singular point, J. of Differential Equations, **107** (1), to appear (1994).

[T2] A. Tovbis, Asymptotics beyond all orders and analytic properties of inverse Laplace transforms of solutions, submitted.

[T3] A. Tovbis, On exponentially small terms of solutions to nonlinear ordinary differential equations, submitted.

[Wa] W. Wasow, "Asymptotic expansions for ordinary differential equations", Dover, N.Y., 1976.

# On the differential equation $y'' + \exp(y^m - t) = 0$

Hans Volkmer
Department of Mathematical Sciences
University of Wisconsin–Milwaukee
P.O. Box 413
Milwaukee, WI 53201
USA

**Abstract**

We derive an asymptotic expansion of the zero of the solution $y(\cdot, a)$ of the nonlinear ordinary differential equation $y'' + \exp(y^m - t) = 0$, $y(t, a) \to a$ as $t \to \infty$. Our results are related to those of Atkinson and Peletier (1986).

## 1 The Problem

The nonlinear partial differential equation

$$-\Delta u = f(u) \text{ in } \mathbf{R}^N \tag{1}$$

arises in many questions of Mathematical Physics such as existence of solitary waves, nonlinear field equations, problems of false vacuum etc. In particular, one is interested in the solution of the corresponding Dirichlet problem on balls. The results are strikingly different for $N \geq 3$ (see for instance [1], [3]) and $N = 2$ (see for instance [2], [4], [5]).

In this talk we study radially symmetric solutions of equation (1) in two dimensions $N = 2$ where $f(u) = \exp(u^m), m \geq 1$.

The substitution

$$t = -\log(\frac{1}{2}r), u(r) = y(t)$$

transforms equation (1) with the initial condition $u(0) = a$ into the ordinary differential equation with initial condition at $t = \infty$ (see [2, (4.1)] with $g(y) = y^m$)

$$y'' + \exp(y^m - t) = 0,\ y(t) \to a > 0 \text{ as } t \to \infty,\ m \geq 1. \tag{2}$$

The study of the solution $y(t) = y(t,a)$ of this problem was initiated in the interesting paper [2]. As $a \to \infty$, the graph of the function $y(t)$ looks like a ray coming in horizontally from infinity which then is reflected at the curve $y^m = t$. For $1 \leq m < 2$ we have only one reflection whereas we have multiple reflection for $m > 2$.

Our goal is to find the asymptotic behavior of the zero $T(a)$ of $y(t,a)$ as $a \to \infty$. The results depend critically on the value of the parameter $m$. If $m = 1$ then $y(t)$ can be given explicitely

$$y(t) = a - 2\log(1 + \frac{1}{2}\exp(a - t))$$

which implies

$$T(a) = \frac{a}{2} - \log 2 - \log(1 - \exp(-\frac{a}{2})) = \frac{a}{2} - \log 2 + \mathcal{O}(\exp(-\frac{a}{2})) \quad \text{as } a \to \infty.$$

If $m > 1$ then we know that $y(t)$ satisfies the inequality [2, Lemma 1]

$$y(t,a) \leq a - \frac{2}{ma^{m-1}} \log\left(1 + \frac{m}{2}a^{m-1}\exp(a^m - t)\right) \quad \text{for } t \geq T(a) \tag{3}$$

which implies that (see [2, Lemma 2])

$$T(a) > \left(1 - \frac{m}{2}\right)a^m + (m-1)\log a + \log\frac{m}{2}. \tag{4}$$

Our two main results given in the next section concern the difference between the two sides of inequality (4). We show that the difference converges to $\frac{3}{2}(m-1)$ as $a \to \infty$ if $1 < m < 2$. We also calculate the limit if $m = 2$.

The asymptotic behavior of $T(a)$ is unknown for $m > 2$ but results were obtained in [4].

## 2 Main results

We state the main results on the asymptotic behavior of the zero $T(a)$ of $y(t,a)$.

**Theorem 2.1** *Let $y(t,a)$ be the solution of (2). Then, if $1 \leq m < 2$, the zero $T(a)$ of $y(\cdot,a)$ satisfies*

$$T(a) = \left(1 - \frac{m}{2}\right)a^m + (m-1)\log a + \log\frac{m}{2} + \frac{3}{2}(m-1) + \mathcal{O}\left(\frac{\log^3 a}{a^m}\right) \quad \textit{as } a \to \infty.$$

In the borderline case $m = 2$, the asymptotic formula for $T(a)$ of Theorem 2.1 is not true anymore, but the following similar result holds.

**Theorem 2.2** *Let $y(t,a)$ be the solution of (2) with $m = 2$. Then the zero $T(a)$ of $y(\cdot, a)$ satisfies*

$$T(a) = \log a + s_0 + \mathcal{O}\left(\frac{\log^3 a}{a^2}\right) \quad \text{as } a \to \infty,$$

*where $s_0 = s = 1.6853749\ldots$ is the solution of*

$$s - 1.5 = \exp(-s).$$

For the proofs, we derive asymptotic expansions for $y(t)$ and $y'(t)$ at an intermediate point $T_1 = T_1(a)$ as in [2]. The intermediate point is close to but smaller than the reflection point of $y(t)$. If $m = 2$ then a second reflection close to $y = 0$ has to be taken into consideration. Details of the proofs of Theorems 2.1 and 2.2 are given in [6].

# References


[1] F. V. Atkinson and L. A. Peletier, Ground states of $-\Delta u = f(u)$ and the Emden-Fowler equation. Arch. Rational Mech. Anal. 93 (1986), 103-127.

[2] F. V. Atkinson and L. A. Peletier, Ground states and Dirichlet problems for $-\Delta u = f(u)$ in $\mathbf{R}^2$. Arch. Rational Mech. Anal. 96 (1986), 147-165.

[3] H. Berestycki and P.-L. Lions, Existence of solutions for nonlinear scalar field equations, Part I, The ground state. Arch. Rational Mech. Anal. 82 (1983), 313–345.

[4] B. McLeod and K. McLeod, The critical Sobolev exponent in two dimensions. Proc. Royal Soc. Edinburgh 109A (1988), 1-15.

[5] S. I. Pohozaev, Eigenfunctions of the equation $\Delta u + \lambda f(u) = 0$. Dokad. Akad. Nauk SSSR 165 (1965), 36-39 (in Russian) and Sov. Math. 6 (1965) 1408–1411 (in English).

[6] H. Volkmer, On the differential equation $y'' + \exp(y^m - t) = 0$, to appear in Asymptotic Analysis.

# On a Generalization of the Levinson Theorem

PO-FANG HSIEH AND FEIPENG XIE

*Department of Mathematics and Statistics*
*Western Michigan University*
*Kalamazoo, MI 49008-5152 USA*

## Abstract

The Levisnson theorem, which plays an important role in the asymptotic study of a system of linear ordinary differential equations, and most of subsequent studies assume the integrability of the coefficients of the equations. In this paper, the same result is obtained by only assume the integrability of the half of coefficients. A successive approximation method similar to Gauss-Seidel itereation is developed to prove the existence of the desired transformation matrix.

**1. Introduction.** The result of N. Levinson [L] plays an important role in the study of the asymptotic behavior of solutions of a linear system of differential equations

$$\frac{dy}{dt} = A(t)y, \tag{1.1}$$

as $t \to \infty$, where $y$ is an $n-$dimensional vector and $A(t)$ is an $n \times n$ matrix continuous on $\mathcal{I}_0 = [t_0, \infty)$ ($t_0$ : finite). In order to state the Levinson theorem we need:

**Assumption 1.1.** The matrix $A(t)$ is in the form

$$A(t) \ = \ \Lambda(t) \ + \ R(t), \tag{1.2}$$

where $\Lambda(t)$ is an $n \times n$ diagonal matrix

$$\Lambda(t) \ = \ diag\{\lambda_1(t), \lambda_2(t), \cdots, \lambda_n(t)\}, \tag{1.3}$$

with $\lambda_j(t)$ $(j = 1, 2, \cdots, n)$ continuous on $\mathcal{I}_0$.

**Assumption 1.2.** Let

$$D_{jk}(t) = \Re\big(\lambda_j(t) - \lambda_k(t)\big), \qquad (j,k = 1,2,\cdots,n). \tag{1.4}$$

For each fixed $j$, the set of positive integers $\{1,2,\cdots,n\}$ is the union of two disjoint subsets $\mathcal{P}_{j1}$ and $\mathcal{P}_{j2}$, where

(i) $k \in \mathcal{P}_{j1}$ if

$$\begin{cases} \lim_{t\to+\infty} \int_{t_0}^{t} D_{jk}(\tau)\, d\tau = -\infty, \\ \int_{s}^{t} D_{jk}(\tau)\, d\tau < K \qquad \text{for} \quad t_0 \le s \le t, \end{cases}$$

for some positive number $K$,

(ii) $k \in \mathcal{P}_{j2}$ if

$$\int_{s}^{t} D_{jk}(\tau)\, d\tau < K \qquad \text{for} \quad s \ge t \ge t_0$$

for some positive number $K$.

**Assumption 1.3.** The matrix $R(t)$ is an $n \times n$ matrix satisfying

$$R(t) \in L_1(\mathcal{I}_0). \tag{1.5}$$

A version of the Levinson theorem can be stated as the following (eg. see E. A. Coddington and N. Levinson [CL] or M. S. P. Eastham [E]).

**Theorem 1.1.** *Under the Assumptions 1.1, 1.2 and 1.3, there exists an $n \times n$ matrix $Q(t)$ such that*

*(1) the derivative $\frac{dQ(t)}{dt}$ exists and the entries of $Q(t)$ and $\frac{dQ(t)}{dt}$ are continuous in $t$ on the interval $\mathcal{I}_0$,*

*(2) $\lim_{t\to+\infty} Q(t) = 0$,*

*(3) the transformation:*

$$y = [\, I_n + Q(t)\,]\, z \tag{1.6}$$

*changes system (1.1) to*

$$\frac{dz}{dt} = \Lambda(t) z \tag{1.7}$$

*on the interval $\mathcal{I}_0$, where $I_n$ is the $n \times n$ identity matrix.*

**Remark 1.1.** Assume that the functions $\lambda_1(t), \cdots, \lambda_n(t)$ are continuous on the interval $\mathcal{I}_0$ and that

$$\lim_{t\to+\infty} \lambda_j(t) = \mu_j \qquad (j = 1,2,\cdots,n)$$

exist. Then, if the real parts of $\mu_1, \cdots, \mu_n$ are mutually distinct, the functions $\lambda_1(t), \cdots, \lambda_n(t)$ satisfy Assumption 1.2.

**Remark 1.2.** Theorem 1.1 has been shown to be the basis of many important results for asymptotic integrations of differential equations. For instances, see W. A. Harris, Jr. and D. A. Lutz [HL3] and M. S. P. Eastham [E].

**Remark 1.3.** A similar result of Theorem 1.1 for the system (1.1) with $A(t, h(t))$ is obtained recently by W. A. Harris, Jr. and Y. Sibuya [HY]. Here $A(t, \epsilon)$ is periodic in $t$ for every fixed vector parametrer $\epsilon$ and $h(t)$ a vector function tending to the zero vector as $t \to +\infty$.

**Remark 1.4.** The conditions required for the existence of $Q(t)$ globally analytic on $\mathcal{I}_0$ and $\overline{\mathcal{I}}_0$ are studied recently by H. Gingold, P. F. Hsieh and Y. Sibuya [GHY].

Theorem 1.1 can be proved in the following manner. From (1.1), (1.2), (1.6) and (1.7), we see that $Q(t)$ satisfies a linear differential equation:

$$\frac{dQ(t)}{dt} = \Lambda(t)Q - Q\Lambda(t) + R(t)[I + Q]. \tag{1.8}$$

As (1.8) is a linear equation, if a solution $Q(t)$ is shown to satisfy condition (2) in an interval $\mathcal{I} = [t_1, \infty)$, for a large $t_1$, then, $Q(t)$ exists on $\mathcal{I}_0$ and satisfies Theorem 1.1. Assumptions 1.2 and 1.3 are employed to show the existence of such $Q(t)$.

P. Hartman and A. Wintner[HW] studied the same problem for (1.1) under slightly different set of assumptions:

**Assumption 1.4.** There exists a positive constant $\delta$ such that for each pair of indices $j$ and $k, (j \neq k)$,

$$|D_{jk}(t)| \geq \delta > 0, \quad \text{for} \quad t \in \mathcal{I}_0. \tag{1.9}$$

**Assumption 1.5.** There is a positive $p, (1 < p \leq 2)$, such that

$$|R(t)| \in L_p(\mathcal{I}_0), \tag{1.10}$$

where $|\cdot|$ denotes the maximum norm.

A version of the Hartman and Wintner Theorem can be given as follows:

**Theorem 1.2.** *Under the Assumptions 1.1, 1.4 and 1.5, there exists an $n \times n$ matrix $Q(t)$ such that*

*(1) the derivative $\frac{dQ(t)}{dt}$ exists and the entries of $Q(t)$ and $\frac{dQ(t)}{dt}$ are continuous in $t$ on the interval $\mathcal{I}_0$,*

*(2)* $Q(t) \in L_p(\mathcal{I}_0)$, $diag Q(t) = 0$ *and* $\lim_{t\to+\infty} Q(t) = 0$,

*(3) the transformation:*

$$y = [\, I_n + Q(t)\,]\, z \tag{1.11}$$

*changes system (1.1) to*

$$\frac{dz}{dt} = [\Lambda(t) + diag R(t)]z \tag{1.12}$$

*on the interval* $\mathcal{I}_0$, *where* $I_n$ *is the* $n \times n$ *identity matrix.*

For a system (1.1) with almost constant coefficient, consider the following:

**Assumption 1.6.** The matrix $A(t)$ is in the form

$$A(t) = A_1 + B(t). \tag{1.13}$$

satisfies the conditions:

(i) the constant matrix $A_1$ is in block-diagonal form

$$A_1 = diag\{A_{11}, A_{12}, \cdots, A_{1m}\}$$

where $A_{1j}$ is an $n_j \times n_j$ matrix in the form

$$A_{1j} = \lambda_j I_{n_j} + E_j + N_j \quad (j = 1, 2, \cdots, m)$$

with $\lambda_j$ a real number, $E_j$ an $n_j \times n_j$ diagonal matrix of purely imaginary diagonal entries, $N_j$ an $n_j \times n_j$ nilpotent matrix and $n_1 + n_2 + \cdots + n_m = n$;

(ii) $\lambda_m < \lambda_{m-1} < \cdots < \lambda_2 < \lambda_1$ ;

(iii) $N_j E_j = E_j N_j \quad (j = 1, 2, \cdots, m)$;

(iv) the matrix $B(t)$ is expressed in the form $(B_{jk}(t))_{j,k=1}^m$ with $B_{jk}(t)$ $n_j \times n_k$ matrices satisfying

$$\lim_{t\to\infty} B_{jk}(t) = 0 \qquad (j, k = 1, \cdots, m).$$

Let

$$y = [y_1 \;\; y_2 \;\; \cdots \;\; y_m]^T, \tag{1.14}$$

where $y_j$ are $n_j$ column vectors. Sibuya [S] proved the following (see also Hartman [H, Ch. X, Lemma 12.1]):

**Theorem 1.3.** *Under the Assumption 1.6, there exists a linear transformation*

$$y_j = z_j + \sum_{j\neq k} T_{jk}(t)\, z_k \qquad (j = 1, 2, \cdots, m) \tag{1.15}$$

*with $n_j \times n_k$ matrices $T_{jk}(t)$ such that*

*(1) for every pair of indices $j$ and $k$, $(j \neq k)$, the derivative $\dfrac{T_{jk}(t)}{dt}$ exists and the entries of $T_{jk}(t)$ and $\dfrac{T_{jk}(t)}{dt}$ are continuous on the intrerval $\mathcal{I}_0$;*

*(2) $\lim_{t\to\infty} T_{jk}(t) = 0, \quad (j \neq k)$;*

*(3) the transformation (1.15) changes the system (1.1) with (1.13) to*

$$\frac{dz_j}{dt} = \left[A_{jj} + B_{jj}(t) + \sum_{h\neq j} B_{jh}(t)T_{hj}(t)\right] z_j \qquad (j = 1, 2, \cdots, m). \tag{1.16}$$

For a more general setting, suppose that the system (1.1) is in the form

$$\frac{dy_j}{dt} = A_j(t)y_j + \sum_{k=1}^{m} B_{jk}(t)y_k, \tag{1.17}$$

where $y_j$ are $n_j$-column vectors given in (1.14), $A_j(t)$ and $B_{jk}(t)$ are $n_j \times n_j$ and $n_j \times n_k$ matrices, respectively. Let $G_j(t,s)$ be the $n_j \times n_k$ matrix such that

$$\begin{cases} \dfrac{dG_j(t,s)}{dt} &= A_j(t)G_j(t,s), \\ G_j(s,s) &= I_{n_j}, \qquad (j = 1, 2, \cdots, m), \end{cases} \tag{1.18}$$

for $t, s \in \mathcal{I}_0$. Assume the following:

**Assumption 1.7.** There exist two positive constants $K$ and $\delta$ such that for any $n_j \times n_k$ matrix $C_{jk}$, we have

$$\|G_j(t,s)C_{jk}G_k(t,s)^{-1}\| \leq Ke^{+\delta(t-s)}\|C_{jk}\|, \quad \text{for} \quad t \leq s, j > k, \tag{1.19}$$

and

$$\|G_j(t,s)C_{jk}G_k(t,s)^{-1}\| \leq Ke^{-\delta(t-s)}\|C_{jk}\|, \quad \text{for} \quad t \geq s, j < k, \tag{1.20}$$

for $s, t \in \mathcal{I}_0$. Here $\|\cdot\|$ denote the Euclidian norm.

**Assumption 1.8.** There exists a function $f(t)$ such that

$$\|B_{jk}(t)\| \leq f(t) \qquad (j = 1, 2, \cdots, m). \tag{1.21}$$

and

$$\sup_{p\geq t}(1 + p - t)^{-1} \int_t^p f(\tau)d\tau \to 0, \tag{1.22}$$

as $t \to +\infty$.

Sibuya [S] proved the following

**Theorem 1.4.** *Under the Assumptions 1.7 and 1.8, there exists a linear transformation*

$$y_j = z_j + \sum_{j \neq k} T_{jk}(t)\, z_k \quad (j = 1, 2, \cdots, m) \tag{1.23}$$

*with $n_j \times n_k$ matrices $T_{jk}(t)$ such that*

*(1) for every pair of indices $j$ and $k$, $(j \neq k)$, the derivative $\frac{T_{jk}(t)}{dt}$ exists and the entries of $T_{jk}(t)$ and $\frac{T_{jk}(t)}{dt}$ are continuous on the interval $\mathcal{I}_0$;*

*(2) $\lim_{t \to +\infty} T_{jk}(t) = 0, \quad (j \neq k)$;*

*(3) the transformation (1.23) reduces the system (1.17) to*

$$\frac{dz_j}{dt} = \left[ A_{jj} + B_{jj}(t) + \sum_{h \neq j} B_{jh}(t) T_{hj}(t) \right] z_j \quad (j = 1, 2, \cdots, m). \tag{1.24}$$

Note that the reduced systems (1.16) and (1.24) involve the entries of the respective transformation matrices, while the reduced systems (1.7) and (1.12) do not. Further study of the asymptotic behavior of (1.1) are made to generalize these theorems and applied to topics such as adiabatic oscillators and deficiency index problems (eg. see K. Chiba and T. Kimura [CK], A. Devinatz [D], M. S. P. Eastham [E] and W. A. Harris and D. A. Lutz [HL1, HL2, HL3]). In these studies, the integrability of $R(t)$ and/or its derivatives is always assumed, except for Theorems 1.3 and 1.4. In this paper, we will generalized these results by assuming only the integrability of entries of $R(t)$ above (or below) the diagonal and replace the other half by zero limit as $t \to \infty$. Furthermore, the reduced equations of our new result do not involve the entries of the transformation matrices. In Section 2, the main theorem is to be stated and examples are to be given to illustrate the features of these theorems. The proof of the main theorem is to be given in Section 3.

**2. The Main Theorem and Examples.** In order to formulate the main theorem, let

$$R(t) = \Big( r_{jk}(t) \Big)_{j,k=1}^{n} \tag{2.1}$$

and

$$r_1(t) = \max_{j>k} |r_{jk}(t)|, \qquad r_2(t) = \max_{j<k} |r_{jk}(t)|. \tag{2.2}$$

We assume the following

**Assumption 2.1.** There exist two positive constants $\delta$ and $K$ and a constant $\alpha$ $(0 \leq \alpha < 1)$, independent of $t$, such that for each pair of indices $j$ and $k$, $(j, k = 1, 2, \cdots, n; j \neq k)$,

$$\exp\left\{\int_t^s D_{jk}(\tau)d\tau\right\} \leq K \exp\left\{-\delta(s^{1-\alpha} - t^{1-\alpha})\right\}, \quad \text{whenever} \quad s \geq t, \tag{2.3}$$

or

$$\exp\left\{\int_t^s D_{jk}(\tau)d\tau\right\} \leq K \exp\left\{-\delta(t^{1-\alpha} - s^{1-\alpha})\right\}, \quad \text{whenever} \quad s \leq t. \tag{2.4}$$

**Assumption 2.2.** The matrix $R(t)$ is continuous and satisfies:

$$|R(t)| = o(t^{-\alpha}), \qquad \text{as} \quad t \to +\infty, \tag{2.5}$$

and either

$$\int_{t_0}^{+\infty} t^{-\alpha} \max_{s \geq t}[s^\alpha r_1(s)]dt < +\infty, \tag{2.6}$$

or

$$\int_{t_0}^{+\infty} t^{-\alpha} \max_{s \geq t}[s^\alpha r_2(s)]dt < +\infty. \tag{2.7}$$

We will establish the following main theorem:

**Theorem 2.1.** *Under the Assumptions 2.1 and 2.2, there exists an $n \times n$ matrix $Q(t)$ such that*

*(1) the derivative $\dfrac{dQ(t)}{dt}$ exists and the entries of $Q(t)$ and $\dfrac{dQ(t)}{dt}$ are continuous in $t$ on the interval $\mathcal{I}_0$,*

*(2) $\lim_{t \to +\infty} Q(t) = 0$,*

*(3) the transformation:*

$$y = [\ I_n + Q(t)\ ]\ z \tag{2.8}$$

*changes system (1.1) to*

$$\frac{dz}{dt} = [\Lambda(t) + diag R(t)]z \tag{2.9}$$

*on the interval $\mathcal{I}_0$, where $I_n$ is the $n \times n$ identity matrix.*

Note that Assumption 2.1 is equivalent to Assumption 1.7 when $\alpha = 0$. On the other hand (2.5) and (2.6) implies Assumption 1.8. Thus Assumption 2.1 is weaker than Assumption 1.7 for $\alpha > 0$ and Assumption 2.2 is stronger than Assumption 1.8. However, the reduced equation (2.9) does not involve the entries of the transformation matrix $Q(t)$ in (2.8).

The proof of this theorem can be found in Hsieh and Xie [HX]. In order to utilize the Assumption 2.2, we will develop a "row-wise" (or "colum-wise") successive approximations method, similar to the Gauss-Seidel iteration (e.g., see [GVl] and [V]), in the proof. In this process, we need only the continuity of $R(t)$ along with Assumption 2.2, and do not require the existence of $R'(t)$ as in many theorems of this type (e.g., see [CK], [E], [HL1] and [L]). Furthermore, as we do not require that $\lim_{t\to+\infty} \Lambda(t)$ has distinct eigenvalues, Theorem 2.1 can be applied to diagonalize some systems which are not diagonalizeable by previous results.

**Remark 2.1.** If $\{t^\alpha r_{jk}(t)\,|\,j>k\}$ (or $\{t^\alpha r_{jk}(t)\,|\,j<k\}$) are all monotonic decreasing, then $t^\alpha r_1(t)$ (or $t^\alpha r_2(t)$) is monotonic decreasing and, consequently, (2.6) (or (2.7)) can be replaced by

$$\int_{t_0}^{+\infty} r_1(s)dt < +\infty, \quad \left(\text{or} \quad \int_{t_0}^{+\infty} r_2(s)dt < +\infty\right). \tag{2.10}$$

The system (1.1) with following $A(t)$ are not applicable by any of Theorems 1.1, 1.2, 1.3 or 1.4, but applicable by Theorem 2.1 on $[t_0, \infty)$.

**Example 1.** Consider

$$A(t) = \begin{bmatrix} 1+\dfrac{1}{t^{\frac{2}{3}}} & \dfrac{1}{t^2} \\ \dfrac{|\sin(t^2)|}{\sqrt{t}\ln t} & 1+\dfrac{1}{\sqrt{t}} \end{bmatrix},$$

for $t \in [2, +\infty)$. We consider $\lambda_1(t) = 1$, $\lambda_2(t) = 1+\dfrac{1}{\sqrt{t}}$, $r_{11}(t) = \dfrac{1}{t^{\frac{2}{3}}}$ and $r_{22}(t) = 0$. The system (1.1) with this $A(t)$ is not diagonalizeable by Theorem 1.1, because Assumption 1.1 is not satisfied. It is not diagonalizeable by Theorem 1.2 or 1.3 because $\lim_{t\to+\infty} \lambda_2(t) = \lim_{t\to+\infty} 1+\dfrac{1}{\sqrt{t}} = 1 = \lambda_1(t)$. It is also not diagonalizeable by Theorem 1.4, because Assumption 1.7 is not satisfied. On the other hand, it is diagonalizeable by Theorem 2.1 with $\alpha = \dfrac{1}{2}$.

**Example 2.** Consider

$$A(t) = \begin{bmatrix} t & \dfrac{\sin t}{t^2} & \dfrac{1}{t^{\frac{2}{3}}} \\ \dfrac{1}{t} & 2 & \dfrac{1}{t^3} \\ \dfrac{\ln t}{t^{\frac{1}{2}}} & \dfrac{-2}{t^{\frac{1}{3}}\ln t} & 2+\dfrac{1}{t^{\frac{1}{3}}} \end{bmatrix},$$

for $t \in [2, +\infty)$. The system (1.1) is not diagonalizeable by Theorem 1.1, 1.2, 1.3 or 1.4 because $\lim_{t \to +\infty} \lambda_3(t) = \lim_{t \to +\infty} \left[ 2 + \frac{1}{t^{\frac{1}{3}}} \right] = 2 = \lambda_2(t)$.. On the other hand, it is diagonalizeable by Theorem 2.1 with $\alpha = \frac{1}{3}$. In this case, $[\max r_2(t)] = \frac{1}{t^{\frac{3}{2}}}$, thus, $[t^{\frac{1}{3}} \max r_2(t)] = \frac{1}{t^{\frac{1}{3}}}$ is monotonic decreasing for $t \in [2, +\infty)$.

**Example 3.** Consider

$$A(t) = \begin{bmatrix} 1 & \frac{1}{t^{\frac{3}{2}}} \\ \frac{1}{\ln(t+2)} & e^{it} \end{bmatrix},$$

for $t \in [1, +\infty)$. The system (1.1) with this $A(t)$ is not diagonalizeable by Theorem 1.1 as Assumption 1.1 is not satisfied. As the diagonal elements coincide at $t = 2k\pi$, ($k$: positive integers), Theorems 1.2 and Theorems 1.3 are not applicable. On the other hand, the system (1.1) with this $A(t)$ is diagonalizeable by Theorem 2.1 with $\alpha = 0$. This system (1.1) is diagonalizeable by Theorem 1.4 , however, the reduced system has different coefficients from that obtained by Theorem 2.1

**Example 4.** Consider

$$A(t) = \begin{bmatrix} 1 & \frac{1}{e^t} \\ \frac{2 + \sin t}{\sqrt{t}} & \sin t \end{bmatrix},$$

for $[1, +\infty)$. The system (1.1) with this $A(t)$ is not diagonalizeable by Theorem 1.1 as Assumption 1.1 is not satisfied. As the diagonal elements coincide at $t = \left(2k + \frac{1}{2}\right)\pi$, ($k$ : nonnegative integers), Theorems 1.2 and Theorems 1.3 are not applicable. This system (1.1) with this $A(t)$ is diagonalizeable by Theorem 2.1 with $\alpha = 0$. Note that in this case, $r_{22}(t) = \frac{2 + \sin t}{\sqrt{t}}$ is not monotonic decreasing. It is also diagonalizeable by Theorem 1.4 with different coefficients in the reduced system.

## References.


[CK] K. Chiba and T. Kimura, On the asymptotic behavior of solutions of a system of linear ordinary differential equations, *Comm. Math. Univ. St. Pauli* 18 (1970), 61 – 80.

[CL] E. A. Coddington and N. Levinson, "Theory of Ordinary Differential Equations", McGraw-Hill, New York, 1955.

[D] A. Devinatz, Asymptotic theory for systems of linear differential equations, *Trans. Amer. Math. Soc.*, **160** (1971), 353-363.

[E] M. S. P. Eastham, "The asymptotic solution of linear differential systems. Applications of the Levinson theorem", Oxford Science Publications, Oxford, 1989.

[GHY] H. Gingold, P. F. Hsieh and Y. Sibuya, Globally analytic simplification and the Levinson Theorem, to appear in *J. Math. Anal. and Appl.*

[GVl] G. H. Golub and C. F. Van Loan, "Matrix Computations", The John Hopkins University Press, Baltimore, 1983.

[HL1] W. A. Harris, Jr. and D. A. Lutz, On the asymptotic integration of linear differential system, *J. Math. Anal. Appl.* 48(1974), 1-16.

[HL2] W. A. Harris, Jr. and D. A. Lutz, Asymptotic integration of adiabatic oscillators, *J. Math. Anal. Appl.* 51(1975), 76-93.

[HL3] W. A. Harris, Jr. and D. A. Lutz, A unified theory of asymptotic integration, *J. Math. Anal. Appl.* 57(1977), 571-586.

[HS] W. A. Harris, Jr. and Y. Sibuya, Asymptotic behavior of solutions of a system of linear ordinary differential equations as $t \to \infty$, in "Delay Differential Equations and Dynamical Systems", Lecture Notes in Math., No. 1475, S. Busenberg and M. Martelli ed., 210–217, Springer-Verlag, 1991.

[H] P. Hartman, "Ordinary Differential Equations", Wiley, New York, 1964.

[HW] P. Hartman and A. Wintner, Asymptotic integration of linear differential equations, *Amer. J. Math.* 77(1955), 45-96, 932.

[HX] P. F. Hsieh and F. Xie, Asymptotic diagonalization of a system of linear differential equations, submitted to *J. Math. Anal. Appl.*

[L] N. Levinson, The asymptotic nature of solutions of linear ordinary differential equations, *Duke Math. J.* 15(1948), 111-126.

[V] R. S. Varga, "Matrix Iterative Analysis", Prentice-Hall, Inc., 1962.

# COMPUTING CONTINUOUS SPECTRUM

Anton Zettl[1]

**Abstract**

The continuous, or essential, spectrum of singular Sturm-Liouville (S-L) problems can be approximated numerically just by computing eigenvalues of regular S-L problems with separated boundary conditions - provided one knows which regular problems to use. These have been identified by Bailey, Everitt, Weidmann and Zettl in [BEWZ]. Since there are highly effective codes, such as SLEIGN and SLEDGE, available for the computation of eigenvalues of regular S-L problems with separated boundary conditions this scheme works surprisingly well for a large class of problems.

The purpose of this note is to show that the results in [BEWZ] by Bailey, Everitt, Weidmann and Zettl, together with the software packages SLEIGN or SLEDGE can be used to approximate the essential spectrum of a large class of singular Sturm-Liouville (S-L) problems. Below we demonstrate this in particular for the unperturbed and perturbed Mathieu equation and for what we call the Plum equation. But it is to be emphasized that these examples were chosen solely because of their interest in the literature and because there are independent checks available.

Consider the S-L equation

$$-(py')' + qy = \lambda wy \tag{1}$$

Let $J = (a, b)$, $-\infty \leq a < b \leq \infty$, be a real interval, let $p$, $q$, $w$ map $(a, b)$ into $\mathbf{R}$, the reals, and assume that

$$\frac{1}{p},\ q, w \in L_{loc}(J),\ p > 0\ a.e.,\ w > 0\ a.e. \tag{2}$$

[1]This report is on joint work with P. B. Bailey and W. N. Everitt on a project partially funded by NSF under grant DMS-9106470.

Let $\{a_r;\ r \in N_0\}$, $\{b_r:\ r \in N_0\}$, $N_0 = 0, 1, 2, 3, \ldots$ be (endpoint) sequences such that

$$-\infty \le a < a_r < b_r < b \le \infty,\ a_r \to a,\ b_r \to b. \tag{3}$$

Conditions (2) ensure that equation (1) is regular on each interval $[a_r, b_r]$, $r \in N_0$ but the endpoints $a$, $b$ may be singular.

If $S_r$ is any self-adjoint realization of (1) on $[a_r, b_r]$ then it is well known [W] that the spectrum $\sigma(S_r)$ of $S_r$ is discrete i.e. consists entirely of isolated eigenvalues and these can be ordered to satisfy

$$-\infty < \lambda_0 \le \lambda_1 \le \lambda_2 \le \ldots,\ \lambda_n \to +\infty \text{ as } n \to \infty, \tag{4}$$

where

$$\sigma(S_r) = \{\lambda_n:\ n \in N_0\},\ \lambda_n = \lambda_n(S_r). \tag{5}$$

If $S$ is a self-adjoint realization of (1) on (a,b) then since one or both endpoints may be singular the spectrum $\sigma(S)$ of $S$ is, in general, quite complicated. It may consist of a discrete and a continuous part. Let

$$\sigma(S) = \sigma_d(S) \cup \sigma_e(S), \tag{6}$$

where $\sigma_d$ denotes the set of all isolated eigenvalues and $\sigma_e$ the rest of the spectrum. Either one of $\sigma_d$ or $\sigma_e$, but not both, may be empty. Let

$$\sigma_0 = inf.\ \sigma_e,\ -\infty \le \sigma_0 \le +\infty. \tag{7}$$

Here $\sigma_0 = +\infty$ means that $\sigma_e = \phi$.

**Definition.** Let $S_r$ be a self-adjoint realization of (1) on $[a_r, b_r]$, $r \in N_0$; and let $S$ be a self-adjoint realization of (1) on $(a, b)$.

(i) The sequence $\{S_r:\ r \in N_0\}$ is spectral included for $S$ if for any $\lambda \in \sigma(S)$ there exists a sequence of eigenvalues

$$\{\lambda_r \in \sigma(S_r):\ r \in N\} \text{ such that } \lambda_0,\ \lambda_1,\ \lambda_2,\ \lambda_3, \ldots \text{ converges to } \lambda. \tag{8}$$

(ii) The sequence $\{S_r : r \in N_0\}$ is spectral exact for $S$ if it is spectral included and, in addition, if $\lambda_r \in \sigma(S_r)$, $r \in N_0$ and the sequence $\{\lambda_r : r \in N_0\}$ converges to $\lambda$, implies that $\lambda \in \sigma(S)$.

**Remark.** Note that in this definition each operator $S_r$ "lives" in a different space i.e. $H_r = L^2((a_r, b_r); w)$, no kind of convergence on the operator sequence $\{S_r : r \in N_0\}$ is assumed and, of course, this definition can be made for abstract operators $S_r$, $S$ not just for S-L operators.

The endpoint $a$ is said to be in the limit-circle case, $LC$ for short, if all solutions of (1) are in $L^2((a,c);w)$ for some $c$, $a < c < b$. Similarly for the endpoint $b$. It is well known that the $LC$ classifications is independent of $\lambda$. An endpoint is said to be in the limit-point or $LP$ case if it is not $LC$.

Next we state the result from [BEWZ] upon which our computations are based. This theorem identifies which regular problems, or more accurately, which sequences of regular problems, approximate a given singular one.

**Theorem 1.** (Bailey, Everitt, Weidmann and Zettl, [BEWZ]). Assume each endpoint $a$, $b$ is in the $LP$ case. Let $\{a_r : r \in N_0\}$, $\{b_r : r \in N_0\}$ be endpoint sequences satisfying (3).

a) Let $S_r$ be any self-adjoint realization of (1) on $[a_r, b_r]$, $r \in N_0$. (Note that $S_r$ can be determined by any self-adjoint boundary conditions, separated or coupled, specified at $a_r$ and $b_r$ with possibly different boundary conditions for different $r$'s.) Then the sequence $\{S_r : r \in N_0\}$ is spectral included for $S$ but, in general, not spectral exact.

b) Suppose that $S$ is bounded below and each $S_r$, is determined by Dirichlet boundary conditions:

$$y(a_r) = 0 = y(b_r), \quad r \in N_0. \tag{9}$$

Then

i) The sequence $\{S_r : r \in N_0\}$ is spectral exact for $S$ below $\sigma_0(S)$. (This means that if (8) holds with $\lambda < \sigma_0(S)$ then $\lambda \in \sigma(S)$.)

ii) If $S$ has no eigenvalue below $\sigma_0(S)$, then the sequence

$$\{\lambda_0(S_r);\ r \in N_0\} \text{ converges to } \sigma_0(S), \quad \text{as } r \to +\infty. \tag{10}$$

iii) If $S$ has exactly $k$ eigenvalues below $\sigma_0(S)$, $0 \le k < \infty$, then

$$\{\lambda_k(S_r);\ r \in N_0\} \to \sigma_0(S), \quad r \to +\infty. \tag{11}$$

(Recall that the eigenvalues are indexed starting with 0 not 1.)

iv) The sequence $\{S_r : r \in N_0\}$ cannot be spectral exact for $S$ above $\sigma_0(S)$ for arbitrary endpoint sequences $\{a_r : r \in N_0\}$ converging to $a$ and $\{b_r : r \in N_0\}$ converging to $b$.

*Proof. This proof is due to W.N.Everitt in a private communication.* Parts (a) and (b) (i), (ii), (iii) follow from Theorems 6.1 and 6.2 in [BEWZ]. To prove (b) (iv) note that for any fixed $n \in N_0$, the $n$-th eigenvalue $\lambda_n(S_r)$ is a continuous function of the endpoints $a_r$, $b_r$. Now consider a problem such as the Mathieu equation where the spectrum of $S$ consists of spectral bands separated by gaps. As an eigenvalue "moves" from the second spectral band into the first it must move through the first gap by continuity. Choose $n_1$ and $r_1$ such that $\lambda_n(S_{r_1})$ is in the first gap of $\sigma(S)$ and not an eigenvalue of $S$. Now move the endpoints $a_r$, $b_r$ further out towards $a$, $b$ and "watch" the eigenvalues moving through the first gap until one of them is exactly equal to $\lambda_{n_1}(S_{r_1})$. Then record the values of the endpoints $a_{r_2}$, $b_{r_2}$ for which this happens. Repeating this process we can construct a sequence of endpoints $a_{r_j}$, $b_{r_j}$ for which there exists a corresponding sequence of eigenvalues converging to $\lambda_{n_1}(S_{r_1})$ which is a point inside the first gap and which is not in $\sigma(S)$.

In [BEWZ] analogues of Theorem 1 for the other endpoint classifications $LP/LC$, $LC/LP$ and $LC/LC$ are also established. The $LC/LC$ case is illustrated in [BEZ].

The $LP/LC$ and $LC/LP$ cases are similar although technically more complicated to the $LP/LP$ case and hence we illustrate only the LP/LP case. The results in [BEWZ] require that "inherited" boundary conditions on the truncated intervals $(a_r, b_r)$ be used. Near an $LP$ endpoint a Dirichlet condition can be used but near an $LC$ endpoint an "inherited" boundary condition is obtained from a "boundary condition function" which determines the singular boundary conditions of $S$. For details see [BEWZ].

## THE MATHIEU EQUATION

$$-y'' + (\sin\ x)y = \lambda y \quad \text{on} \quad (-\infty, \infty) \ \& \ [0, 2\pi]$$

$$\left\{ \begin{array}{l} y(0) = y(2\pi) \\ y'(0) = y'(2\pi) \end{array} \right\} \quad -\infty < \lambda_0 < \lambda_1 \le \lambda_2 \le \ldots \to +\infty$$

$$\left\{ \begin{array}{l} y(0) = -y(2\pi) \\ y'(0) = -y'(2\pi) \end{array} \right\} \quad -\infty < \mu_0 \le \mu_1 \le \mu_2 \le \ldots \to \infty$$

SLEIGN2: These eigenvalues were computed by P.B. Bailey using SLEIGN2, a code under development by Bailey, Everitt and Zettl, partially funded by NSF grant: DMS - 9106470.

$$\lambda_0 = -0.378492; \quad \lambda_1 = 0.918058; \quad \lambda_2 = 1.29315; \quad \lambda_3 = 4.03192; \quad \lambda_4 = 4.053$$

$$\mu_0 = -0.34767; \quad \mu_1 = 0.594797; \mu_2 = 2.28515; \quad \mu_3 = 2.3425$$

$\sigma(-\infty, \infty)$ :

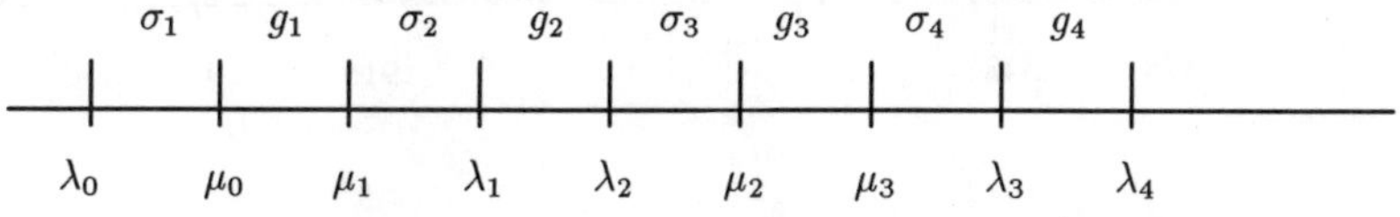


$\sim 1$ : .03 1. .3 .4 1. .05 .7 .02

$b_r, \quad a_r = -b_r$

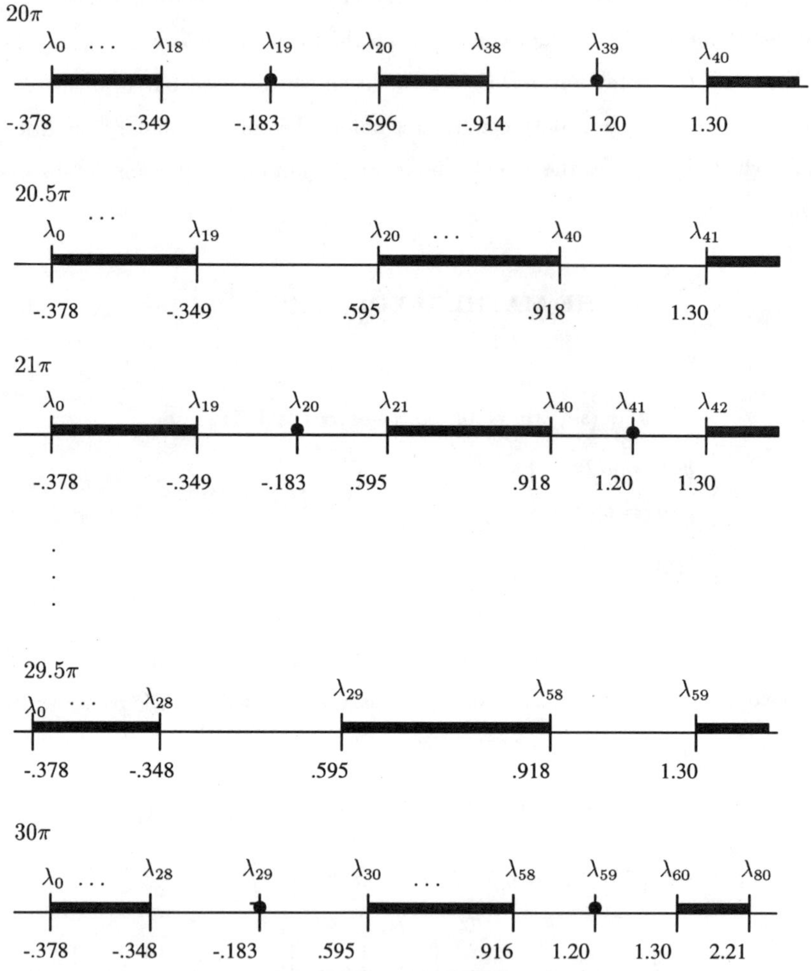
20π
λ0 . . . λ18 λ19 λ20 λ38 λ39 λ40
-.378 -.349 -.183 -.596 -.914 1.20 1.30
20.5π
. . .
λ0 λ19 λ20 . . . λ40 λ41
-.378 -.349 .595 .918 1.30
21π
λ0 λ19 λ20 λ21 λ40 λ41 λ42
-.378 -.349 -.183 .595 .918 1.20 1.30
29.5π
λ0 . . . λ28 λ29 λ58 λ59
-.378 -.348 .595 .918 1.30
30π
λ0 . . . λ28 λ29 λ30 . . . λ58 λ59 λ60 λ80
-.378 -.348 -.183 .595 .916 1.20 1.30 2.21

$$-y'' + q(x)y = \lambda y \quad \text{on} \quad [-b, b], \quad q(x) = \sin(x) + \frac{1}{x^2+1}$$

$b = 29.5\pi$

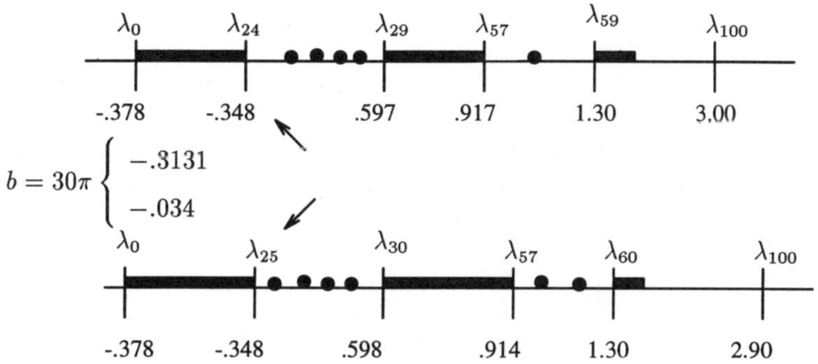


<u>Remark:</u>

1. The numerical evidence suggests that there are no eigenvalues below the essential spectrum which starts at approximately $\sigma_0 \sim -.378$.

2. There are two eigenvalues, roughly located at -.3131 and at -.034 both in the first gap and both appearing (with the same values to three decimal places) for both BEWZ approximations on the intervals $[-29.5\pi, 29.5\pi]$ and on $[-30\pi, 30\pi]$. These seem to be further illustrations of the "trapping and cascading" phenomenon.

3. On the interval $[-30\pi, 30\pi]$ the first 100 eigenvalues of the Dirichlet problem are all less than 2.90 . Thus the asymptotic behavior has not yet begun to assert its influence.

$$-y'' + q(x)y = \lambda y, \quad [-b, b], \; q(x) = \sin \; x + \frac{1}{1+x^2}$$

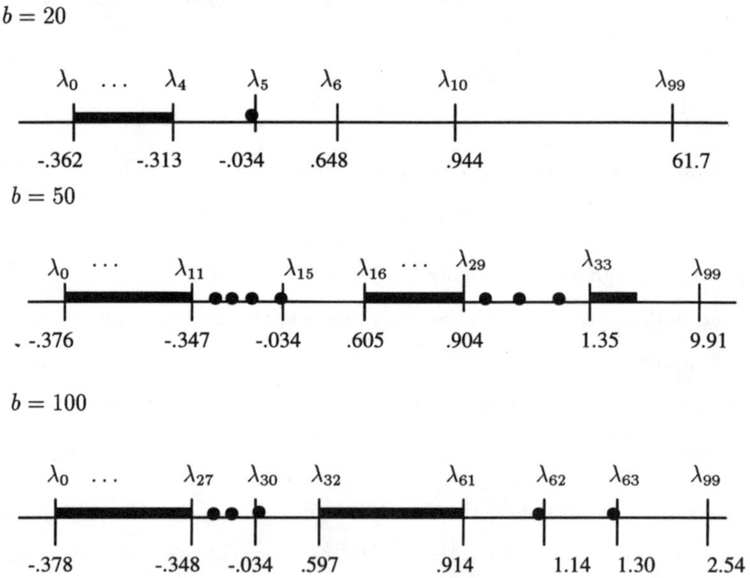

$$-y'' + q(x)y = \lambda y \quad \text{on} \quad [-b, b], \quad y(-b) = 0 = y(b), \quad q(x) = \sin(x + \frac{1}{1+x^2})$$

$b = 92.6770 \sim 2.95\pi$

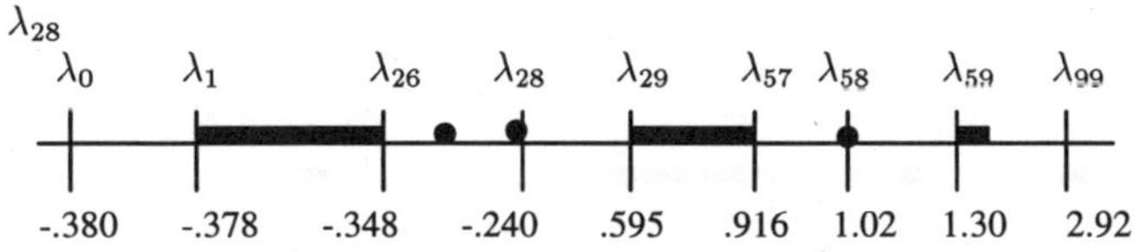


$b = 94.2478 \sim 30\pi$

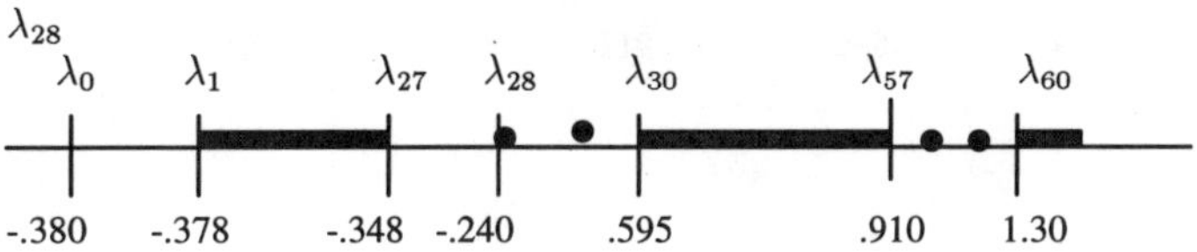


Remark.

1. There is strong numerical evidence for the existence of an eigenvalue near $-.380 < \sigma_0 \sim -.378$.
2. None of the gaps are "clear".
3. For many truncated intervals of the BEWZ approximations the eigenvalue near -.240 persists in the first gap. Is this an indication of the existence of an eigenvalue in the first gap of the singular problem or merely another illustration of "trapping and cascading"?

$$-y'' + q(x)y = \lambda y, \quad [-b, b], \quad q(x) = \sin(x + \tfrac{1}{1+x^2}), \quad y(-b) = 0 = y(b)$$

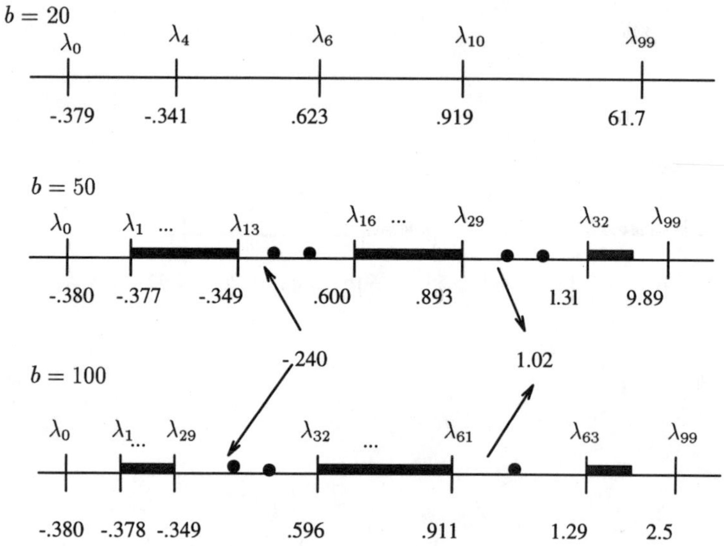


Remark.

1. There is strong numerical evidence for the existence of an eigenvalue near $-.380 < \sigma_0 \sim -.378$.

2. None of the gaps are "clear".

3. For many truncated intervals of the BEWZ approximations the eigenvalue near -.240 persists in the first gap. Is this an indication of the existence of an eigenvalue in the first gap of the singular problem or merely another illustration of "trapping and cascading"?

## Plum Problem

$$-y'' + 100\ \cos^2(x)\ r = \lambda y, \text{ on } [0, 3.14159]:$$

Let the periodic, semi-periodic and Dirichlet eigenvalues be denoted by $\lambda_n^P$, $\lambda_n^S$, $\lambda_n^D$, respectively. These were computed with SLEIGN2. The following inequalities are well known [W].

$$-\infty < \lambda_0^P < \lambda_0^S \leq \left\{ \begin{array}{c} \lambda_0^D \\ \\ \lambda_1^N \end{array} \right\} \leq \lambda_1^S < \lambda_1^P \leq \lambda_1^D \leq \lambda_2^P < \lambda_2^S \leq \lambda_2^D.$$

$$\lambda_0^P = 9.74321; \quad \lambda_1^P = 28.6851$$

$$\lambda_0^S = 9.74321; \quad \lambda_1^S = 28.6851$$

$$\lambda_0^D = 9.74322; \quad \lambda_1^D = 28.6851$$

On $(-\infty, \infty)$ :

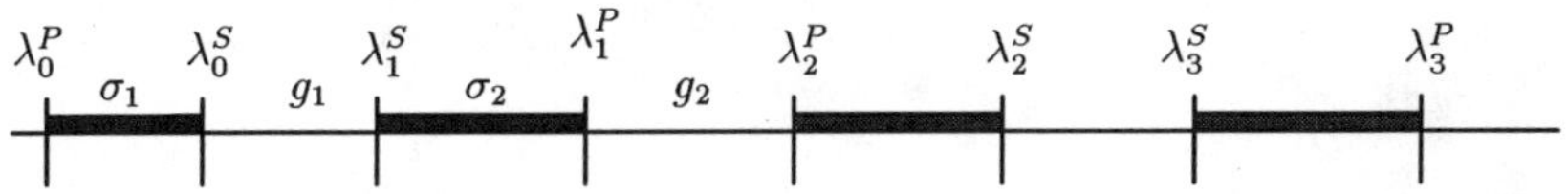


BEWZ appr. on [-50,50] : y(-50) = 0 y(50) = 0.

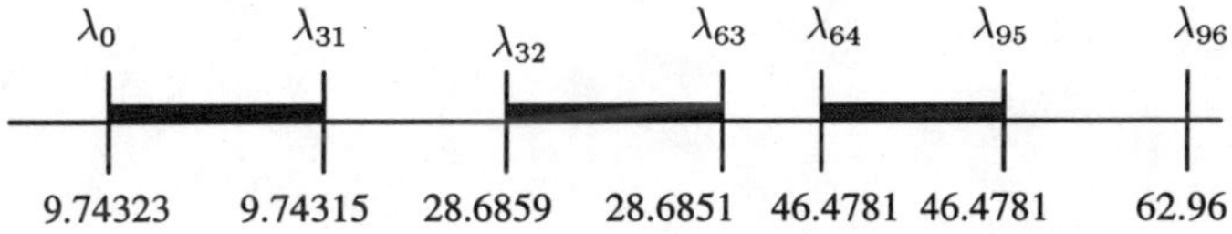

$y(-60) = 0 = y(60)$

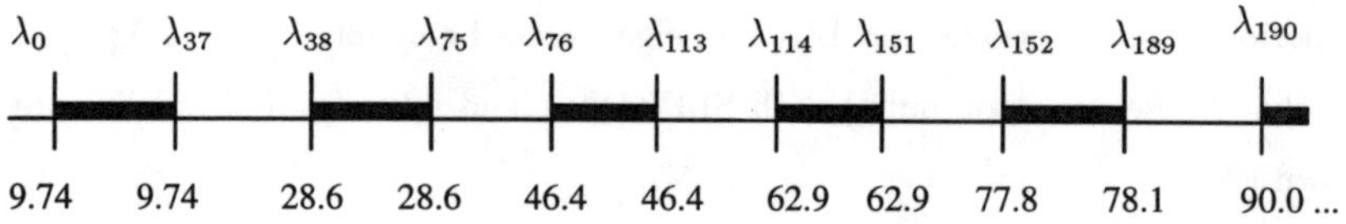


$\lambda_0 = 9.7432; \quad \lambda_{38} = 28.6855$

$\lambda_{37} = 9.7464; \quad \lambda_{75} = 28.685(1)$

$\lambda_{114} = 62.9651; \quad \lambda_{76} = 46.4780$

$\lambda_{151} = 62.9862; \quad \lambda_{113} = 46.4789$

$\lambda_{152} = 77.8057; \quad \lambda_{190} = 90.0525$

$\lambda_{189} = 78.0623;$

|– – – –|

All gaps are clear for both interval $[-b, b]$, $b = 50$, $b = 6$.

**Comments and questions:**

1. For the Mathieu equation: Is the sequence $\{S_r : r \in N_0\}$ spectral exact for $S$ when $a_r = -b_r$ and $b_r = (r+1/2)\pi$ and not spectral exact when $a_r = -b_r$ and $b_r = r\pi$?

   Note that for all the cases we checked the first few gaps are clear when $a_r = -b_r$ and $b_r$ is a half-integer multiple of $\pi$ but not when $b_r$ is an integer multiple of $\pi$.

2. Note the numerical detection of an eigenvalue $\lambda_0 \sim -.380$ below the essential spectrum $\sigma_0 \sim -.378$ for the potential $q(x) = \sin(x + \frac{1}{(x^2+1)})$. (Theorem 1 guarantees spectral exactness below $\sigma_0$.)

3. For the potential $q(x) = 100\ \cos^2 x$ note the extremely thin first few bands and the extraordinary corresponding distribution of the eigenvalues of the BEWZ approximations. All this for the truncated intervals [-50,50] and [-60,60] approximating $(-\infty, \infty)$. It is also interesting that the first few gaps are clear in both cases.

   The first few bands appear to have a width $< 10^{-5}$. (Each band is known to have a positive width [W].)

   These eigenvalue distributions, remarkable as they are, seem even more remarkable in view of the fact that for each problem $S_r$ the eigenvalues are asymptotically like $n^2$ i.e.

$$\lambda_n(S_r) \sim C_r\, n^2, \quad \text{as } n \to \infty. \tag{12}$$

   Of course $C_r \to 0$ as $r \to \infty$ but still it is interesting to note that the continuous spectrum of $S$ is detected by the low eigenvalues before the asymptotic distribution (12) takes effect.

## References

[BEWZ] P. B. Bailey, W. N. Everitt, J. Weidmann and A. Zettl, "Regular approximations of singular Sturm-Liouville problems", Results in Mathematics, v. 23 (1993), 3-22.

[BEZ] P. B. Bailey, W. N. Everitt and A. Zettl, "Computing eigenvalues of singular Sturm-Liouville problems", Results in Mathematics, v. 20 (1991), 391-423.

[W] J. Weidmann, "Spectral theory of ordinary differential operators", Lecture Notes in Mathematics '1258, Springer-Verlag, Heidelberg, 1987.